Tailoring of Nanocomposite Dielectrics

Tailoring of Nanocomposite Dielectrics

From Fundamentals to Devices and Applications

edited by

Toshikatsu Tanaka

Alun S. Vaughan

PAN STANFORD PUBLISHING

Published by

Pan Stanford Publishing Pte. Ltd.
Penthouse Level, Suntec Tower 3
8 Temasek Boulevard
Singapore 038988

Email: editorial@panstanford.com
Web: www.panstanford.com

British Library Cataloguing-in-Publication Data
A catalogue record for this book is available from the British Library.

Tailoring of Nanocomposite Dielectrics: From Fundamentals to Devices and Applications

ISBN 978-981-4669-80-1 (Hardcover)
ISBN 978-981-4669-81-8 (eBook)

Printed in the USA

Contents

Preface

The inclusion of nanoscale particulates within a host matrix has been exploited by mankind for millennia as a means of producing materials with unique properties. Notable examples of this include the Lycurgus Cup and Damascus steel, albeit that the origins of the effects were not understood at the time. More recently, the inclusion of nanometric carbon—carbon black—in rubber was found to be an effective means of producing improved automobile tires. As such, the topic that we now know as nanocomposites has a long history. In the electrical area, such materials have a rather shorter history, with many authors citing John Lewis' 1994 paper, "Nanometric Dielectrics," as catalyzing the current extensive and global interest in the field of so-called nanodielectrics, although it is clear that some pioneers had been researching this topic well before the 1990s. So what has caused the massive growth in interest in this class of materials over the last two decades? A reasonable answer to this is that experimental evidence emerged to demonstrate that introducing nanoparticles at a low loading level can be an effective means of improving a range of electrically relevant properties of bulk commodity polymers. While the potential of the approach progressively evolved during the first decade or so of empirical study, understanding of the origins of the effects that were reported lagged behind and this, combined with issues of scale-up and reproducibility, have meant that the clear laboratory promise of nanodielectrics has not been transferred into extensive technological impact. However, recently, this has begun to change and, therefore, the publication of this book is most timely.

The aim of this book is to bring together contributions from leading researchers who are active in topics pertinent to the electrical applications of nanocomposite systems. While the topics addressed are covered in detail, it is never possible to provide an entirely comprehensive account in a concise manner and, therefore, extensive reference lists are provided within each chapter, in order to facilitate further inquiry. The book is divided into four sections that address, in order: material preparation; experimental characterization and techniques for computer simulation; properties, both bulk and surface, that are relevant to the use of nanocomposites in dielectric application; a range of potential devices and applications. In this way, the book aims to provide a multidisciplinary overview of nanodielectrics from a range of different perspectives.

Finally, the editors wish to express their appreciation to all the contributors who have each brought their own perspective of the topic to this book. In addition, we would also like to thank Pan Stanford Publishing and, in particular, Stanford Chong and Arvind Kanswal for their hard work in pulling together the various contributions and for their patience and enthusiasm throughout the project.

Chapter 1

Introduction

Toshikatsu Tanaka
IPS Research Center, Waseda University,
Kitakyushu-shi, Fukuoka 808-0135, Japan
t-tanaka@waseda.jp

Nanotailoring is the theme of this book. In this way, you can formulate whatever materials you want. Such an era is coming near and polymer nanocomposites are an example of this. By adding a small amount of nanofiller into conventional polymers, you can actively design the very materials you might need for potential application. These materials have ever appeared as advanced dielectric and electrical insulating materials since the year 2000. They include inorganic fillers of nanometer scale and surprisingly are characterized by their interfaces that are internally formed between nanofillers and their surrounding polymermatrices. Such inner-interface dominant materials can exhibit much improved performance compared with neat polymers. Properly tailored interfaces will result in significant improvements in dielectric properties and, consequently, various applications of such materials are under R&D consideration. Indeed, some of these materials have been utilized in reality. This book is intended to give a comprehensive explanation of how

Tailoring of Nanocomposite Dielectrics: From Fundamentals to Devices and Applications
Edited by Toshikatsu Tanaka and Alun S. Vaughan

ISBN 978-981-4669-80-1 (Hardcover), 978-981-4669-81-8 (eBook)
www.panstanford.com

we should tailor polymer materials, what characteristics such tailored materials exhibit, how anti-aging they are against ambient conditions and, finally, how they can be used in the electrical power and microelectronics sectors. Many leading international experts participated in contribution to this book. Each of the chapters has a special touch that is characteristic of each author's talent and experience, but as a whole, the book is rather well balanced. Thereby, it has a strong appeal that will guide you to enjoy and understand such a nanodielectric world.

1.1 Invitation to the Nanodielectric World

Welcome to the nanodielectric world. Nanocomposites are composed of host and guest materials in general. They should be fabricated so that they may be endowed with the superb performance of the guest filler materials, while keeping the original performance of the host materials. In general, inorganic materials exhibit excellent optical, electrical, mechanical and thermal properties, while organic materials are superb in terms of light weight, flexibility, and processability. Such characteristics for both materials can be transferred, complementarily, to nanocomposites. Furthermore, novel performances that neither of them holds by nature will possibly appear in newly fabricated nanocomposites. There are three kinds of combination as for hosts and guests, i.e. inorganic-inorganic, inorganic-organic, and organic-organic composite systems. The materials that attract most attention to-date are nanocomposites that consist of organic polymers as host and inorganic substances as guest. This field was pioneered by the successful invention of polyamide/organic clay nanocomposites in the 1990s. Much attention was directed toward mechanical and optical properties in the beginning. Furthermore, recent investigations have, in addition, been directed to characteristics including gas barrier, lubrication, thermal endurance, heat radiation, electrical conductivity, and electrical insulation, with a view to producing drastic enhancements by controlling material structures in the nanometer scale through self-assembly. The investigation of various polymer nanocomposites has been intensively made concerning permittivity, loss tangent, electrical conductivity, space charge, thermally stimulated current (TSC), dielectric breakdown, treeing breakdown (short time breakdown and treeing *V-t* characteristics), partial discharge (PD) resistance, electroluminescence and many other relevant characteristics. It is now recognized that all the characteristics indicated above reveal the superiority of nanocomposites over conventional composites; the improvement of PD resistance, the prolongation of treeing lifetime, and the suppression of space charge formation are most prominent among these. New application innovation for power apparatus is certainly expected through the use of polymer nanocomposites.

1.2 Interfaces

Interface science and colloid science have dealt with nanometer size substances long before nanotechnology emerged. Gold sols and other nanomaterials with sizes down to 10 nm were studied in the 1920s, while Langmuir–Blodgett thin films consisting of one or more monolayers of organic material have attracted considered interest since the early 1950s. The first measurement of surface forces was made in the early 1950s; the characteristics of interfaces between mutually immiscible substances in colloids were also clarified. Colloid science reached a significant level of maturity during the second half of the 20th century. It is important to note that in colloid science the forces acting between particles and solvent are well understood for both dilute and condensed phases. Various forces act between colloidal particles and include excluded volume repulsion; electrostatic forces; van der Waals' force; entropic effects; steric forces. Excluded volume repulsion, sometimes called Born repulsion, is caused by the impossibility of overlap between hard particles. Electrostatic forces surrounding a colloidal particle were analyzed in detail and clarified by Derjaguin, Landau, Verwey, and Overbeek, resulting in the DLVO theory. There is a deep minimum in the potential (attractive force) near the particle surface, a shallow minimum (attractive force) far from the particle surface, with a high maximum (repulsive force) in between. Entropic effects act to increase the entropy until thermal equilibrium is reached. Steric forces arise between polymer-covered surfaces, or in solutions containing non-adsorbing polymer, producing an additional steric repulsive force (which is predominantly conformational and thereby entropic in origin) or an attractive force due to depletion of counter ions. There are also other interfacial phenomena associated with the interaction between particles and their environment. Physical and chemical adsorption is one of them. The former is related to hydrogen, hydrophobic or van der Waals bonding. The latter is caused by chemical bonding between adsorbate and adsorbent. Surface tension appears on a soft surface, generated by the difference between the free energies of the surface and the bulk. Wetting is the contact of a solid surface by a liquid and includes adhesion, spreading and immersion wetting. Knowledge of interfacial phenomena in colloids has proved extremely useful in developing polymer nanocomposites.

Interfaces in dielectric nanocomposites are now discussed in terms of the following subjects:

(1) Bound polymers
- Silane coupling—bonding between organic and inorganic substances
- Wettability and affinity between inorganic filler and polymer
- Bound polymer involving covalent bonding
- Evidence for bound polymers

(2) Interface models
- Wilkes' model
- Conceptual illustration of interfaces
- Evidence for far-distance interaction
- Charge tails in interface: Stern–Gouy–Chapman counter-charge layer and triboelectricity charge
- Multicore model
- Water shell model

We come to the stage at which we can tailor interfaces. Surprisingly we can control more than two characteristics at the same time, and thereby simultaneously improve both dielectric and thermal properties, for instance. Double core shell and core satellite structures are some of the examples that can be tailored by reversible addition-fragmentation chain-transfer (RAFT) and atom-transfer radical-polymerization (ATRP) techniques. Bimodal polymer brush structures can be formed by click chemistry such as "grafting from" and "grafting to" techniques. Then we can control dispersion and dielectric performances separately. In this way, we can seek to create the second generation of nanocomposites.

1.3 Improvement in Materials Performances

As polymer nanocomposites are expected to appear as advanced dielectric materials, significant investigations have been undertaken in connection with both fundamental characteristics and application specific characteristics. These include (i) electrical and dielectric properties, such as dielectric spectroscopy, dc conductivity at low and high electric field, space charge, dielectric breakdown, treeing resistance, partial discharge resistance, tracking resistance, and flame retardancy, and (ii) closely associated physical properties, such as thermal endurance, thermal conductivity, and glass transition temperatures. Overall, better properties have been obtained for polymer nanocomposites than for their base resins and conventional micro filler filled resins. However, the production of materials with excellent properties requires excellent material processing skills. This involves achieving not only a good dispersion with sound binding between the filler particles and the matrices, but also a freedom from inclusion of unwanted impurities. It is not possible to

Table 1.1 Bird's-eye view of the characteristics of polymer nanocomposites

Properties	Improvement	Remarks
PD resistance	Big enhancement	Nano segmentation of base polymers
Treeing lifetime	Big prolongation	Especially under moderate field
Space charge	Big reduction	Elevated threshold for charge injection
Tracking resistance	Enhancement	Similar to PD resistance
BD strength	Increase or same	Coupling and impurity inclusion effects
Permittivity	Increase or decrease	Decrease in limited conditions
Dielectric loss	Complicated	Complicated structures of interfaces
DC conductivity	Increase or decrease	Roles of ion traps and impurity inclusion

make a comprehensive description of all the data available for polymer nanocomposites. Basic properties have been obtained mainly through dielectric properties and dc conductivity to correlate with interfacial problems that are very important for nanocomposites. From the material application point of view, resistance against partial discharges, treeing lifetime and reduction of space charge formation are outstanding and extremely promising characteristics. Tracking resistance also seems to be positive. There are supportive data available for an increase of dielectric breakdown strength by nanofiller addition. Physical properties such as thermal endurance, thermal conductivity, and glass transition temperatures are used to explain why dielectric and electrical properties change when polymers are nanostructured. Additionally, one kind of polymer nanocomposite such as LDPE/MgO nanocomposite was assessed by cooperative individual groups to demonstrate its feasibility for application.

1.4 Expectation to Computer Simulation

Dielectric Breakdown of Solids and *Theory of Dielectrics* were published by Stanley Whitehead in 1953 and by Herbert Fröhlich in 1958, respectively. To our surprise, such theories include quantum mechanics. Evidently, these theoretical treatments were completed in the middle of the 20th century, and for more than half a century since then, engineering problems have been solved on the basis of such approaches, in connection with practical applications, materials aging, etc. In the 1940s during the Second World War, computer technology began to appear, especially to calculate ballistic trajectories and, since then, has rapidly evolved with the development of semiconductor devices and memory technologies in microelectronics. Moore's law (1965) tells that the number of components per unit area in integrated circuits would double every year and, now, we live in an era in which we cannot survive without computers. In the dielectrics world, on the other hand, computer simulation has only recently emerged, but its role is crucial now.

Although dielectric phenomena have been simulated by a variety of methods, thus far, insufficient reviews and discussions have been made to provide comprehensive understanding. It is now understood from a preliminary survey that simulation works in dielectrics can be represented by the following methods:

(1) Ab initio with density-functional theory
(2) Molecular dynamics (coarse-grained MD)
(3) Monte Carlo simulation
(4) Cellar automata with finite element analysis
(5) Fractal analysis
(6) Stochastic analysis
(7) Direct solution from governing formulae

Computer simulation of the mechanical properties of composites is popular. It includes molecular dynamics simulation, Monte Carlo simulation and ab initio computation. It would also be useful for clarifying dynamic behavior such as the formation of interfaces and dispersion of nanoparticles in polymer matrices. Some phenomena and characteristics have been qualitatively explained, but the technique is still maturing in the context of nanocomposites. Molecular dynamics simulation of thermosetting resins incorporating nanofillers has demonstrated several features. First, the distribution of created bonds is changed by the asymmetrical interaction between monomer and surface, and the affinity of the hardener brings about an increase in the number of near-surface bonds. Second, the segmental motion of reacted resins is not affected by the surface affinity in regions far from the surface, but it changes drastically near the surface. It was suggested that the state of interfaces is correlated with the multicore interfacial model. In an ab initio computational study, the theory of local dielectric permittivity was applied in order to understand variations in the static and optical dielectric constant across idealized Si-SiO_2, Si-HfO_2, and SiO_2-polymer interfaces over interatomic distances. These variations have been correlated with the chemistry of the interfaces, e.g., dangling bonds and multiple oxidation states. The layer-decomposed density of states approach has been used to compute the electronic structure variation across Si-HfO_2 and SiO_2–polymer interfaces and to elucidate band bending, band offsets and the creation of trap states at interfaces. Filler dispersion in a polyamide-imide/silica nanocomposite system has been studied by molecular dynamics simulation; adhesive fracture energy at the interfaces between silica and solvents was calculated and Brownian motion of the silica particles was simulated to clarify their dispersion. It was found that the colloidal state of silica is maintained by covering the silica surface with a new low-hygroscopicity solvent and the chemical structure of the polymer influences the dispersion of the silica. It was claimed that the method is useful when attempting to determine the optimum conditions for good filler dispersion in nanocomposite systems.

These techniques are essentially beneficial to the R&D of nanocomposites. This approach is linked with the exploration of nanocomposites with new possible structures that it is rather difficult to obtain experimentally.

1.5 Applications

Some of the possible applications of nanocomposites lie in motor windings (synthetic enamel), generator windings (epoxy), switchgear spacers (epoxy), capacitors (polypropylene), ac and dc cable insulation (polyethylene and cross-linked polyethylene (XLPE)), silicone elastomers for outdoor use, microelectronics PWB (printed wiring board) and packaging (epoxy). Plausible applications of nanocomposites are under R&D in the electric power field. They seem to offer advantages of higher efficiency, longer service life and reduced size of equipment. At present, energy consumption is increasing around the world and global warming is becoming more serious every year. Energy saving and environmental friendliness are international concerns. Nanocomposites have the potential to help solve these problems. When the use of nanotechnology and nanocomposites is widespread, one might reasonably be concerned about potential nanohazards in work places and in homes. Important as it is, that topic is beyond the scope of this book.

Part 1
Preparation

Chapter 2

Preparation of Nanoparticles

Thomas Andritsch,[a] Matthew Praeger,[a] Alun Vaughan,[a] and Mikimasa Iwata[b]

[a]*ECS, University of Southampton, Southampton, SO22 4HZ, UK*
[b]*Central Research Institute of Electric Power Industry (CRIEPI), 2-6-1 Nagasaka, Yokosuka-shi, Kanagawa-ken 240-0196, Japan*

asv@ecs.soton.ac.uk, m-iwata@criepi.denken.or.jp

2.1 Introduction

Nanoparticles have been produced from many different materials and may take many different forms, including platelets (e.g., nanoclays and graphene), rods (e.g., carbon nanotubes and metal oxide nanofibers) and isometric particles (e.g., colloidal metals and metal nitrides). Although it is often convenient to focus on the dimensions of such structures as being the primary characteristic, in reality, it is likely that many other factors will also play a critical role. In materials science, properties are dependent upon structure and structure is depend upon processing, such that the crystallinity, impurity levels, surface structure, morphology, surface chemistry, amongst others, are all likely to influence the properties of nanoparticulate systems and, hence, the behavior of nanocomposites produced using them. This chapter provides

Tailoring of Nanocomposite Dielectrics: From Fundamentals to Devices and Applications
Edited by Toshikatsu Tanaka and Alun S. Vaughan

ISBN 978-981-4669-80-1 (Hardcover), 978-981-4669-81-8 (eBook)
www.panstanford.com

a brief overview of various nanoparticle synthesis routes, both to illustrate the range of different procedures that can be used and to highlight how processing affects the final products.

2.2 Sol-Gel Synthesis of Nanoparticles

Sol-gel chemistry has a very long history and, according to Cushing et al. [1], can be traced back to the mid-1840s. More recently, the utility of the approach is evinced by the number of papers that have been published on the subject, with a search of the ISI Web of Science database in 2015 yielding more than 200,000 publications. As such, a comprehensive account of this topic is not feasible and, consequently, this section of this chapter will address just the following areas. First, it presents an overview of the fundamentals of the sol-gel process and, then, describes a number of ways in which the basic methodology has been modified to exploit different chemistries as a means of producing different products. Finally, the in situ production of nanocomposites is briefly considered.

2.2.1 The Fundamentals of the Sol-Gel Approach

The account that follows is based upon the reviews by Cushing et al. [1] and Hench and West [2] and, consequently, for more detail, reference to these articles is recommended. In general, the sol-gel process can be divided into two types: the gelation of a suspension of colloidal powders; the hydrolysis and polycondensation of some appropriate precursor compounds, which are commonly alkoxides or nitrates, followed by drying. A common precursor for the production of sol-gel silica (SiO_2) is tetraethyl orthosilicate or tetraethoxysilane (TEOS), which has the chemical formula $Si(OC_2H_5)_4$ and is used to produce silicas in various forms. These include sols, that is, dispersions of colloidal particles typically ranging in size from 1 nm to 1 μm, to highly porous monolithic solids. Although this article is primarily concerned with the synthesis of nanoparticles, it is important to note that the sol-gel process concerns a spectrum of products and that a key requirement for the generation of a porous monolithic mass is that the gel does not collapse on drying, which will consequently lead to a fragmented product.

To illustrate the sol-gel production of nanoparticles, consider the aqueous hydrolysis and condensation reactions of a generic silicon-based alkoxide, which will here be represented $Si(OR)_4$. In the presence of water, the hydrolysis reaction shown below in Fig. 2.1a occurs. Hydrolysis can occur under both acidic and basic condition and is also influenced by the amount of water present. Since silicon is a relatively electronegative atom, this weakens the O-H bonds shown, which leads to condensation. This is influenced by both the pH of the system and by the number of alkoxy groups that are substituted. The product that results from condensation can therefore by dimers (f = 1), polymers (f = 2) or ring structures and network (f > 2). Figure 2.1b shows, for illustration, the idealized case where f = 2, such that linear chains are formed.

(a) $Si(OR)_4 + fH_2O \rightarrow Si(OR)_{4-f}(OH)_f + fROH$

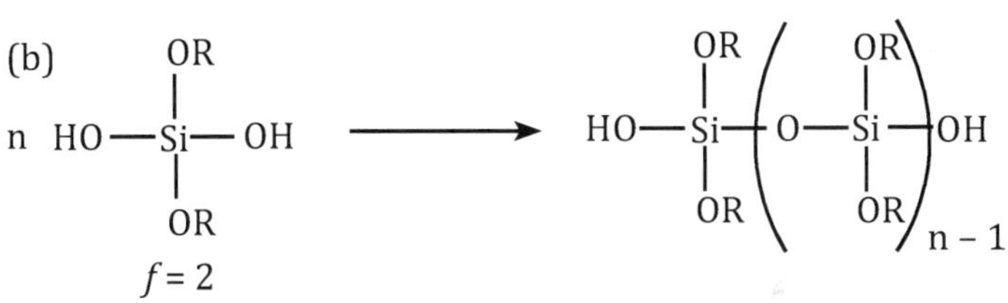

Figure 2.1 Typical reactions that occur in the sol-gel reaction of a silicon alkoxide precursor: (a) hydrolysis; (b) condensation. Reprinted with permission from Cushing, B. L., Kolesnichenko, V. L., and O'Connor, C. J. (2004). Recent advances in the liquid-phase syntheses of inorganic nanoparticles, *Chem. Rev.*, **104**, pp. 3893–3946. Copyright (2004) American Chemical Society.

The relative rates of the above reactions are dependent upon whether they occur under acidic or basic conditions. For example, when the pH < 4, the rate of hydrolysis will occur more rapidly than the rate of condensation. The evolution of sol-gel silica structures as a function of time and pH is shown schematically in Fig. 2.2.

Although the above account has focused on the aqueous synthesis of sol-gel silica, other oxide systems can be synthesized in a comparable way [4]. Also, the use of non-aqueous solvents has been shown to be a flexible alternative, as discussed by Niederberger [5].

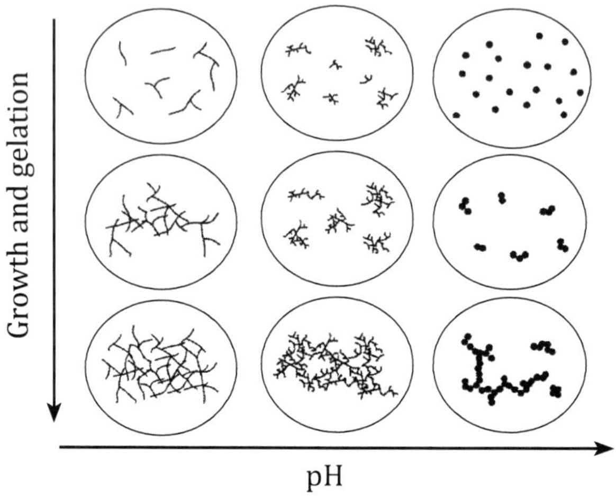

Figure 2.2 Schematic representation of the evolution of sol-gel produced silica structures as a function of time and pH.

2.2.2 Processing of Sol-Gel Materials

The preceding section has focused on the basic chemistry of sol-gel synthesis, as applied to one exemplar system. To supplement factors such as solvent composition and pH, a number of additional parameters can also be used to control the final product.

Although the precise mechanisms are not well understood, it has been stated [6] that microwave irradiation can be used to increase the rate of reaction, reduce reaction times and also improve the quality of the final product (e.g., narrow particle size distribution). In support of this, Panda et al. [7] reported on the synthesis of nanorods and nanowires of samarium (Sm_2O_3), neodymium (Nd_2O_3) and gadolinium (Gd_2O_3) oxides using microwave irradiation and compared the results of this with control samples prepared using direct heating. Transmission electron microscopy (TEM) of the products of both processes revealed that the nanorods produced with microwave irradiation were more uniform; the required reaction time was also significantly reduced. Another approach that has been used to enhance the production of nanostructured materials is through the application of high intensity ultrasound, a topic that is considered by Bang and Suslick in their review [8]. Within the context of this chapter, the most significant modus operandi of sonication is, however, through nebulization; that is, the creation of fine droplets. This use of ultrasound therefore relates to nanoparticle synthesis in a confined geometry and, as such, is addressed below along with other related techniques.

Post-production modification of sol-gel nanoparticles is commonly accomplished by an appropriate thermal treatment. While heating is likely to increase crystallinity, it can also lead to undesirable agglomeration if the chosen temperature is too high. Wu et al. [9] used a nitrate precursor to prepare a zirconia aerogel, consisting of small, highly porous, high surface area particles, and studied the effect of subsequent annealing (calcination) in the temperature range 500–1000 °C. A combination of TEM and wide angle X-ray diffraction (XRD) were subsequently used to study the effect of the thermal treatment on the zirconia structure. Both the specific surface area and pore volume were found to decrease as the annealing temperature was increased but, even after being treated at 1000 °C, the primary particle size was still found to be no larger than 30 nm. Elsewhere, the effect of synthesis conditions and subsequent calcination temperature on the structure and properties of zinc oxide nanoparticles has been described, in which a linear relationship between the calcination temperature and particle size was reported. In this, the average particle size increased from 20 nm to 41 nm for calcining temperature of 400 and 600 °C, respectively [10].

2.2.3 Non-Oxide Nanoparticles

All the exemplars considered above have concerned oxide systems, but examination of the published literature indicates that sol-gel approaches have also been used to synthesis fluorides, where a number of review articles exist [11, 12]. For example, Fujihara et al. [13] described variants on the process described above in which silica/lanthanum trifluoride (LaF_3) composites were prepared using tetramethylorthosilicate, combined with trifluoroacetic acid (TFA) and lanthanum acetate as chemical sources for LaF_3. This study examined the effect of both the initial composition and post-processing on the final products. Other fluorine-containing compounds that have been generated by the sol-gel method include nanosized magnesium fluoride and amorphous high surface area aluminum trifluoride [11]. All of these various compounds were synthesized using particular chemical routes and, consequently, reference to the primary sources is required to appreciate the details.

In addition to fluorides, carbides and nitrides can also be synthesized by sol-gel routes [14]. An example of this is the hydrazide sol-gel process reported by Kim and Kumta [15] for the synthesis of nanostructured titanium nitride. Here, a titanium alkoxy hydrazide precursor was generated by reacting titanium isopropoxide with anhydrous hydrazine in the presence of anhydrous acetonitrile. Characterization, after heat treatment at 1200 °C in an inert atmosphere, revealed the production of titanium nitride nanoparticles 10–20 nm in size.

2.2.4 In situ Sol-Gel Synthesis

A major issue with the reliable production of nanocomposites concerns aggregation and agglomeration of nanoparticles. Consequently, the production of nanoparticles within a matrix, under conditions where they are unable to form larger structures, is highly attractive. Sol-gel chemistry is compatible with various polymers, which facilitates the formation of inorganic structures in the presence of an organic host [16, 17]. Consequently, two different strategies can be devised, whereby inorganic nanostructures either form in the presence of pre-existing polymer or where nanoparticle formation occurs in parallel with matrix polymerization.

As discussed above, sol-gel, reactions can occur in both aqueous and non-aqueous media but, to illustrate the principles, first consider the former. The in situ production of silica from a tetra-alkoxysilane precursor involves the formation of silanol groups and, consequently, synthesis in the presence of a polymer that can undergo hydrogen bonding will lead to strong interactions between the evolving nanosilica and the polymer matrix. A polymer that exhibits the required characteristics is poly(ethylene oxide) (PEO) and, consequently, the behavior of such systems has been widely studied in connection with, for example, the influence of the polymer on the development of the silica phase [18, 19].

A key concept in the in situ sol-gel production of nanoparticles is compatibility between the polymer and the evolving silica system, such that a homogenous product results. In non-aqueous media, this can be facilitated by the use of solvents with an amphiphilic character, through direct covalent bonding between the organic and inorganic components or by the inclusion of appropriate coupling agents. *N,N*-dimethylformamide (DMF) and *N,N*-dimethylacetamide (DMAc) are solvents that can hydrogen bond with silanol groups and which are compatible with many organic polymers; examples where this strategy has been employed include organic–inorganic hybrids based upon a range of polymers, including polystyrene (PS) and poly(vinyl chloride) (PVC) [20]. A widely exploited strategy for increasing favorable interactions between nanofillers and the host matrix, and hence improving dispersion, is through control of surface chemistry. Conducting sol-gel synthesis of inorganic nanoparticles in the presence of polymers that include appropriately reactive moieties within their structure constitutes a means of controlling the structure of the inorganic phase during its evolution. An approach that has exploited this involves the copolymerization of the monomer of interest with a monomer such as 3-(methacryloxypropyl)-trimethoxysilane and examples of polymers that have been modified in this way include PS, [21] and poly(methyl methacrylate) (PMMA) [22]. Alternatively, the required functionality can be incorporated into the inorganic phase, through the inclusion of an appropriate coupling agent [23].

The parallel synthesis of sol-gel nanoparticles and polymer necessarily requires that different chemical reactions occur simultaneously. Consequently, for this to be viable, first, the two reactions need to be non-interacting and, second, as the reactions proceed, macroscopic phase separation should not occur. From the perspective of electrical insulation, epoxy-based materials are of considerable practical interest and, in epoxy/silica systems, a number of variables can be used to control the parallel production of sol-gel silica and the curing of the polymer. For example, amine hardeners act both as epoxy crosslinking agents and as catalysts in the sol-gel process [24], while the addition of a suitably functionalized silane can help to control phase separation. Evidently, such strategies mean that the formation of the silica and the curing of the polymer are, however, coupled processes [25]. Elsewhere, similar methodologies have also been reported in connection with polyacrylates [26] and materials based upon cyclic alkenyl monomers [27]. For a general overview of this topic, see the review by Pomogailo [28].

2.3 Nanoparticle Synthesis within Confined Geometries

Mini-emulsions utilize traditional synthetic methods, like sol-gel chemistry, but within confined geometries in the form of solvent droplets. These are therefore heterophase systems, containing droplets with diameters of about 100 nm, which provide a unique reaction environment. They are not thermodynamically stable, but stabilized against droplet collisions and diffusional degradation. Mixing of two micro-emulsions leads to immediate exchange of the components, while larger mini-emulsions need stimuli in the form of, for example, ultrasound to initiate the reaction.

There are a number of reasons why mini-emulsion techniques are gaining attraction. Due to the synthesis being restricted to the amount of material in the immediate vicinity, the size of the resulting particles is restricted. At the same time, multiple reactions can occur within the same droplet, allowing for the creation of mixed crystals and virtually unlimited potential for hybrid structures. These generated structures can then be stabilized with surfactants, in order to prevent subsequent large scale agglomeration [29]. Mini-emulsions can be generated in a variety of different solvents, allowing for reaction temperatures up to approximately 200 °C.

Mini-emulsions are usually generated with ultrasonication on the lab scale. For industrial use, high-pressure homogenization can be used. A number of different types of emulsions can be generated depending on the liquids and surfactants used. The direct route results in oil-in-water emulsions, which can also be inverted to achieve water-in-oil emulsions. By substituting water with, for example, a polar, organic solvent, non-aqueous mini-emulsions can be generated [30].

A number of review articles exist that address elements of this approach, including ex-situ prepared composites [31, 32], inorganic particles in mini-emulsion [33, 34] to metal oxides in supercritical fluids [35].

2.3.1 Preparation of Inorganic Nanoparticles in Mini-Emulsion

Porous oxides are one of the most conveniently produced materials by sol-gel mini-emulsions. Sol-gel processes are generally conducted at temperatures below 100 °C, starting from hydrolyzable metal compounds or salts, as described above. Surfactants can be used to control the size of micelles in the framework [36, 37]. Inverse mini-emulsions are typically used for direct preparation of nanoparticles, which tend to be based on aqueous solutions and water-soluble precursors, dispersed in an inert hydrocarbon. Such solutions can range from glycolates of titanium or silicon to hydrophobic alkoxides or ionic precursors. The use of titanium or silicon glycolate can be beneficial due to the lyotropic templating with the surfactant happens in one phase [38]. This technique allows for porous nanoparticles of silica [39], ceria [40], titania [41] and others. For insulating dielectrics, this approach poses some issues, since porous particles can result in a degradation of dielectric performance [42]. But they can be created with properties that make them suitable for use in smart materials. Ang et al. [43] generated paramagnetic maghemite nanoparticles and silica by dispersing the maghemite nanoparticles into the silica xerogel. Field emission gun scanning electron microscopy (FEGSEM) revealed that the silica gel formed a porous network structure with an average pore size of 15 nm, while energy dispersive X-ray spectroscopy (EDS) indicated that the iron oxide particles were embedded in this silica structure along with maghemite nanoparticles, quoted as being 5 and 4.4 nm in size on average, respectively.

In order to prepare mesoporous silica nanoparticles, the energy input during homogenization of a mini-emulsion is vital. Schiller et al. [39] showed one of the earliest implementations of lyotropic templating of silica in mini-emulsions and demonstrated that the duration and intensity of the ultrasonication, as well as the temperature of the system during homogenization, affect the nanostructure. An increase in ultrasonic intensity prevents porosity, resulting in non-porous silica structures instead [39]. As discussed above, another factor is the pH, which determines the balance between hydrolysis of the silica precursor and condensation. Since hydrolysis is catalyzed by acids or bases, the reaction rate is the lowest at a neutral pH. Longer reaction times on the other hand yield porous, flake-like material [44]. This shows that sol-gel reactions can easily be performed within mini-emulsions, albeit that the usability for dielectric insulation currently appears limited.

Titania is a crystalline material that can occur in several forms: anatase, brookite, or rutile. It has found application as the anode material in lithium-ion batteries, mostly in the anatase form, due to its crystalline structure and band gap. Here, key factors to consider are specific surface area and crystallite size. Smaller crystals offer improved charging behavior, due to short diffusion times, while increased specific surface area improves contact with the electrolyte in the battery. This can be achieved by an inverse mini-emulsion, as shown by Rossmanith et al. [41]. With reaction temperatures above 100 °C, single-phase anatase can be obtained with a crystalline size estimated to be 5 nm, which is to be expected since anatase crystals below 14 nm are considered thermodynamically stable [45]. Kubiak et al. established that anatase with a specific surface area of 150 m^2 g^{-1}, crystallite size of 10 nm and a pore size distribution between 7 and 15 nm, has a capacity retention of 90% after 20 charge/discharge cycles at up to 3 V. Lithium insertion indicated charge capability of 88 mAh g^{-1} [46].

Another way of forming nanoparticles is by precipitation within each of the droplets, which can be induced by factors such as temperature stimulus, solvent evaporation, or by applying a co-homogenization procedure using a second mini-emulsion in concert [47]. Inverse mini-emulsion techniques can also be used to obtain inorganic crystalline films of red, green or blue phosphor [48]. These particles show high stability with amphiphilic block copolymers used as non-ionic surfactants. In the case that a system does not dissolve or melt easily, a multi-stage approach is necessary, whereby precursors transform the compound particle-by-particle [47]. The preparation of silver nanoparticles from silver nitrate was used as a model system, in order to show the feasibility of using polymer-enclosed nanoreactors [49]. First, silver nitrate is dissolved in the dispersed phase, with subsequent polymerization of the capsule. Hydrazine, which can diffuse through the capsule walls, is then used to reduce the metal salt.

Li et al. [50] reported on the use of oil-in-water micro-emulsions for the synthesis of various types of inorganic nanoparticles. The method, termed "oil-water interface controlled reaction", used inorganic salts as precursors. Unlike regular oil-in-water emulsion methods, in which the metallic precursors and the precipitating agent are confined inside the micro-droplets, in this case, they are dissolved into the continuous phase. After the solution forms and the metal cations have been added, they are adsorbed at the interface between the water and the oil. The ions prefer to be located at said interfaces and are, in turn, stabilized by both water and surfactant molecules. Once the precipitating agent is introduced, the balance is destroyed and reaction occurs within seconds [50]. The anions can then react with the cations close to the interface, or with cations from the interfaces of two colliding nanodroplets. A continuous and rapid "reaction transfer" was proposed, after which hexane was added to induce phase separation. The nanoparticles remain in the oil phase, where they can be washed and dispersed in a non-polar solvent. Alternatively, thermal treatment can be used to increase the crystallinity. With this method, small particles ranging from 2 to 13 nm could be produced from a range of metals (Ag, Cu), semiconductors, fluorides, phosphates and chromates. Subsequent research resulted in nanoparticles, nanocubes, nanorods, and multipods [51–53]. The main strategy for controlling particle shape, size and type was hereby the concentration variation of ethanol and cyclohexane, as well as the alkalinity. In the work of Liu et al. [52], hexane was omitted and the co-solvent altered (ethanol or DMF), which resulted in other particle sizes, without the need for additional thermal treatment.

2.3.2 Polymers Prepared in Mini-Emulsion

Inspired by natural growth processes, material scientists have applied lessons learned from biominerals in order to use polymers to assist the formation of inorganic crystals. Synthesized like examples of biological origin, these polymers are used as additives to control nucleation and growth processes. Mini-emulsions can be used in order to prepare functionalized polymer particles with a narrow size distribution. However, challenges nevertheless exist when trying to utilize mini-emulsion polymerization on an industrial scale [54], which explains why even 40 years after the work of Ugelstad et al. [55] demonstrated the viability of such approaches, their use in the preparation of commercial products is still scarce. In general, there are two approaches to polymer synthesis in mini-emulsion: polymers as additives in the mini-emulsion or they are used as templates. What follows is just a glance at what is possible; much more in-depth accounts exist elsewhere [56, 57].

When using polymers as additives, mini-emulsion latexes have attracted significant attention and have been used as controlling agents in the crystallization of different systems since the early 2000s [58]. They can be used with a multitude of different systems, with zinc oxide often being used as a model system, since it crystallizes in aqueous media in a single hexagonal phase. These crystals reach micrometric dimensions and incorporation of polymeric particle can be easily observed [59]. The surface chemistry of the latex particles has hereby shown a strong effect on the morphology of the resulting crystals [60]. However, the inclusion of latex can result in point defects, after the polymer was dissolved or removed via thermolysis, resulting in a "Swiss cheese" morphology [61].

If the process is inversed, by moving away from nanoscaled inorganic materials, inorganic nanoparticles will grow on the surface of the polymer beads; the polymers thereby act as templates for inorganic crystallization. Most of this work has not been conducted using mini-emulsions, rather, alternative emulsion techniques have been employed [62]. When considering mini-emulsions, the formation of calcium phosphate has been thoroughly explored [63], but other nanoparticles can be obtained in a "one-pot" reaction of styrene with a suitable comonomer [64]. When it comes to surfactants, both ionic and non-ionic types have their drawbacks. Ionic surfactants introduce charged groups onto the particle surfaces; these make it difficult to determine the amount of functional groups on the surface and negatively impact the space charge behavior and the complex permittivity of insulating dielectrics. Non-ionic surfactants can also have a negative effect if their concentration is too high [65], with pH and surface chemistry playing a vital role in crystal formation. However, the use of ionic as well as non-ionic surfactants can be avoided outright by the use of surface-active co-monomers, which are able to act as surfactants as well as a monomers. This versatile approach only introduces the desired functional moiety [66]. Apart from serving as templates, latex particles can also act as scaffolds that confine the crystallization process [67].

In the examples cited so far, polymer templates were created in a first step, with subsequent formation of the inorganic compound on the pre-existing polymeric skeleton. However, as Lu et al. [68] showed, with asymmetric PS/SiO_2 nanocomposites, it is possible to achieve the entire process in one step. The same laboratory showed later the formation of PS/TiO_2 hybrids, with polymerization and the formation of the inorganic particles via the sol-gel route happening in the same step [69].

2.3.3 Hybrid Nanoparticle Precursors

This approach depends on oxidation or reduction of the organic component of polymer/inorganic hybrid nanostructures, in order to form inorganic nanoparticles. As an example, metal nanoparticles can be obtained from mini-emulsions by removing the organic component via plasma processing, before reduction of the remaining particles [70]. The resulting particles are arranged in highly regular patterns on solid substrates. The final size of the inorganic features seems to depend on the initial metal loading of the hybrid particles, while their interparticle distance is determined by the dimensions of the initial colloids [71].

Inverse mini-emulsions use metal salts; loading levels >50 wt.% are achievable, together with a hydrophilic polymer. Such loading levels are much higher than achievable by encapsulation of hydrophilic metal precursors in hydrophobic polymers like PS, using the direct mini-emulsion process as described above [72]. However, one drawback of inverse mini-emulsions compared with the direct approach is the wider size distribution of the resulting particles. The inverse approach tends to produce particles 100 to 150 nm in size; larger particles are possible, but then, the size distribution will be broader [73].

To deposit the particles on a substrate, several techniques can be applied [74, 75], resulting in controlled, single-crystalline domains several hundred μm^2 in size. Subsequent plasma assisted reduction is performed, during which the plasma parameters need to be tightly controlled in order to prevent loss of symmetry by thermal effects, uncontrolled size reduction or sputtering effects [76].

2.3.4 Microfluidic Synthesis

Another approach that exploits geometrical effects is the production of sol-gel nanoparticles within microfluidic devices. In one approach, highly crystalline iron oxide–based nanoparticles with a narrow size distribution (mean diameter 3.6 nm; standard deviation 0.8 nm) were synthesized within a capillary-based droplet reactor [77]. In this work, the required reactants in aqueous solution were injected at a controlled rate, into flowing octadecene, which constituted an immiscible carrier. The inclusion of dextran within the reactant solution formulation, which co-ordinates to the surface of the particles, stabilizes them within the nanometer size range. Elsewhere [78], the synthesis of gold nanoparticles with average sizes ranging from 19 nm to 58 nm within a microfluidic reaction chip was described. In addition to offering the ability to tailor nanoparticle dimensions, the approach also resulted in a dramatic (83%) reduction in the synthesis duration. For more details concerning the synthesis of nanostructures within microfluidic systems, reference to the review of Marrea and Jensen is [79] recommended.

2.4 Arc Plasma Methods for Aluminum Nitride Nanoparticles

Downsizing of electrical equipment has meant that such equipment requires a high thermal conductivity. In order to develop polymer insulating materials with high thermal conductivities, it is necessary to add large quantities of ceramic powder of high thermal conductivity to organic insulating materials and mixing small-grained particles with large-grained particles is considered to be a good way of achieving a high ceramic particle content.

Aluminum nitride (AlN) is a ceramic with a high thermal conductivity. Commercially available AlN particles have a size of a few microns and, therefore, AlN nanoparticles are required to complement these. However, the low production capacity of nanoparticles makes it extremely expensive. Although there are many methods of producing nanoparticles, methods that utilize arc plasma are a good choice in terms of raising the production rate because they allow plasma power to be upgraded with ease. There are many reports on the synthesis of AlN nanoparticles using an arc plasma [80–87]. The production rate will increase if the raw materials are injected into a high-power arc plasma continuously in order to produce much vapor from the raw materials for synthesizing nanoparticles. The transferred-type arc plasma has a long high-temperature zone and the advantage of making it easy to upgrade plasma power. Consequently, this approach has the potential to reduce production costs [88]. This section of this chapter describes the synthesis of AlN nanoparticles from aluminum (Al) particles using transferred-type arc plasmas [89].

2.4.1 Experimental Setup and Conditions

Figure 2.3 shows the experimental setup. A transferred-type arc plasma was generated between the tungsten cathode of a plasma torch and a graphite anode in the chamber. The distance between the two electrodes was 75 mm. The cathode consisted of a tube-shaped electrode, the anode was a rod-shaped electrode and the two electrodes were water-cooled. Al particles were injected into the arc plasma through the cathode using a plasma gas (N_2). The size (diameter) of the Al particles was 20–25 μm, with a feed rate of 0.5 g min^{-1} and a feed time of 1–2 min. The dc arc current was 140 A, and the arc voltage was 100–150 V. A suction pipe was positioned downstream from the anode. The distance between the anode and the tip of the suction pipe was 75 mm. Ammonia (NH_3) gas was blown into the suction pipe as a reacting/quenching gas. The Al particles melted and evaporated in the arc plasma and, then, the Al gas reacted with and was quenched by the NH_3 gas. The off-gas was exhausted through the suction pipe, a tank and a filter, using a vacuum pump. The tank contained ethanol. The synthesized particles were collected in the filter. The composition of the synthesized particles was analyzed by X-ray diffraction

(XRD). The diameter and shape of the powder was analyzed by transmission electron microscopic (TEM).

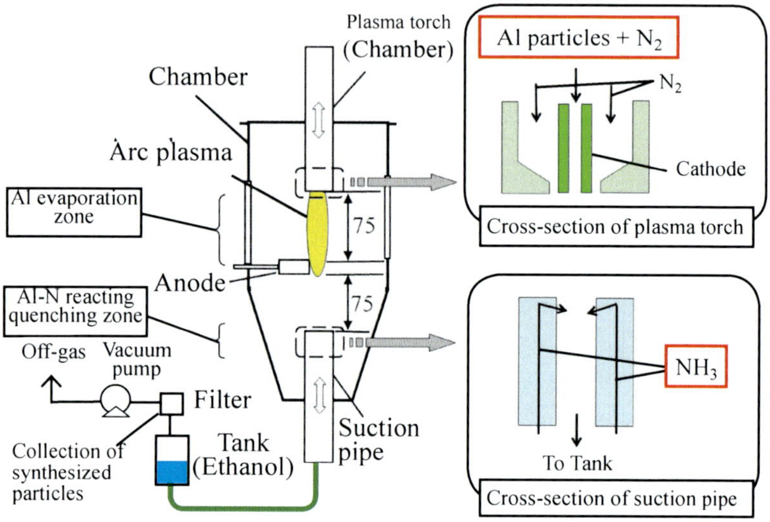

Figure 2.3 Experimental setup. Based upon [89]. From Iwata, M., Adachi, K., Furukawa, S., and Amakawa, T. (2004). Synthesis of purified AlN nano powder by transferred type arc plasma, *J. Phys. D: Appl. Phys.*, **37**, pp. 1041–1047. © IOP Publishing. Reproduced with permission. All rights reserved.

2.4.2 Experimental Results

Figure 2.4 shows an XRD profile of the particles collected in the filter. Only AlN peaks were observed; no Al peak was present. Figure 2.5 shows a TEM micrograph of the synthesized particles collected in the filter. Many particles of 10–30 nm were observed. As the flow rate of the reacting/quenching gas increased, the average diameter of the synthesized particles decreased. A reason for this decrease in diameter was considered to be that the temperature of the particle synthesis space decreased [90, 91] and the AlN vapor density and time spent for growing nanoparticles in the space decreased [91].

By using a similar technique to that described above, spherical nanostructured composite particles of AlN were synthesized [92]. In this method, the particles injected into the arc plasma were crushed, commercial AlN particles. The injected particles melted and vaporized partially in the arc plasma. The synthesized particles consisted of small-grained particles (nanoparticles) dispersed on the surface of large-grained particles (micro-particles). The shape of the micro-particles was approximately spherical. The AlN composite particles were mixed with epoxy resin, and nanoparticles were dispersed in the epoxy resin. As a result, the electric breakdown strength of the epoxy resin with the composite particles was higher than that of the epoxy resin containing commercial, crushed particles [93]. It was considered that the spherical particles of the composite particles restrained the concentration of the electric field at the edge of particles in the epoxy resin.

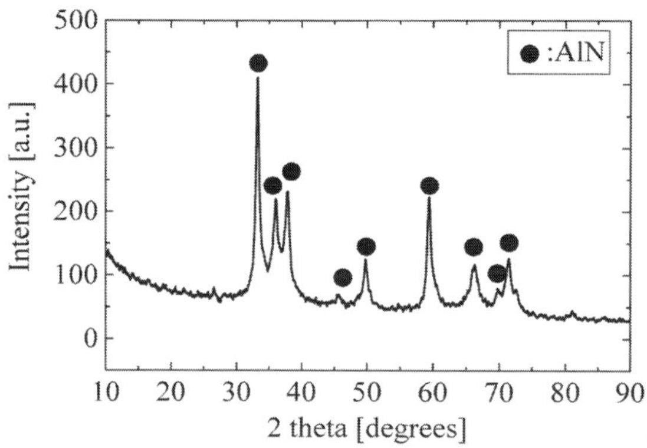

Figure 2.4 XRD profile of the synthesized AlN particles. Based upon [89].

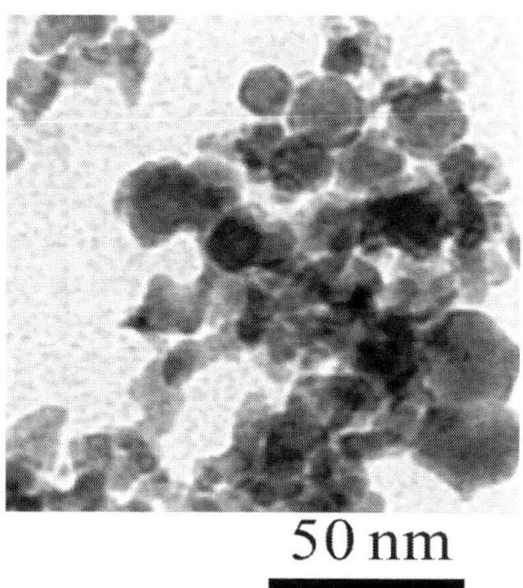

Figure 2.5 TEM micrograph of the synthesized AlN particles.

2.5 Production of Nanoparticles by Flame Hydrolysis

In flame hydrolysis, precursor materials are burned in a high temperature, hydrogen-oxygen flame; hydrolysis reactions occur that consolidate molecules of the precursor into small particles in the resulting soot. This method can be used as a means of producing nanoparticles [94] (original patent filed in Germany by Kloepfer [95]) or, if the material is collected and annealed on a surface, it can be used as a means of depositing layers (termed flame hydrolysis deposition FHD [96] or vapor-phase axial deposition VAD [97]). Nanoparticles produced by flame hydrolysis may be labeled as fumed or pyrogenic particles. It should be noted, however, that the term pyrogenic is a general one and encompasses any production method that employs a high-temperature flame or arc (it would apply equally to particles produced by the arc plasma method described in Section 2.4). Since the method of production may affect the properties of the nanoparticles, it is important to establish this piece of information and to consider its impact (particularly in relation to the nanoparticle surface chemistry) when interpreting results. The following section will offer fumed silica (one of the most widely used nanofillers) as an example of nanoparticle production by flame hydrolysis.

2.5.1 Flame Hydrolysis (Fumed Silica)

The precursor for production of fumed silica is silicon tetrachloride ($SiCl_4$). The boiling point of this material is approximately 57.6 °C and so, at room temperature, it is a volatile liquid. For flame hydrolysis, a stream of $SiCl_4$ vapor is typically produced using a bubbler flask. In this device, liquid $SiCl_4$ is heated slightly to increase its vapor pressure and a carrier gas (i.e., dry nitrogen, which is ignored in the hydrolysis reactions) is blown through the liquid. The $SiCl_4$ vapor is then burned in a hydrogen oxygen flame to produce the fumed silica soot (see Fig. 2.6). The primary reactions taking place in the flame (shown below) are similar in form to those of the sol-gel method described in Section 2.2.1; however, rather than being mediated by a liquid solvent, the reactions take place within a gaseous flame. In the hydrogen-oxygen flame:

$$2H_2 + O_2 \rightarrow H_2O$$

such that hydrolysis of the $SiCl_4$ occurs:

$$SiCl_4 + 2H_2O \rightarrow SiO_2 + 4HCl$$

The gaseous hydrogen chloride (HCl) byproduct can be dissolved into water for treatment by passing the gas extract through a scrubber consisting of numerous baffles and sprayed water curtains.

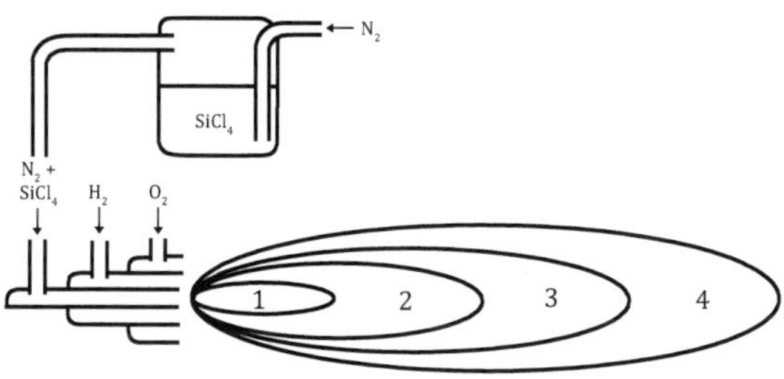

Flame zone	Zone 1	Zone 2	Zone 3	Zone 4
Temperature	$T > 2950$ °C	$T < 2950$ °C	$T > 1700$ °C	$T < 1700$ °C
SiO_2 phase	Vapour	Liquid (low viscosity)	Liquid (high viscosity)	Solid
Particles produced		Spherical droplets	Fused clusters	Loosely bound clumps

Figure 2.6 Diagram of flame zones and table of properties in each zone.

The resultant silica may be generated in different phases, depending upon the local temperature within the flame. The highest temperature expected within a hydrogen-oxygen flame is of the order of 3200 °C, but this can be varied to some extent via the mass flow rate of the precursor gasses and the carrier gas; in general the peak temperature will be somewhat lower than this extreme value. The flame temperature varies both radially and with distance from the torch (see Fig. 2.6). A region exists within the flame (zone 2), where the local temperature falls below the boiling point of SiO_2 (~2950 °C) but exceeds its melting point (~1700 °C). In this zone, the SiO_2 molecules produced by the hydrolysis reaction condense and, because they are in a molten state with low viscosity, subsequent collisions lead to the production of spherical nanoscale droplets. Downstream in the flame, the temperature is reduced and the viscosity of the SiO_2 droplets increases rapidly. In zone 3 the surface tension of the SiO_2 is unable to overcome its viscosity and so collisions here result in the production of granular clusters of quasi-spherical particles, which are permanently fused together. In still cooler regions of the flame (zone 4, below the melting point of SiO_2) particles and/or clusters of particles may become entangled to form loosely bound agglomerates.

2.5.2 Fumed Silica: Surface Properties

As a result of being created in an environment that is saturated with hydrogen, oxygen, water and associated moieties, the surface of fumed silica is in a hydroxylated condition [98]. That is, a large number of silicon atoms on the surface of the nanoparticles will be terminated with hydroxyl groups (OH). If surface treatment is being undertaken, for example, to generate covalent bonds between nanoparticles and a matrix in order to improve mechanical performance, then this can be beneficial. Many functionalizing agents (i.e., those based on silane) are designed to react chemically with and bond to these surface hydroxyl groups. However, on untreated or partially treated fumed silica, surface hydroxyls have the effect of making the particles hydrophilic. The OH groups act as sites for the adsorption of water molecules and promote absorption of water into composite materials. Electrically, this can be highly undesirable, as water content is now recognized as one of the most critical parameters in determining the electrical properties of nanoparticles and nanocomposites.

2.5.3 Calcination of Fumed Silica

Surface hydroxyl groups can be removed from fumed silica nanoparticles by calcination, which changes their surface character from hydrophilic to hydrophobic. The calcination process involves holding the nanoparticles at high temperatures >1000 °C for several hours under an inert atmosphere such as dry nitrogen or under vacuum. The various surface groups present on fumed silica require different temperatures for removal [99, 100]. Surface adsorbed water will be completely removed by 200 °C, whilst vicinal groups will have condensed by about 500 °C and may form strained siloxane groups. Heat treatment up to ~500 °C produces a reversible dihydroxylation and upon exposure to atmospheric humidity the surface will rapidly rehydroxylate. Elevating the temperature to >1000 °C allows the strained siloxane groups to relax, forming stable silanol bridges. This creates a permanent dehydroxylation and leads to hydrophobic properties being observed. When incorporated into a nanocomposite, specimens filled with calcined fumed silica exhibit markedly reduced water uptake than those prepared with untreated fumed silica. Whilst addition of conventional fumed silica often tends to decrease electrical breakdown strength (especially in a humid environment), enhanced electrical breakdown strength (relative to an unfilled polymer control sample) has been demonstrated with calcined fumed silica nanocomposites [101].

2.5.4 Flame Hydrolysis with Variable Composition

Flame hydrolysis is possible with a range of halide precursors other than silicon tetrachloride ($SiCl_4$) and, for example, germanium tetrachloride ($GeCl_4$), phosphorus trichloride (PCl_3) and boron trichloride (BCl_3) can all be used [94]. By incorporating additional bubbler flasks it is possible to introduce mixtures of precursors into the flame. Similar techniques have been applied in the fabrication of optical fiber preforms, where adjusting the precursor flow rates produces phosphosilicate or borophosphosilicate glasses with variable composition for refractive index control [102]. By controlling the deposition conditions, it is even possible to affect the morphology of the resulting material, determining whether it is amorphous or crystalline [103]. It is becoming increasingly clear that the surface chemistry of nanofillers plays a crucial role in determining the electrical properties of nanodielectrics; therefore, direct control of nanoparticle composition offers an intriguing alternative to surface functionalization.

References

1. Cushing, B. L., Kolesnichenko, V. L., and O'Connor, C. J. (2004). Recent advances in the liquid-phase syntheses of inorganic nanoparticles, *Chem. Rev.*, **104**, pp. 3893–3946.

2. Hench, L. L., and West, J. K. (1990). The sol-gel process, *Chem. Rev.*, **90**, pp. 33–72.
3. Gallagher, D., and Ring, T. A. (1989). Sol-gel processing of ceramic films, *Chimia*, **43**, pp. 298–304.
4. Gupta, S. M., and Tripathi, M. (2012). A review on the synthesis of TiO_2 nanoparticles by solution route, *Cent. Eur. J. Chem.*, **10**, pp. 279–294.
5. Niederberger, M. (2007). Nonaqueous sol-gel routes to metal oxide nanoparticles, *Acc. Chem. Res.*, **40**, pp. 793–800.
6. Li, W., and Lee, J. (2008). Microwave-assisted sol-gel synthesis and photoluminescence characterization of $LaPO_4:Eu^{3+}$, Li^+ nanophosphors, *J. Phys. Chem. C*, **112**, pp. 11679–11684.
7. Panda, A. B., Glaspell, G., and El-Shall, M. S. (2007). Microwave synthesis and optical properties of uniform nanorods and nanoplates of rare earth oxides, *J. Phys. Chem. C*, **111**, pp. 1861–1864.
8. Bang, J. H., and Suslick, K. S. (2010). Applications of ultrasound to the synthesis of nanostructured materials, *Adv. Mater.*, **22**, pp. 1039–1059.
9. Wu, Z. G., Zhao, Y. X., Xu, L. P., Liu, D. S. (2003). Preparation of zirconia aerogel by heating of alcohol–aqueous salt solution, *J. Non-Crystal. Solids*, **330**, pp. 274–277.
10. Ba-Abbad, M. M., Kadhum, A. A. H., Mohamad, A. B., Takriff, M. S., and Sopian, K. (2013). The effect of process parameters on the size of ZnO nanoparticles synthesized via the sol-gel technique, *J. Alloys Compounds*, **550**, pp. 63–70.
11. Fedorov, P. P., Luginina, A. A., Kuznetsov, S. V., and Osiko, V. V. (2011). Nanofluorides, *J. Fluorine Chem.*, **132**, pp. 1012–1039.
12. Rüdiger, S., Groß, U., and Kemnitz, E. (2007). Non-aqueous sol-gel synthesis of nano-structured metal fluorides, *J. Fluorine Chem.*, **128**, pp. 353–368.
13. Fujihara, S., Kato, T., and Kimura, T. (2000). Influence of solution composition on the formation of SiO_2/LaF_3 composites in the sol-gel process, *J. Mater. Sci.*, **35**, pp. 2763–2767.
14. Giordano, C., and Antonietti, M. (2011). Synthesis of crystalline metal nitride and metal carbide nanostructures by sol-gel chemistry, *Nano Today*, **6**, pp. 366–380.
15. Kim, I.-S., and Kumta, P. N. (2003). Hydrazide sol-gel synthesis of nanostructured titanium nitride: precursor chemistry and phase evolution, *J. Mater. Chem.*, **13**, pp. 2028–2035.

16. Schottner, G. (2001). Hybrid sol-gel-derived polymers: applications of multifunctional materials, *Chem. Mater.*, **13**, pp. 3422–3435.

17. Kickelbick, G. (2002). Concepts for the incorporation of inorganic building blocks into organic polymers on a nanoscale, *Prog. Polym. Sci.*, **28**, pp. 83–114.

18. Vong, M. S. W., Bazin, N., and Sermon, P. A. (1997). Chemical modification of silica gels, *J. Sol-Gel Sci. Technol.*, **8**, pp. 499–505.

19. Takahashi, R., Nakanishi, K., and Soga, N. (2000). Aggregation behavior of alkoxide-derived silica in sol-gel process in presence of poly(ethylene oxide), *J. Sol-Gel Sci. Technol.*, **17**, pp. 7–18.

20. Ogoshi, T., and Chujo, Y. (2003). Synthesis of organic–inorganic polymer hybrids utilizing amphiphilic solvent as a compatibilizer, *Bull. Chem. Soc. Jpn.*, **76**, pp. 1865–1871.

21. Wei, Y., Yang, D. C., Tang, L. G., and Hutchins, M. G. K. (1993). Synthesis, characterization, and properties of new polystyrene-SiO_2 hybrid sol-gel materials, *J. Mater. Res.*, **8**, pp. 1143–1152.

22. Wei, Y., Jin, D., Brennan, D. J., Rivera, D. N., Zhuang, Q., DiNardo, N. J., and Qiu, K. (1998). Atomic force microscopy study of organic-inorganic hybrid materials, *Chem. Mater.*, **10**, pp. 769–772.

23. Bonilla, G., Martinez, M., Mendoza, A. M., and Widmaier, J.-M. (2006). Ternary interpenetrating networks of polyurethane-poly(methyl methacrylate)-silica: preparation by the sol-gel process and characterization of films, *Eur. Polym. J.*, **42**, pp. 2977–2986.

24. Matejka, L., Plestil, J., and Dusek, K. (1998). Structure evolution in epoxy–silica hybrids: sol-gel process, *J. Non-Cryst. Solids*, **226**, pp. 114–121.

25. Davis, S. R., Brough, A. R., and Atkinson, A. (2003). Formation of silica/epoxy hybrid network polymers, *J. Non-Crystal. Solids*, **315**, pp. 197–205.

26. Wei, Y., Jin, D., Yang, C., and Wei, G. (1996). A fast convenient method to prepare hybrid sol-gel materials with low volume-shrinkages, *J. Sol-Gel Sci. Technol.*, **7**, pp. 191–201.

27. Ellsworth, M. W., and Novak, B. M. (1991). Mutually interpenetrating inorganic-organic networks. New routes into nonshrinking sol-gel composite materials, *J. Am. Chem. Soc.*, **113**, pp. 2756–2758.

28. Pomogailo, A. D. (2005). Polymer sol-gel synthesis of hybrid nanocomposites, *Colloid J.*, **67**, pp. 658–677.

29. Maier, M., Kotman, N., Friedrichs, C., Andrieu, J., Wagner, M., Graf, R., Strauss, W. S. L., Mailander, V., Weiss, C. K., and Landfester, K.

(2011) Highly site specific, protease cleavable, hydrophobic peptide-polymer nanoparticles, *Macromolecules*, **50**, pp. 6258–6267.

30. Crespy, D., and Landfester, K. (2011). Making dry fertile: a practical tour of non-aqueous emulsions and miniemulsions, their preparation and some applications, *Soft Matter*, **7**, pp. 11054–11064.
31. Hu, J., Chen, M., and Wu, L. M. (2011). Organic-inorganic nanocomposites synthesized via miniemulsion polymerization, *Polym. Chem.*, **2**, pp. 760–772.
32. Weiss, C. K., and Landfester, K. (2010). Miniemulsion polymerization as a means to encapsulate organic and inorganic material, *Adv. Polym. Sci.*, **233**, pp. 185–236.
33. Munoz-Espi, R., Weiss, C. K., and Landfester, K. (2012). Inorganic nanoparticles prepared in miniemulsion, *Curr. Opin. Colloid Interface Sci.*, **17**, pp. 212–224.
34. Sanchez-Dominguez, M., Pemartin, K., and Boutonnet, M. (2012). Preparation of inorganic nanoparticles in oil-in-water microemulsions: a soft and versatile approach, *Curr. Opin. Colloid Interface Sci.*, **17**, pp. 297–305.
35. Sui, R., and Charpentier, P. (2012). Synthesis of metal oxide nanostructures by direct sol-gel chemistry in supercritical fluids, *Chem. Rev.*, **112**, pp. 3057–3082.
36. Garcia-Bennett, A. E. (2011), Synthesis toxicology and potential of ordered mesoporous materials in nanomedicine, *Nanomedicine*, **6**, pp. 867–877.
37. Kumar, P., and Guliants, V. V. (2010). Periodoc mesoporous organic-inorganic hybrid materials: applications in membrane separations and adsorption, *Microporous Mesoporous Mater.*, **132**, pp. 1–14.
38. Hartmann, S., Brandhuber, D., and Husing, L. (2007). Glycol-modified silanes: novel possibilities for the synthesis of hierarchically organized (hybrid) porous materials, *Acc. Chem. Res.*, **40**, pp. 885–894.
39. Schiller, R., Weiss, C. K., Geserick, J., Husing, N., and Landfester, K. (2009). Synthesis of mesoporous silica particles and capsules by miniemulsion technique, *Chem. Mater.*, **21**, pp. 5088–5098.
40. Nabih, N., Schiller, R., Lieberwirth, I., Kockrick, E., Frind, R., Kaskel, S., Weiss, C. K., and Landfester, K. (2011). Mesoporous CeO_2 nanoparticles sythesized by an inverse miniemulsion technique and their catalytic properties in methane oxidation, *Nanotechnology*, **22**, 135606.
41. Rossmanith, R., Weiss, C. K., Geserick, J., Husing, N., Hormann, U., Kaiser, U., and Landfester, K. (2008). Porous anatase nanoparticles

with high specific surface area prepared by miniemulsion technique, *Chem. Mater.*, **20**, pp. 5768–5780.

42. Andritsch, T. (2010). Epoxy based nanodielectrics for high voltage DC applications–synthesis, dielectric properties and space charge dynamics, ISBN 978-90-5335-331-8.

43. Ang, B. C., Yaacob, I. I., and Nurdin, I. (2013). Investigation of Fe_2O_3/SiO_2 nanocomposite by FESEM and TEM, *J. Nanomat.*, **2013**, 980390.

44. Beck, J. S., Vartuli, J. C., Roth, W. K., Leonowicz, M. E., Kresge, C. T., Schmitt, K. D., Chu, C. T. W., Olson, D. H., Sheppard, E. W., McCullen, S. B., Higgins, J. B., and Schlenker, J. L. (1992). A new family of mesoporous molecular-sieves, prepared with liquid-crystal templates, *J. Am. Chem. Soc.*, **114**, pp. 10834–10843.

45. Zhang, H. Z., and Banfield, J. F. (1998). Thermodynamic analysis of phase stability of nanocrystalline titania, *J. Mater. Chem.*, **8**, pp. 2073–2076.

46. Kubiak, P., Froschl, T., Husing, N., Hormann, U., Kaiser, U., Schiller, R., Weiss, C. K., Landfester, K., and Wolfahrt-Mehrens, M. (2011). TiO_2 anatase nanoparticle networks: synthesis, structure, and electrochemical performance, *Small*, **7**, pp. 1690–1696.

47. Willert, M., Rothe, R., Landfester, K., and Antonielli, M. (2001). Synthesis of inorganic and metallic nanoparticles by miniemulsification of molten salts and metals, *Chem. Mater.*, **13**, pp. 4681–4685.

48. Taden, A., Antonietti, M., Heilig, A., and Landfester, K. (2004). Inorganic films from three different phosphors via a liquid coating route from inverse miniemulsions, *Chem. Mater.*, **16**, pp. 5081–5087.

49. Crespy, D., Stark, M., Hoffmann-Richter, C., Ziener, U., and Landfester, K. (2007). Polymeric nanoreactors for hydrophilic reagents synthesized by interfacial polycondensation on miniemulsion droplets, *Macromolecules*, **40**, pp. 3122–3135.

50. Ge, J. P., Chen, W., Liu, Y. D., and Li, Y. D. (2006). Formation of disperse nanoparticles in oil/water interface in normal microemulsions, *Chem. Eur. J.*, **12**, pp. 6552–6558.

51. Zhuang, Z., Peng, Q., Liu, J. F., Wang, X., and Li, Y. D. (2007). Indium hydroxides, oxyhydroxides, and oxide nanocrystals series, *Inorg. Chem.*, **46**, pp. 5179–5187.

52. Liu, J., Chen, W., Liu, X. W., Zhou, K. B., and Li, Y. D. (2008). Au/LaVO(4) nanocomposite: preparation, characterization and catalytic activity for CO oxidation, *Nano Res.*, **1**, pp. 46–55.

53. Shan, S. N., Wang, X. Y., and Jia, N. Q. (2011). Synthesis of $NaYF_4$: Yb^{3+}, Er^{3+} upconversion nanoparticles in normal microemulsions, *Nanoscale Res. Lett.*, **6**, pp. 539–543.

54. Asua, J. M. (2014). Challenges for industrialization of miniemulsion polymerization, *Prog. Polym. Sci.*, **39**, pp. 1797–1826.

55. Ugelstadt, J., El-Aasser, M. S., Vanderhoff, J. W. (1973). Emulsion polymerization: initiation of polymerization in monomer droplets, *J. Polym. Sci. Polym. Lett. Ed.*, **11**, pp. 503–513.

56. Asua, J. M. (2002). Miniemulsion polymerization, *Prog. Polym. Sci.*, **27**, pp. 1283–1346.

57. Nicolas, J., Guillaneuf, Y., Lefay, C., Bertin, D., Gigmes, D., and Charleux, B. (2013). Nitroxide-mediated polymerization, *Prog. Polym. Sci.*, **38**, pp. 63–235.

58. Wegner, G., Baum, P., Müller, M., Norwig, J., and Landfester, K. (2001). Polymers designed to control nucleation and growth of inorganic crystals from aqueous media, *Macromol. Symp.*, **175**, pp. 349–355.

59. Gorna, K., Munoz-Espi, R., Grohn, F., and Wegner, G. (2007). Bioinspired mineralization of inorganics from aqueous media controlled by synthetic polymers, *Macromol. Biosci.*, **7**, pp. 167–173.

60. Munoz-Espi, R., Qi, Y., Lieberwirth, I., Gomez, C. M., and Wegner, G. (2006). Surface-functionalized latex particles as controlling agents for the mineralization of zinc oxide in aqueous medium, *Chem. Eur. J.*, **12**, pp. 118–129.

61. Wegner, J., Demir, M. M., Faatz, M., Gorna, K., Munoz-Espi, R., Guillemet, B., and Grohn, F. (2007). Polymers and inorganics: a happy marriage? *Macromol. Res.*, **15**, pp. 95–99.

62. Tamai, H., and Yasuda, H. (1999). Preparation of polymer particles coated with hydroxapatite, *J. Colloid Interface Sci.*, **212**, pp. 585–588.

63. Ethirajan, A., and Landfester K. (2010). Functional hybrid materials with polymer nanoparticles as templates, *Chem. Eur. J.*, **16**, pp. 9398–9412.

64. Schöller, K., Ethirajan, A., Zeller, A., and Landfester, K. (2011). Biomimetic route to calcium phosphate coated polymeric nanoparticles: influence of different functional groups and pH, *Macromol. Chem. Phys.*, **212**, pp. 1165–1175.

65. Ethirajan, A., Ziener, U., and Landfester, K. (2009). Surface-functionalized polymeric nanoparticles as templates for biomimetic mineralization of hydroxyapatite, *Chem. Mater.*, **21**, pp. 2218–2225.

66. Fischer, V., Lieberwirth, I., Jakob, G., Landfester, K., and Munoz-Espi, R. (2013). Metal oxide/polymer hybrid nanoparticles with versatile functionality prepared by controlled surface crystallization, *Adv. Func. Mater.*, **23**, pp. 451–466.

67. Ethirajan, A., Ziener, U., Chuvilin, A., Kaiser, U., Colfer, H., and Landfester, K. (2008). Biomimetic hydroxyapatite crystallization in gelatin nanoparticles synthesized using a miniemulsion process, *Adv. Func. Mat.*, **18**, pp. 2221–2227.

68. Lu, W., Chen, M., and Wu, L. M. (2008). One-step synthesis of organic-inorganic hybrid asymmetric dimer particles via miniemulsion polymerization and functionalization with silver, *J. Colloid Interface Sci.*, **328**, pp. 98–102.

69. Wu, Y. F., Zhang, Y., Xu, J. X., Chen, M., Wu, L. M. (2010). One-step preparation of PS/TiO_2 nanocomposite particles via miniemulsion polymerization, *J. Colloid Interface Sci.*, **343**, pp. 18–24.

70. Manzke, A., Pfahler, C., Dubbers, O., Plettl, A., Ziemann, P., Crespy, D., Schreiber, E., Ziener, U., and Landfester, K. (2007). Etching masks based on miniemulsions: a novel route towards ordered arrays of surface nanostructures, *Adv. Mater.*, **19**, pp. 1337–1341.

71. Manzke, A., Vogel, N., Weiss, C. K., Ziener, U., Plettl, A., Landfester, K., and Ziemann, P. (2011). Arrays of size and distance controlled platinum nanoparticles fabricated by a colloidal method, *Nanoscale*, **3**, pp. 2523–2528.

72. Vogel, N., Ziener, U., Manzke, A., Plettl, A., Ziemann, P., Biskupek, J., Weiss, C. K., and Landfester, K. (2011). Platinum nanoparticles from size adjustment functional colloidal particles generated by a seeded emulsion polymerization process, *Beilstein J. Nanotechnol.*, **2**, pp. 459–472.

73. Aguirre, M., Paulis, M., and Leiza, J. R. (2014). Particle nucleation and growth in seeded semibath miniemulsion polymerization of hybrid CeO_2/acrylic latexes, *Polymer*, **55**, pp. 752–761.

74. Retsch, M., Zhou, Z. C., Rivera, S., Kappl, M., Zhao, X. S., Jonas, U., and Li, Q. (2009). Fabrication of large-area, transferable colloidal monolayers utilizing self-assembly at the air/water interface, *Macromol. Chem. Phys.*, **210**, pp. 230–241.

75. Vogel, N., Weiss, C. K., and Landfester, K. (2012). From soft to hard: the generation of functional and complex colloidal monolayers from nanolithography, *Soft Matter*, **8**, pp. 4044–4061.

76. Plettl, A., Enderle, F., Saitner, M., Manzke, A., Pfahler, C., Wiedemann, S., and Ziemann, P. (2009). Non-close-packed crystals from self-

assembled polystyrene spheres by isotropic plasma etching: adding flexibility to colloid lithography, *Adv. Funct. Mater.*, **19**, pp. 3279–3284.

77. Kumar, K., Nightingale, A. M., Krishnadasan, S. H., Kamaly, N., Wylenzinska-Arridge, M., Zeissler, K., Branford, W. R., Ware, E., deMello, A. J., and deMello, J. C. (2012). Direct synthesis of dextran-coated superparamagnetic iron oxide nanoparticles in a capillary-based droplet reactor, *J. Mater. Chem.*, **22**, pp. 4704–4708.

78. Yang, S.-Y., Cheng, F.-Y., Yeh, C.-S., and Lee, G.-B. (2010). Size-controlled synthesis of gold nanoparticles using a micro-mixing system, *Microfluid Nanofluid*, **8**, pp. 303–311.

79. Marrea, S., and Jensen, K. F. (2010). Synthesis of micro and nanostructures in microfluidic systems, *Chem. Soc. Rev.*, **39**, pp. 1183–1202.

80. Uda, M., Ohsaki, K., and Morita, Y. (1989). Preparation of mixed ultrafine (AL + ALN) powders and their nitridation, *Proceedings of the International Symposium on Physical Chemistry of Powders Metals Production Processing*, pp. 261–269.

81. Lu, Z. P., and Pfender, E. (1990). DC plasma synthesis of aluminum nitride ceramic powders, *Proceedings of the Material Research Society Symposium*, pp. 857–860.

82. Kikukawa, N., and Makino, M. (1990). Synthesis of material using arc plasma reactor with a secondary anode, *Bull. Nat. Res. Inst. Pollut. Resources*, **20**, pp. 19–26 (in Japanese).

83. Inoue, A., Kim, B. G., Nosaki, K., Yamaguchi, T., and Masumoto, T. (1992). Production of ultrafine aluminum nitride particles by plasma-alloy reaction and their microstructure and morphology, *J. Appl. Phys.*, **71**, pp. 4025–4029.

84. Ageorges, H., Megy, S., Chang, K., Baronnet, J. M., Williams, J. K., and Chapman, C. (1993). Synthesis of aluminum nitride in transferred arc plasma furnaces, *Plasma Chem. Plasma Process.*, **13**, pp. 613–632.

85. Oh, S. M., and Park, D. W. (1998). Preparation of AlN fine powder by thermal plasma processing, *Thin Solid Films*, **316**, pp. 189–194.

86. Munz, R. J., Addona, T., and Cruz, A. C. (1999). Application of transferred arcs to the production of nanoparticles, *Pure Appl. Chem.*, **71**, pp. 1889–1997.

87. Cruz, A. C., Bender, O. W., Thomaz, O., Yoshimura, H. N., and Kiyohara, P. K. (2001). Production of Al and AlN powders in a new version of transferred-arc reactor, *Proceedings of the 15th International Symposium on Plasma Chemistry*, pp. 2729–2734.

88. Iwata, M., and Adachi, K. (2001). Synthesis of alumina ultra fine powder by transferred type arc plasma, *Proceedings of the 15th International Symposium on Plasma Chemistry*, pp. 2769–2774.

89. Iwata, M., Adachi, K., Furukawa, S., and Amakawa, T. (2004). Synthesis of purified AlN nano powder by transferred type arc plasma, *J. Phys. D: Appl. Phys.*, **37**, pp. 1041–1047.

90. Ulrich, G. D. (1971). Theory of particle formation and growth in oxide synthesis flames, *Combustion Sci. Technol.*, **4**, pp. 47–57.

91. Cruz, A. C., Bender, O. W., and Yoshimura, H. N. (2001). Modeling study of the production of AlN nanoparticles in a new version of a plasma synthesis reactor, *Proceedings of the 15th International Symposium on Plasma Chemistry*, pp. 2723–2728.

92. Iwata, M., Furukawa, S., Amakawa, T., and Adachi, K. (2006). A novel method of synthesizing spherical nano-structured composite particles of aluminum nitride using transferred type dc arc plasma, *IEEJ Trans. Fundamentals Mater.*, **126**(11), pp. 1050–1056.

93. Iwata, M., Furukawa, S., Mizutani, Y., Adachi, K., and Amakawa, T. (2006). Design and evaluation of all solid transformer (Part 4): thermal conductivity and breakdown strength of epoxy resin with spherical nano-structured composite particles of aluminum nitride-, *CRIEPI Report*, H05008 (in Japanese).

94. Kriechbaum, G. W., and Kleinschmit, P. (1989). Superfine oxide powders: flame hydrolysis and hydrothermal synthesis, *Angew. Chem.-Int. Ed. Eng.*, **28**, pp. 1416–1423.

95. Kloepfer, H. (1942). Verfahren zum reinigen von hochdispersen oxyden von metallen bzw metalloxiden, DE patent 762723A.

96. Kawachi, M., Yasu, M., and Edahiro, T. (1983). Fabrication of SiO_2-TiO_2 glass planar optical-waveguides by flame hydrolysis deposition, *Electron. Lett.*, **19**, pp. 583–584.

97. Izawa, T., and Inagaki, N. (1980). Materials and processes for fiber preform fabrication: vapor-phase axial deposition, *Proc. IEEE*, **68**, pp. 1184–1187.

98. Dugas, V., and Chevalier, Y. (2003). Surface hydroxylation and silane grafting on fumed and thermal silica, *J. Colloid Interface Sci.*, **264**, pp. 354–361.

99. Zhuravlev, L. T. (2000). The surface chemistry of amorphous silica. Zhuravlev model, *Colloids Surfaces A-Physicochem. Eng. Aspects*, **173**, pp. 1–38.

100. Zhuravlev, L. T. (1987). Concentration of hydroxyl-groups on the surface of amorphous silicas, *Langmuir*, **3**, pp. 316–318.

101. Praeger, M., Hosier, I. L., Vaughan, A. S., Swingler, S. G. (2015). The effects of surface hydroxyl groups in polyethylene-silica nanocomposites, *Electrical Insulation Conference (EIC), Seattle USA, June 7–10*, pp. 201–204.

102. Inada, K. (1982). Recent progress in fiber fabrication techniques by vapor-phase axial deposition, *IEEE J. Quantum Electron.*, **18**(10), pp. 1424–1431.

103. Sekiya, E. H., et al. (2000). Crystalline and amorphous phases of GeO_2 in VAD silica-germania soot preform, *J. Non-Crystal. Solids*, **273**(1–3), pp. 228–232.

Chapter 3

Nano-Filler Dispersion for Tailoring of Nanocomposite Dielectrics

Takahiro Imai

Power and Industrial Systems R&D Center, Toshiba Corporation, 1,Toshiba-cho, Fuchu-shi, Tokyo 183-8511, Japan

takahiro2.imai@toshiba.co.jp

"Everything should be made as simple as possible, but not simpler," Albert Einstein said. Polymer nanocomposite dielectrics, which are composed of a polymer and inorganic fillers with nanometric dimension (nano-filler), are extremely simple materials. However, they have a deceptive simplicity. For the same level of filler loading, nano-fillers have much more extensive interfaces than micro-fillers (inorganic fillers with micrometric dimension) in the polymer. These large interfaces dominate the mesoscopic characteristics and improve macroscopic properties, including electrical, mechanical and thermal behavior. Moreover, an ingenious technique is needed such that the nano-fillers are homogenously dispersed in the polymer without aggregation or agglomeration. Therefore, polymer nanocomposite dielectrics are not simple materials. This chapter focuses on the three processing methodologies: sol-gel, direct dispersion and clay dispersion,

Tailoring of Nanocomposite Dielectrics: From Fundamentals to Devices and Applications
Edited by Toshikatsu Tanaka and Alun S. Vaughan

ISBN 978-981-4669-80-1 (Hardcover), 978-981-4669-81-8 (eBook)
www.panstanford.com

each of which has its advantages. It also considers methods of evaluating the nano-filler dispersion state in these materials.

3.1 Introduction to the Fabrication of Polymer Nanocomposite Dielectrics

The most significant characteristic of nanocomposites is the large interface between the polymer and nano-fillers, which strongly affects the properties of these materials. Moreover, many combinations of polymer and nano-fillers provide many kinds of nanocomposites. As an introduction to the fabrication of nanocomposites, this section compares nano-filler and micro-filler interfaces within a polymer and examines how polymers and nano-fillers can be combined to improve properties.

3.1.1 Large Interface of Nano-Fillers

Nano-fillers have large interfaces with the surrounding polymer compared, to micro-filled systems. For example, when nano-fillers or micro-fillers with the same filler content are disperse into a polymer, the polymer/filler interface area per unit volume, calculated according to the following formula, are very different [1]:

$$\text{Polymer/filler interface area per volume unit} = \frac{6D}{f},$$

where D is the diameter of the filler and f the volume fraction of the filler.

For example, a nanocomposite containing 5 wt% of nano-fillers with 50 nm diameter has approximately 1000 times more interfacial area than a typical microcomposite, as shown in Fig. 3.1. Therefore, the macroscopic properties seem to reflect the mesoscopic characteristics of the large interface area in nanocomposites.

Homogenous dispersion of nano-fillers with diameter below 100 nm is, however, very difficult. The van der Waals force (attractive forces) and the electrostatic repulsion attributed to electric double layer influence inter-particle interactions according to the DLVO (Derjaguin, Landau, Verwey, and Overbeek) theory. When both these forces are considered, the potential peak due to electrostatic repulsion is located just a few nanometers from the surface of

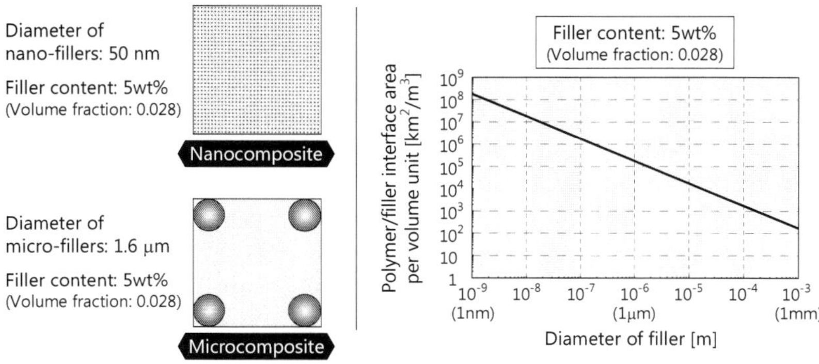

Figure 3.1 Comparison of polymer/filler interface in nanocomposite and microcomposite.

the nano-filler. When the surface distance between the fillers becomes smaller than the potential peak (a few nano-meters), the nano-fillers aggregate with each other due, then, to dominant effect of the van der Waals force. Woodcock et al. proposed the following formula regarding the average inter-particle separation between fillers in a dispersed filler system [2].

$$H = D\left[\left(\frac{1}{3\pi C} + \frac{5}{6}\right)\right]^{0.5}$$

where H is the average surface distance, D the diameter of fillers, and C the filler concentration (volume fraction).

From this, the inter-particle separation in a nanocomposite containing 5 wt% of nano-filler of diameter 50 nm is 107 nm. However, in a microcomposite containing 5 wt% of micro-filler with a diameter of 1.6 μm, the average separation between the surface of neighboring particles is 3.4 μm. This calculation demonstrates that nano-fillers aggregate more easily than micro-fillers.

3.1.2 Combinations of Nano-Fillers and Polymers

Many kinds of nano-fillers are commercially available, such as silica and alumina. Moreover, many kinds of polymer are widely used as insulation materials. Therefore, by combining different nano-fillers with different polymers, it possible to tailor properties.

Epoxy resin, silicone resin, polyamide (PA), polyimide (PI), low-density polyethylene (LDPE), cross-linked polyethylene (XLPE), polyamideimide (PAI), polyester (PE), polyesterimide, polypropylene (PP) and ethylene-vinyl acetate (EVA), amongst others, have all been used as the polymeric matrix, while clay, silica (SiO_2), titania (TiO_2), alumina (Al_2O_3), boehmite alumina (AlOOH), magnesia (MgO), zinc oxide (ZnO), boron nitride (BN), barium titanate ($BaTiO_3$), polyhedral oligomeric silsebquioxane (POSS), silicon carbide (SiC), zirconia (ZrO_2), iron oxide (Fe_2O_3) and calcium carbonate ($CaCO_3$), carbon nanotube (CNT) have been used as nano-fillers for nanocomposite insulation materials. Many combinations of polymers and nano-fillers are evidently possible. Therefore, the selection of the optimal combination, based on the application of the nanocomposite, is very important. Typical combinations of polymers and nano-fillers are summarized in Table 3.1. Typical improved electrical properties of these combinations are also shown.

Table 3.1 Typical combination of polymers and nano-fillers

Polymer	Nano-filler[a]	Typical improved electrical properties	Ref[b]
Epoxy resin	Clay	– ac breakdown strength – Electrical treeing resistance – Partial discharge resistance	[3] [4] [5, 6]
	SiO_2	– ac breakdown strength – Electrical treeing resistance – Partial discharge resistance	[7] [8] [6, 7, 9]
	TiO_2	– Electrical treeing resistance – Partial discharge resistance – Permittivity (decrease)	[10] [6, 9] [11]
	Al_2O_3	– Permittivity (decrease)	[12, 13]
	MgO	– Permittivity (decrease)	[13]
	POSS	– Partial discharge resistance	[14]
Silicone resin	SiO_2	– Tracking/erosion resistance	[15]
	AlOOH	– Partial discharge resistance	[6]
PA	Clay	– Partial discharge resistance	[16]
PI	SiO_2	– Partial discharge resistance	[17]
LDPE XLPE	SiO_2	– Partial discharge resistance – Water treeing resistance	[18] [19]

Polymer	Nano-filler[a]	Typical improved electrical properties	Ref[b]
	MgO	- Space charge accumulation (decrease) - Water treeing resistance	[20] [21]
Polyamideimide	SiO_2	- Partial discharge resistance	[22]
PP	POSS	- Space charge accumulation (decrease) - ac breakdown strength	[23]
EVA	POSS	- Space charge accumulation (decrease) - ac breakdown strength	[23]

[a]Filler content is 1–10 wt%, and average filler diameter is 1–1000 nm.
[b]Typical papers are shown as references.

3.2 Nano-Filler Dispersion Using Sol-Gel and Direct Dispersion Methods

Sol-gel and direct dispersion methods have encouraged the development of nanocomposites, because both methods have crucial advantages. The sol-gel method provides the most homogeneous nanocomposites. The direct dispersion method enables the simplest fabrication of nanocomposites, and has a cost advantage. Therefore, the selection of methods based on the application of the nanocomposites is also important. This section describes the fabrication of nanocomposites using the sol-gel and direct dispersion methods.

3.2.1 Nano-Filler Formation in Polymers by Sol-Gel Process

Generally, in sol-gel methods, the gel is derived from a solution of inorganic compounds via the sol. After that, a solidified oxidation product, such as a glass, is obtained by heating the gel. Metallic alkoxides are usually used as the precursor inorganic compounds and have the chemical structure $M(OR)_n$, where M (metal) indicates Si, Al, Ti, or Zr. Tetraethoxysilane ($Si(OC_2H_5)_4$) is the compound that is most commonly used in the sol-gel method. Figure 3.2a shows a typical process for the formation of SiO_2 particles in fused quartz preparation. The sol-gel reaction needs an acid or alkali catalyst, and hydrochloric acid is frequently used; Figure 3.2b shows the fundamental reactions. When this reaction occurs in the presence of a polymer, it leads to a nanocomposite.

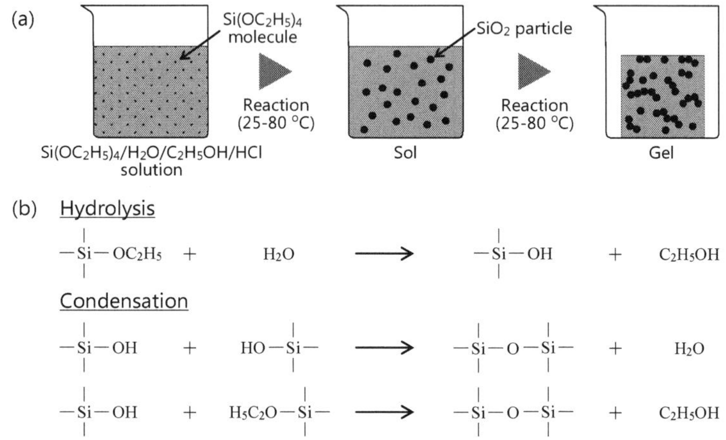

Figure 3.2 Sol-gel method. (a) Formation of SiO_2 particles from tetraethoxysilane via sol and gel. (b) Fundamental sol gel reactions.

For example, polyimide is made by the reaction of pyromellitic dianhydride (PMDA) and 4,4′-oxydianiline (ODA) via polyamic acid (PAA) in *N*,*N*-dimethylacetamide (DMAc) solvent, as shown in Fig. 3.3a. When both this reaction and the sol-gel reaction occur simultaneously, a polyimide/SiO_2 nanocomposite is obtained; Figure 3.3b shows the scheme for fabricating such a nanocomposite [17]. PMDA and tetraethoxysilane (TEOS) are added to the ODA/DMAc solution and the PAA/SiO_2 solution is obtained by stirring at room temperature. A silane coupling agent (γ-aminopropyl triethoxysilane) and H_2O are added to this solution, and this mixture is stirred at room temperature for 16 h. The mixture is subsequently cast onto a glass substrate, and the polyimide/SiO_2 nanocomposite is obtained by gradual heating at 80 °C for 2 h, 120 °C for 1 h, 250 °C for 1 h and 350 °C for 1 h.

Although the sol-gel method has the disadvantages that a large amount of H_2O and alcohol is needed and the fabrication procedure is a little complicated, SEM images demonstrate that the resulting SiO_2 nano-fillers are dispersed homogenously. With this method, the resulting nanocomposites are characterized by a more homogenous structure than is attainable with alternative approaches.

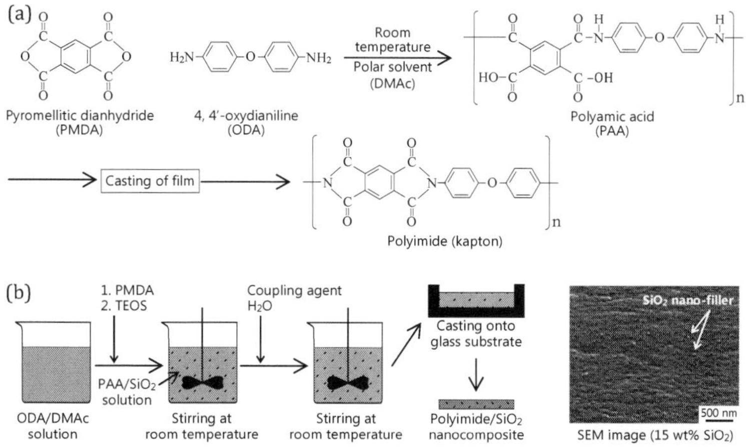

Figure 3.3 Fabrication of polyimide/SiO_2 nanocomposite film. (a) Synthesis of polyimide via polyamic acid. (b) Fabrication scheme of polyimide/SiO_2 nanocomposite [17].

Moreover, some nanocomposites containing silica nano-fillers produced using sol-gel methods have been commercialized [24]. NANOCRYL is based upon acrylate monomers and a large amount (50 wt%) of colloidal silica, while NANOPOX is an epoxy resin containing a large amount (40 wt%) of colloidal silica.

3.2.2 Surface Treatment of Nano-Fillers in Direct Dispersion

Direct dispersion of nano-fillers into polymers is a simple and versatile approach for the fabrication of nanocomposites. Many kinds of nanocomposites can be prepared by the direct dispersion method and their properties evaluated. In the direct dispersion method, the nano-filler and the polymer are directly compounded, as described in the literature. However, since nano-fillers have a huge surface energy because of their nano-scale diameter, insufficient surface treatment tends to degrade the properties of the resulting nanocomposites, due to aggregation or agglomeration of the nano-fillers.

Nano-fillers aggregate and agglomerate as shown in Fig. 3.4, which reduces the effective nano-filler/polymer interface, while a typical nanocomposite fabrication scheme using in the direct dispersion method is shown in Fig. 3.5. In this, surface (silane) treatment of the nano-filler and high shear force compounding are important elements that directly affect the structure and properties of the nanocomposites.

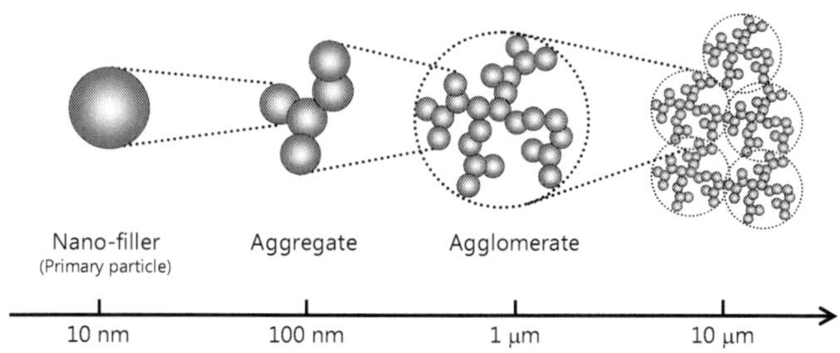

Figure 3.4 Aggregate and agglomerate of the nano-fillers.

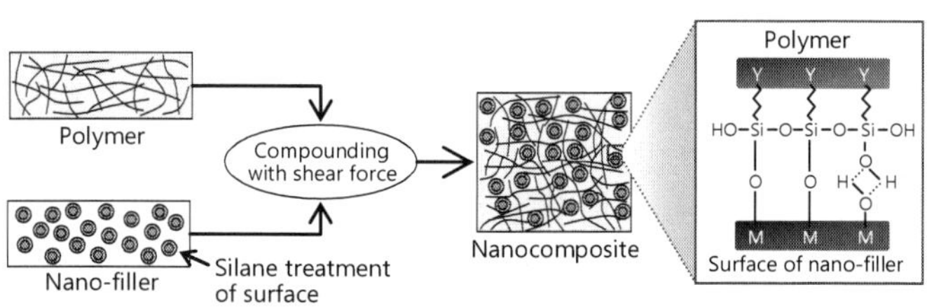

Figure 3.5 Typical fabrication scheme in the direct dispersion method.

Silane coupling agents are often used for the surface treatment of nano-fillers, as part of the fabrication of the nanocomposites, where they have the following roles:

(1) wettability improvement of nano-fillers for polymers
(2) compatibility improvement of nano-fillers and polymers
(3) hydrogen bonding between nano-fillers and polymers
(4) chemical bonding between nano-fillers and polymers

Generally, wettability, compatibility, and hydrogen bonding strength affect the dispersion of the nano-fillers in the polymers. Silane treatment is the most commonly used procedure to improve the wettability and compatibility of the nano-fillers. Many kinds of silane coupling agents have been commercialized, and representative examples are summarized in Table 3.2. Silane coupling agents have the general structure of R-Si-(OR′)$_3$, where the R group determines the nature of the interaction with the polymer and the OR' group results in chemical bonding with the nano-filler. Consequently, the silane coupling agent should be selected based on the relevant combination of polymer and nano-filler. The applicable polymers for each silane coupling agent are shown in Table 3.2 and the effectiveness of the silane coupling agents for a range of nano-fillers are shown in Table 3.3.

Silane treatments are divided into two methods as follows:

(1) pre-treatment of nano-fillers
   - wet method
   - dry method

(2) in situ treatment of nano-fillers

Table 3.2 Representative silane coupling agents

R	OR′	Chemical formula	Chemical name	Applicable polymer
$CH_2=CH$-	-OC_2H_5	$CH_2=$ CH-Si(OC_2H_5)$_3$	Vinyltriethoxysilane	- Polyethylene - Polypropylene
CH_3 $CH_2=C$-$COCH_2CH_2CH_2$- O	-OCH_3	CH_3 $CH_2=C$-$COCH_2CH_2CH_2$-Si(OCH_3)$_3$ O	γ-Ethacryloxypropyl trimethoxysilane	- Urethane
$NH_2CH_2CH_2CH_2$-	-OC_2H_5	$NH_2CH_2CH_2$ CH_2-Si(OC_2H_5)$_3$	γ-Aminopropyl triethoxysilane	- Phenol-formaldehyde - Epoxy resin - Melamine resin - Polyamide - Polyimide - Polycarbonate
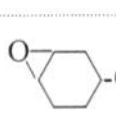-CH_2CH_2-	-OCH_3	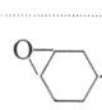-CH_2CH_2-Si(OCH_3)$_3$	β-(3,4-Epoxy cyclohexyl)ethyl trimethoxysilane	- Butyl rubber - Phenol-formaldehyde - Epoxy resin
CH_2-$CHCH_2OCH_2CH_2CH_2$- O	-OCH_3	CH_2-$CHCH_2OCH_2CH_2CH_2$-Si(OCH_3)$_3$ O	γ-Glycidoxypropyl trimethoxysilane	- Polyvinyl chloride - Polyethylene
$HSCH_2CH_2CH_2$-	-OCH_3	$HSCH_2CH_2CH_2$- Si(OCH_3)$_3$	γ-Mercaptopropyl trimethoxysilane	- Polychloroprene - Polybutadiene - Urethane

Table 3.3 Effectiveness of silane coupling agents for inorganic fillers

Effectiveness	Inorganic nano-filler
Very effective	Silica (SiO_2), alumina (Al_2O_3), etc.
Effective	Clay, talc, iron, aluminum hydroxide ($Al(OH)_3$), etc.
Slightly effective	Titania (TiO_2), iron oxide (Fe_2O_3), etc.
Hardly effective	Graphite, carbon black, calcium carbonate($CaCO_3$), Boron nitride (BN), etc.

In the pre-treatment method, the nano-fillers are treated before compounding with the polymers and two methods for this exist: the wet method and the dry method. In the wet method, the nano-fillers are dispersed in a dilute silane solution and, after sufficient mixing, the nano-fillers are separated by filtration, decanting or centrifugation. Drying by heating finally leads to a silane treated nano-filler. In the dry method, the silane coupling agent or the diluted silane solution is added to the nano-filler and mixed at high speed. The separation and drying process is not needed in this approach.

In the in situ treatment method, the nano-fillers, silane coupling agent and polymer are compounded together. This method has the advantage of low cost, and is widely used in industry, but the effectiveness of the approach is inferior to that of the pre-treatment. Also, the addition of a silane coupling agent such as γ-aminopropyltriethoxysilane causes gelation of epoxy resins; the stability of the polymer in the presence of the silane coupling agent is an important factor when considering in situ treatment.

Finally, titanate, aluminate, and phosphate coupling agents have a similar function to the silane coupling agents.

3.2.3 Direct Compounding of Nano-Fillers and Polymers

Nano-fillers usually exist in the form of aggregates and agglomerates before dispersion in the polymer and, consequently, in direct dispersion methods, excellent compounding techniques are essential to break up aggregate and agglomerate of nano-fillers.

The effects of different mixers on nano-filler dispersion have been reported [25]. A boehmite alumina (AlOOH) nano-filler with a primary filler diameter of about 50 nm was mixed with silicone rubber using a high-speed planetary mixer, high pressure homogenizer or triple roll mill, as shown in Fig. 3.6. All mixers are available for compounding with high shear force. Analysis of SEM images of the resulting nanocomposite samples led to estimates of the dispersed filler size (diameter) and the area fraction occupied by the fillers. Nanocomposite prepared using the triple roll mill was found to exhibit a smaller filler size than nanocomposites prepared using the other kinds of mixers. This comparison demonstrates that the triple roll mill seems to be the most effective means of dispersing boehmite alumina nano-fillers into silicone rubber. Nevertheless, some agglomerates up to 20 μm in size possibly remain in the nanocomposite.

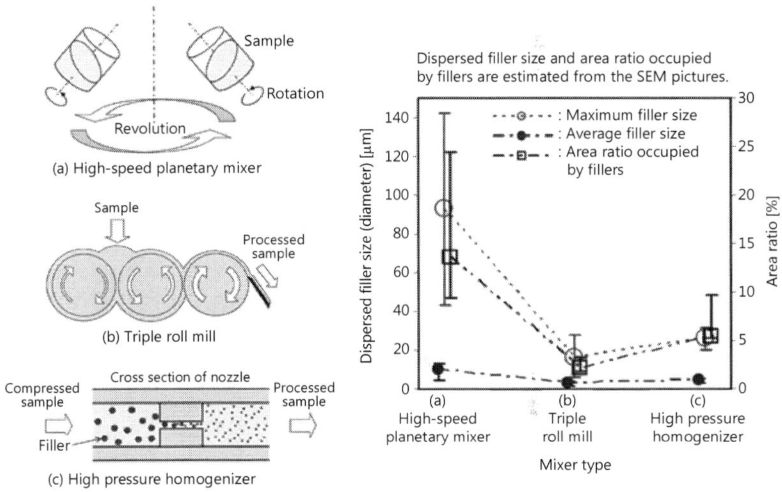

Figure 3.6 Comparison of dispersed filler size in different mixers [25].

It is very difficult fully to break up aggregate and agglomerate of the nano-fillers and, thereby, compound nanocomposites without any residual remaining aggregates and agglomerates. Therefore, a procedure in which centrifugal force removes the agglomerates from the nanocomposite has been reported [26]. An alumina nano-filler with an average filler diameter of about 31 nm was initially dispersed into an epoxy resin using ultrasonic waves (USW), as shown in Fig. 3.7a, after which, centrifugal force (CF) was used to remove agglomerates of the alumina nano-filler. Figure 3.7b shows that the diameter of the remaining agglomerates decreases in the specimens treated with the CF. Therefore, a procedure in which both USW and CF controls the diameter and distribution of the nano-fillers is proposed as shown in Fig. 3.7c.

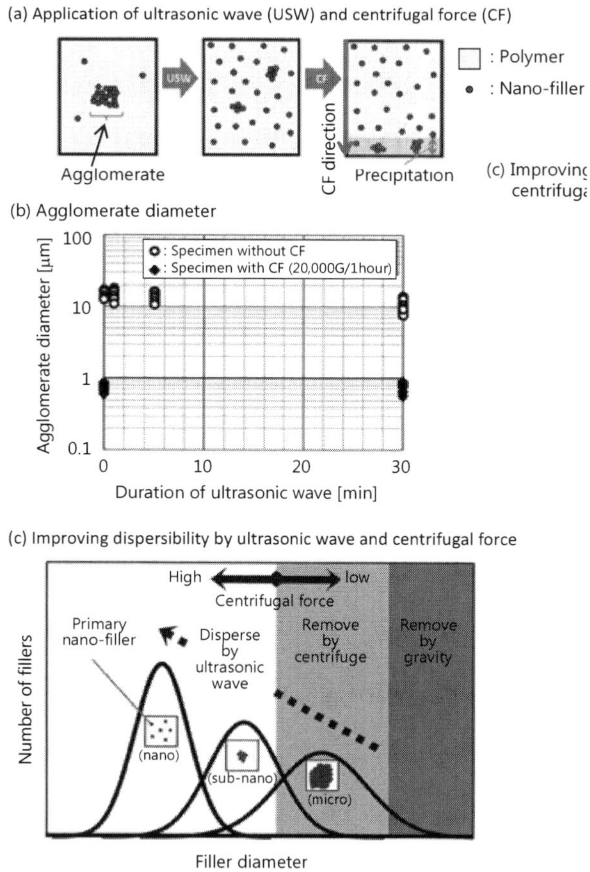

Figure 3.7 Control of nano-filler distribution using ultrasonic wave and centrifugal force [26].

Moreover, a new approach using electrospinning for fabrication of nanocomposite has been reported recently [27]. In this, an ethanolic solution of the silicone rubber containing silica nano-filler with an average diameter of 7 nm was used in the electrospinning apparatus, as shown in Fig. 3.8a. Figure 3.8b shows some stages in the electrospinning process. When the appropriate external voltage is applied to the metal needle, the electrospun fibers erupt from the tip of the needle. The resulting electrospun silicone rubber nanofibers containing the silica nano-filler are then mixed with silicone rubber and a curing agent, as shown in Fig. 3.8c, such that, after degassing and curing, a silicone rubber nanocomposite results. This approach facilitates homogenous dispersion of silica nano-fillers and leads to a significant improvement in the thermal stability of the resulting silicone rubber nanocomposite.

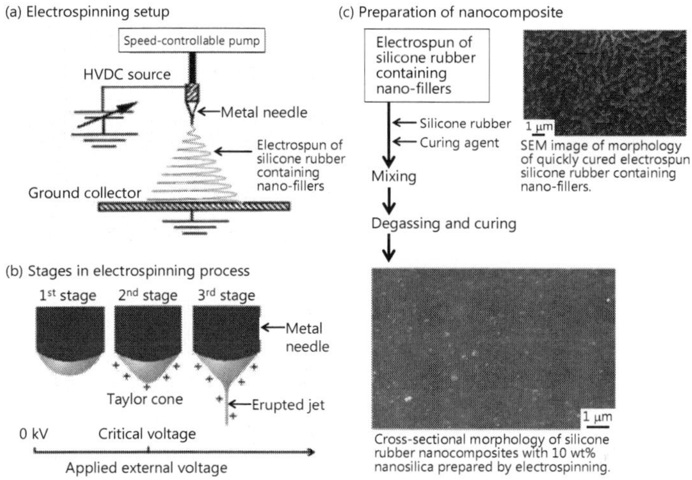

Figure 3.8 Fabrication of nanocomposite using electrospinning procedure [27].

3.3 Modification and Exfoliation of Clay Nano-Fillers

Clays, which are termed layered silicates, are perhaps the most important materials in the fabrication of nanocomposites. Indeed, the first nanocomposites pioneered by the Toyota group used clays as the nano-filler. Clays are characterized by laminar structures, between which, metal ions can be substituted with organic modifiers such as alkylammonium ions. These characteristics are essential to the production of clay-based nanocomposites. This section describes the organic modification, exfoliation, and dispersion of clays for the fabrication of nanocomposites.

3.3.1 The Structure and Properties of Clays

Individual clay layers, which are comprised of corner sharing SiO_4 tetrahedral sheets and $Al(OH)_6$ octahedral sheets, are hydrous aluminum phyllosilicates. They are classified into four typical groups: smectite, kaolin, chlorite, and illite. In particular, the smectite group, which includes montmorillonite, nontronite and saponite, is suitable for reinforcement in clay nanocomposites; the structure of the clay montmorillonite is shown in Fig. 3.9 [28]. This has a laminar structure composed of a base unit (silicate layer) ~100 nm long, ~100 nm wide and about 1 nm thick. This primary layer is composed of SiO_4 tetrahedral sheets and $Al(OH)_6$ octahedral sheets stacked in the order of tetrahedral sheet/octahedral sheet/tetrahedral sheet.

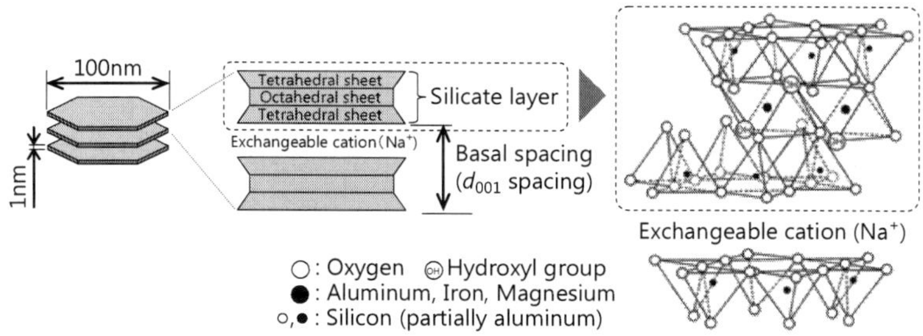

Figure 3.9 Structure of clay (montmorillonite) [28].

Table 3.4 Properties of clay (montmorillonite) [29]

Properties	Unit	Value
Swelling	[ml/2g]	65
pH (2% aqueous dispersion)	—	10.2
Electric conductivity	[μS/cm]	675
Viscosity (4% aqueous dispersion)	[mPa·s]	280
Visible light transmittance (1% aqueous dispersion)	[%]	1
Particle diameter	[nm]	100–2000
Specific surface (N_2, BET)	[m^2/g]	20
MB (methylene blue) absorbed amount	[mmol/100g]	130

Properties		Unit	Value
Cation exchange capacity	—	[meq/100 g]	108.6
Precipitation cation amount	Na^+	[meq/100 g]	114.1
	K^+	[meq/100 g]	2.8
	Mg^{2+}	[meq/100 g]	3.4
	Ca^{2+}	[meq/100 g]	18.2
Chemical composition	SiO_2	[%]	64.4
	Al_2O_3	[%]	25.9
	Fe_2O_3	[%]	3.5
	MgO	[%]	2.4
	CaO	[%]	0.7
	Na_2O	[%]	2.3
	K_2O	[%]	0.1

Substitution of Al^{3+} atoms for Si^{4+} atoms in a tetrahedral sheet, or substitution of divalent metal ion for trivalent metal ion in an octahedral sheet, generates a negative charge in the smectite. Cations such as sodium ions exist between the silicate layers to compensate for this negative charge. This structure imparts cation exchange ability to the montmorillonite and the cation exchange reaction enables the inclusion of an organic modifier (e.g. alkylammonium ion) between the silicate layers. The properties of montmorillonite are summarized in Table 3.4 [29].

3.3.2 Organic Modification of Clay

Organic modification plays an important role in the fabrication of clay-based nanocomposites. Original, unmodified clays with metal ions between the clay layers exhibit hydrophilic properties and lack affinity for most polymers. Therefore, organic modification is required for homogeneous dispersion within the vast majority of polymers.

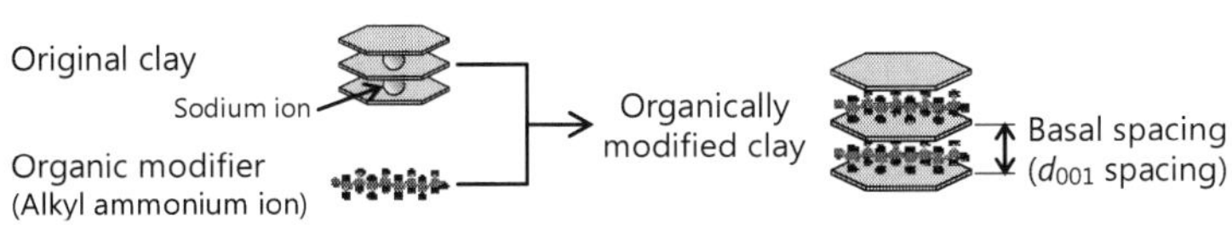

Figure 3.10 Organic modification of clay.

Generally, in the original, unmodified clays, the bonding force between the negatively charged silicate layer surfaces and the interlayer cations is weak. If a solution containing the original clays is mixed with a solution containing alternative cations with the required character, substitution of the original interlayer cations with cations of the required character occurs. The quantity of charge involved in this ion exchange reaction is termed the cation exchange capacity (CEC) of the clay and systems with approximately 100 meq/100 g CEC are suitable for fabrication of clay nanocomposites; the CEC of montmorillonite is 108.6 meq/ 100 g. Figure 3.10 represents this process schematically and shows a clay organically modified by ion exchange between metal ions (sodium ions) and alkylammonium ions. The resulting alkylammonium-modified clay has hydrophobic character and, consequently, exhibits good affinity with many polymers.

Alkylammonium ions are the most commonly used ions to modify clays and typical alkylammonium ions used for this purpose are summarized in Table 3.5. In particular, quaternary alkylammonium ions are suitable as modifiers of clays for nanocomposites and some clays modified with quaternary alkylammonium ions have been commercialized by clay manufacturers. Selection of an organic modifier for clays is very important for homogeneous dispersion within the polymer and, therefore, it is necessary to find a modifier that is compatible with the polarity of the polymer. Organic modifiers containing hydroxyl groups are generally suited to highly polar polymers, such as polyamides and polylactic acid, and organic modifiers containing long alkyl chains are suited to non-polar polymers such as polyolefin.

X-ray diffraction (XRD) analysis and thermal gravimetric analysis (TGA) provide information regarding the organic modification of clays. XRD measures the distance between silicate layers in the clays, a parameter termed the basal spacing (d_{001} spacing). Figure 3.11a compares the basal spacing of the unmodified and the organically modified clays and shows that the basal spacing of expands when sodium ions in the interlayer galleries are replaced alkylammonium ions. Moreover, Fig. 3.11b compares thermogravimetric curves obtained from unmodified and the modified clays, which shows that the weight loss of the modified clay increases markedly compared to that of the original clay. This is attributed to loss of the organic modifier on heating. Consequently, the fraction of organic modifier in the modified clay can be estimated from the change in weight loss in the TGA.

Table 3.5 Organic modifier of clays

Organic modifiers		Examples
Primary alkylammonium ion	$H_2N^+\text{-}R$ (R: Alkyl group)	$H_2N^+\text{-}(CH_2)_{n-1}\text{-}CH_3$ (n = 8, 11, 12, 18, etc.) $H_2N^+\text{-}(CH_2)_{n-1}\text{-}COOH$ (n = 8, 11, 12, 18, etc.)
Tertiary alkylammonium ion	R_1 $HN^+\text{–}R_2$ R_3 (R: Alkyl group)	CH_3 $HN^+\text{–}(CH_2)_{11}\text{-}CH_3$ CH_3 CH_3 $HN^+\text{–}(CH_2)_{17}\text{-}CH_3$ CH_3
Quaternary alkylammonium ion	R_1 $R_4\text{–}N^+\text{–}R_2$ R_3 (R: Alkyl group)	CH_3 $CH_3\text{–}N^+\text{–}(CH_2)_{17}\text{-}CH_3$ $(CH_2)_{15}\text{-}CH_3$ $(CH_2)_7\text{-}CH_3$ $CH_3\text{–}N^+\text{–}(CH_2)_7\text{-}CH_3$ $(CH_2)_7\text{-}CH_3$ T $CH_3\text{–}N^+\text{–}T$ CH_3 $(CH_2)_2\text{-}OH$ $CH_3\text{–}N^+\text{–}T$ $(CH_2)_2\text{-}OH$ CH_3 $CH_3\text{–}N^+\text{–}CH_2\text{–}C_6H_5$ HT (T: Tallow, HT: Hydrogenated tallow)

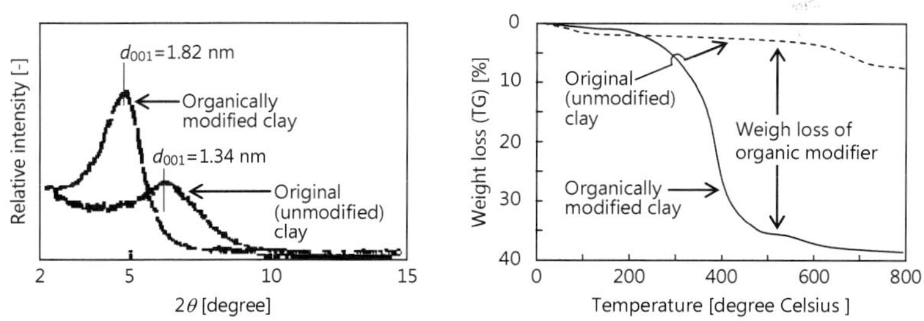

(a) X-ray diffraction (XRD) spectra of original (unmodified) clay and organically modified clay.

(b) Thermal gravimetric curves of original (unmodified) clay and organically modified clay.

Figure 3.11 Analysis of organically modified clay.

3.3.3 Exfoliation of Organically Modified Clays in Nanocomposite Fabrications

The first clay nanocomposite made by the Toyota research group exploited in situ polymerization [30], as shown in Fig. 3.12 [31].

In the first step, the original clays are organically modified using 12-amino-dodecanoic acid (H_2N-(CH_{11})-COOH), through a cation exchange reaction in acidic solution. In the second step, melt mixing of the modified clay and ε-caprolactam (melting point 70 °C) causes the ε-caprolactam to be inserted in the clay interlayer galleries. In the third step, the ε-caprolactam is polymerized by a ring-opening reaction at 250 °C; the basal spacing of the clay increases as polymerization progresses. Finally, the polyamide-based clay nanocomposite, in which the exfoliated clays are dispersed, is obtained.

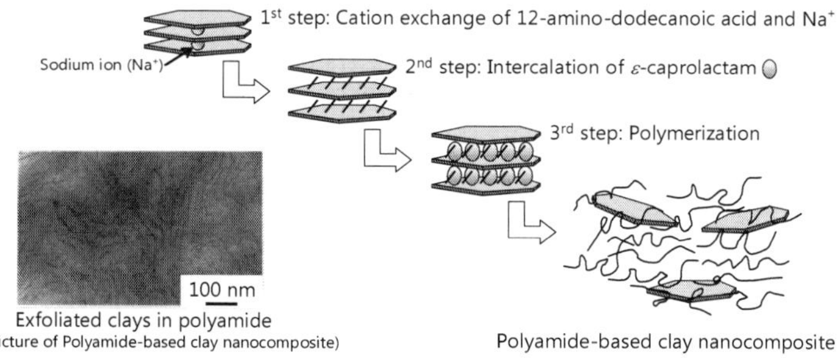

Figure 3.12 Fabrication method of polyamide-based clay nanocomposite [31].

Two approaches for the exfoliation of clays can be considered: the polymerization method and the melt blending method, the latter of which is shown in Fig. 3.13. As described above for the case of polyamide-based clay nanocomposite, in the polymerization method, organically modified clays are first dispersed in the relevant monomer, after which, the derived clay nanocomposite is obtained by polymerization. In this process, the polymerization reaction is the driving force for the exfoliation of the clay. In the melt blending method, clay nanocomposites are obtained by blending an organically modified clay with the required molten polymer, whereupon, dispersion or exfoliation of the clay is promoted by mechanical shear during mixing. Generally, the polymerization method requires a chemical plant to polymerize the monomer dispersed with organically modified clays, while the melt blending method has the advantage that the clay nanocomposites can be made relatively easily using thermoplastic resins. However, it is more difficult fully to exfoliate the clay using the melt blending method. For example, the mechanical properties and thermal resistance of polyamide-based clay nanocomposites made by these two methods are compared in Table 3.6 [29]. This table demonstrates that the polyamide-based clay nanocomposite produced by the polymerization method has superior flexural strength, flexural modulus and heat distortion temperature (HDT) compared with the system processed using the melt blending method. These results are attributed to differences in the exfoliation state of the clays within the polyamide.

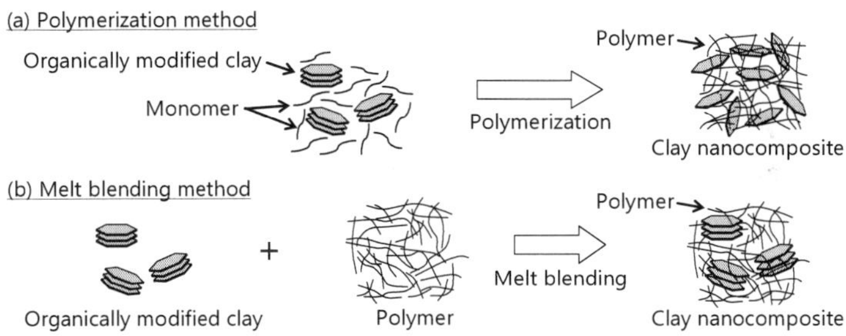

Figure 3.13 Exfoliation methods of organically modified clays in polymers.

Table 3.6 Comparison of polymerization method and melt blending method in polyamide-based clay nanocomposites [29]

Properties		Polyamide[a] (PA6)	PA6/clay nanocomposites Polymerization method	Melt blending method
Amount of clays (Nanomer 1.24TC [b])	[wt%]	0	5.5	5.5
Flexural strength	[MPa]	97.5	143.3	124.3
Flexural modulus	[MPa]	2420	4247	3740
Heat distortion temperature (HDT, 264 psi)	[°C]	59.8	131.9	116.4

[a]Capron 8202 35FAV manufactured by Allied Sigma company.
[b]Organically modified clay manufactured by Nanocor company.

Many factors concerning the exfoliation method affect the exfoliation and dispersion of the clay during the fabrication of nanocomposites. The major factors are summarized in Table 3.7.

It is important to fabricate clay-based nanocomposites with careful consideration of these factors, because exfoliation and homogenous dispersion can improve the properties of the resulting system.

Table 3.7 Major factor that influences exfoliation and dispersion of clays

Category		Major factor
Material	Monomer	– Chemical structure, etc.
	Polymer	– Chemical structure – Molecular weight – Polarity – Functional group, etc.
	Clay	– Shape – Size – Aspect ratio – Cation exchange capacity (CEC) – Purity, etc.
	Organic modifier (alkylammonium ions)	– Kinds of alkylammonium ions (Primary, tertiary and quaternary) – Length and branch of alkyl chain – Rate of the organic modifier in clay (Rate of residual Na^+), etc.
	Swelling solvent	– Polarity – Solubility parameter, etc.
Exfoliation method	Polymerization procedure	– Polymerization catalyst – Kinds of hardener, etc.
	Melt blending procedure	– Mixing equipment – Mixing time – Mixing temperature, etc.

3.3.4 Specific Exfoliation and Dispersion Methods for Clays

Exfoliation and dispersion methods for clays are ever-improving. A brief introduction to four specific exfoliation and dispersion methods are presented herein.

Organically modified clays can be swollen by exposure to an appropriate solvent; Figure 3.14a shows exfoliation methods using swollen clays and a curing accelerator. An organic solvent (polar solvent) such as *N, N*-dimethylacetamide (DMAc) is inserted into the galley between the clay layers, which expands the basal spacing of the clays [32]. Generally, the curing reaction between the epoxy resin and the acid anhydride hardener is slow, and an accelerator is sometimes used to speed it up. The tertiary ammonium ion, which is one of the accelerators, is inserted between the clay layers and the clay modified by the tertiary ammonium ions is dispersed into the epoxy resin. The clay nanocomposite is then obtained by curing in the presence of the hardener. In this system, the curing reaction initiates between the clay layers in which the accelerators exist, such that the cross-linked network of the epoxy resin starts to form from inside the clay layer. The beginning of the curing reaction between the clay layers promotes exfoliation of the clays.

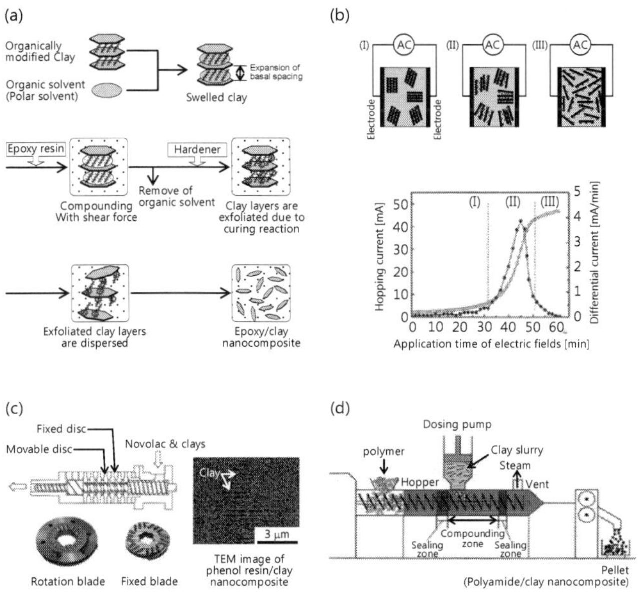

Figure 3.14 Specific exfoliation and dispersion methods for clays. (a) Method using swelled clays and curing accelerator. (b) Method using ac voltage application [33]. (c) Method using high-shear compounding in solid state [34]. (d) Method using unmodified clays [35].

Figure 3.14b shows another approach, which involves the application of an ac voltage [33]. 1 phr (parts per hundred parts of resin) of a clay modified with dimethyl-benzyl-hydrogenated tallow quaternary ammonium ions is first dispersed within the epoxy resin. Then, an ac voltage is applied to the sample using parallel electrodes with a 40 mm gap, which promotes the exfoliation of the clay. The hopping current increases rapidly and the exfoliation of the clays occurs in the process (II); the exfoliated clays are dispersed in the epoxy resin in the process (III).

Figure 3.14c exemplifies methods that exploit high-shear compounding in solid phase [34]. When the shear rate increases in the compounding process, the polymer containing the clays tends to show non-Newtonian behavior associated with decreasing polymer viscosity. Therefore, high-shear compounding in the solid phase using an extruder is an effective means of dispersing the clays homogenously. A Novolac resin (M_n = 1600, M_w = 7400) plus 4 phr organically modified clay are mixed in the solid phase with high shear force in the extruder. The fourth compounding (passing through the extruder) disperses the clays homogenously. After that, the phenolic resin/clay nanocomposite is obtained by curing in the presence of a hardener.

Figure 3.14d shows a processing method that exploits unmodified clays [35]. Although organic modification of clays is generally considered an essential element in the fabrication of nanocomposites, a low-cost fabrication method using unmodified clays has been developed. The water slurry technique disperses the unmodified clays in polyamide. In this, a polyamide resin is supplied from the hopper of the extruder, and the clay slurry is injected from the dosing pump. The clays are exfoliated during the compounding process and evolved steam is exhausted from the vent at the end of the extruder. The polyamide/clay nanocomposite is finally processed into pellets. Although nanocomposites produced using unmodified clays are inferior to those that use organically modified clays, this fabrication method without organic modification of clays seems to be an innovative method from the view-point of industrial manufacturing.

3.4 Blending of Nano-Fillers and Micro-Fillers for Dual-Functional Composites

Insulation materials usually need to exhibit additional characteristics, beyond their electrical properties. An effective approach to realizing insulation materials with multi-functionality is through the use of nano-micro composite. Composites that contain dispersed nano-fillers and micro-fillers within the polymer have superior electrical and other properties, such as low thermal expansion and improved thermal conductivity. This section describes the combined selection of nano- and micro-fillers and the fabrication method for the production of nano-micro composites.

3.4.1 Selection of Combinations of Nano- and Micro-Fillers

Insulation materials that only exhibit excellent electrical properties are often insufficient for practical use. For example, epoxy-based insulation moldings around metal conductors in switchgear need low thermal expansion as well as good electrical properties. The thermal expansion of epoxy resin without fillers is much higher than that of metal conductors, which leads to interfacial failure between the epoxy resin and the metal conductor during thermal-cycling. Moreover, epoxy-based encapsulation materials for power devices and insulation materials for solid molded transformers need high thermal conductivity as well as superior electrical properties, since such devices generate heat during operation that needs to be dissipated. Therefore, insulation materials with dual functions are necessary in such practical applications.

Since the first epoxy-based nano-micro composites with low thermal expansion and superior electrical properties were reported in 2005 [36, 37], many types of composites with dual functionality have appeared. Generally, nano-micro composites are composed of epoxy resin, a few wt% of nano-fillers with average diameter 1–100 nm and 50–70 wt% of micro-fillers with average diameter 1–100 μm; a round-robin study, in which the same epoxy-based nano-micro composite specimens were evaluated by different researchers, was conducted in 2012 [7]. The combination of nano-fillers and micro-fillers is important to obtain the desired functionality and some typical combinations are listed in Table 3.8.

Table 3.8 Typical combinations of nano-filler and micro-filler in epoxy-based nano-micro composites

Nano-filler[a]	Micro-filler[b]	Additional property	Improved electrical properties	Ref[d]
Clay	SiO_2	Low thermal expansion	- ac breakdown strength - Electrical treeing resistance - Partial discharge resistance	[36]
Clay	SiO_2	No measurement	- Partial discharge resistance	[37]
Clay	SiO_2	High tensile and flexural strength	- ac breakdown strength	[38]
SiO_2	SiO_2	No measurement	- ac breakdown strength - Partial discharge resistance - Permittivity - Space charge - dc conductivity	[7]
SiO_2	SiO_2	Glass transition temperature	- ac breakdown strength - Permittivity	[39]
TiO_2	SiO_2	Low thermal expansion	- ac breakdown strength - Electrical treeing resistance - Partial discharge resistance	[40]
SiO_2	BN & AlN[c]	High thermal conductivity	- ac breakdown strength - Partial discharge resistance	[41]
Al_2O_3	Al_2O_3	High thermal conductivity	- ac breakdown strength - Partial discharge resistance	[42]

[a]Filler content is a few wt%, and average filler diameter is 1–100 nm.
[b]Filler content is 50–70 wt%, and average filler diameter is 1–100 μm.
[c]BN means Boron Nitride, and AlN means Aluminum Nitride.
[d]Typical papers are shown as references.

3.4.2 Fabrication Methods for Nano-Micro Composites

As in the case of nanocomposites, the production of nano-micro composites relies critically on the homogenous dispersion of the nano-filler components. Therefore, the first step in fabrication is the dispersion of the nano-fillers according to the methods described in previous sections.

Figure 3.15 shows a typical fabrication scheme for epoxy-based nano-micro composites. In the first step, the nano-fillers are dispersed in the epoxy resin by mixing with shear force. In the second step, compounding of the nano-filler dispersed epoxy resin with the micro-fillers is performed. As described elsewhere, surface treatment using silane coupling agents improves the dispersibility of the nano-fillers and micro-fillers in the epoxy resin. After this, the nano-micro composite is produced by degassing, casting and curing in the presence of a hardener. The order of dispersion of nano-fillers and micro-fillers is important, since dispersion of nano-fillers into an existing micro-filled resin can lead to retention of aggregated and agglomerated nano-fillers, due to a lack of shear force during compounding. Therefore, nano-micro composites produced using the reverse order of filler dispersion to that described above does not seem to lead to final materials with superior electrical properties.

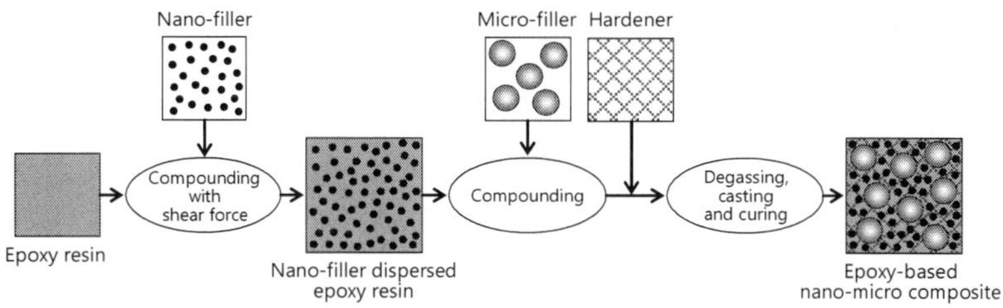

Figure 3.15 Typical fabrication scheme of epoxy-based nano-micro composite.

SEM and TEM images of epoxy-based nano-micro composites containing a few wt% of nano-fillers and 65 wt% of micro-fillers are shown in Fig. 3.16 [40]. In these composites, flake-like shaped clay with dimensions of about 100 × 100 × 1 nm, spherical silica (SiO_2) with an average filler diameter of about 12 nm and cylinder shaped titania (TiO_2) with dimensions of about 15 nm in diameter and 20–100 nm in length are used as nano-fillers, and silica (SiO_2) with an average filler diameter of about 17 μm is used as the micro-filler. These images show that the areas surrounding the micro-fillers are filled with nano-fillers, and demonstrate that these composites have structures with densely packed fillers. Moreover, these nano-micro composites have the dual function of low thermal expansion and superior electrical properties. All these composites have low coefficients of thermal expansion (α_1), comparable to that of aluminum, which is often used as the metallic conductor in heavy electric apparatus. In addition, all these composites have superior resistance to electrical treeing to that of composites containing only 65 wt% of micro-fillers. In particular, the addition of clay nano-fillers enables nano-micro composite to have more than 20 times greater resistance to electrical treeing than composite containing 65 wt% of micro-filler alone.

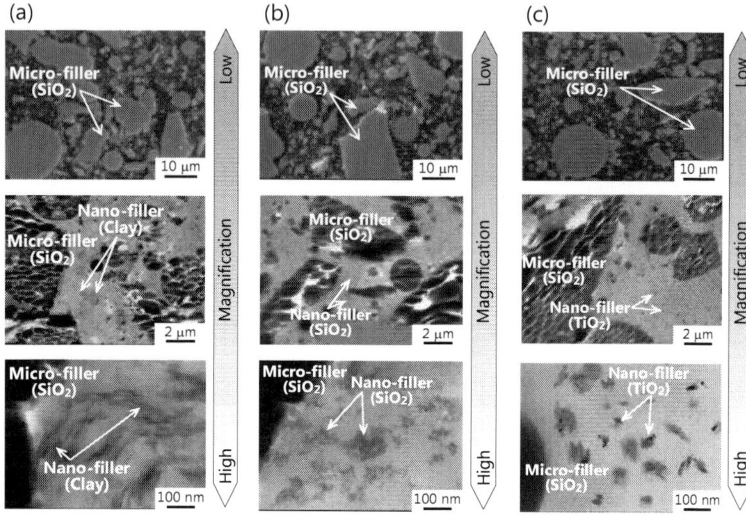

Figure 3.16 SEM and TEM images of nano-micro composites [40]. (a) Nano-micro composite containing clay nano-fillers and SiO_2 micro-fillers. (b) Nano-micro composite containing SiO_2 nano-fillers and SiO_2 micro-fillers. (c) Nano-micro composite containing TiO_2 nano-fillers and SiO_2 micro-fillers.

3.5 Evaluation of the Nano-Filler Dispersion State in Nanocomposites

The nano-filler dispersion state has an impact on the mechanical, thermal and electrical properties of nanocomposites and, consequently, the reliable evaluation of the nano-filler dispersion states is important in the fabrication of nanocomposites. Analysis instruments have greatly improved and rapidly advanced in recent years and can now provide detailed information concerning nano-filler dispersion states. This section therefore describes evaluation examples of nano-filler dispersion states using SEM, TEM, EDX, WAXD, and SAXS.

3.5.1 Scanning Electron Microscopy

Scanning electron microscopy (SEM) is the most readily available evaluation method for nano-filler dispersion. Moreover, a field emission scanning electron microscopy (FE-SEM) provides images with higher resolution than conventional SEM.

Figure 3.17a shows a cross-section image of an epoxy/clay nanocomposite [40], which shows that, in this system, the clay nano-fillers are dispersed homogenously. Figure 3.17b shows a fracture surface image of a sample of silicone rubber/boehmite alumina nanocomposite [25]. In this example, image analysis has facilitated the quantitative evaluation of binarized images derived from the original SEM images, using image analysis software. The filler size distribution is estimated based on the binarized images.

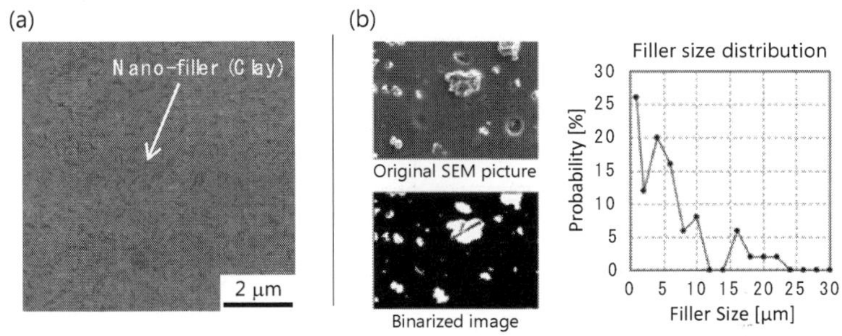

Figure 3.17 SEM images of nano-filler dispersion state in nanocomposites. (a) Epoxy/clay nanocomposite [40]. (b) Silicone rubber/boehmite alumina nanocomposite [25].

3.5.2 Transmission Electron Microscopy

Transmission electron microscopy (TEM) provides images with much higher resolution than SEM. However, this requires the production of thin sections for TEM observation, which is most usually accomplished using an ultramicrotome.

Figures 3.18a,b show cross-sectional images of an epoxy/clay nanocomposite and a polyester-imide/silica nanocomposite, respectively [43, 44]. The dispersion states of the clay and silica nano-fillers are directly observed in the TEM images. Moreover, TEM tomography (three-dimensional TEM) has recently become available for the evaluation of nano-filler dispersion states.

Figure 3.18c shows a TEM tomography image of an epoxy/silica nanocomposite [45]. The 1284 silica nano-filler particles exist within the observed volume, which has dimension of 594 nm × 596 nm × 104 nm dimension, and their three-dimensional configuration is confirmed.

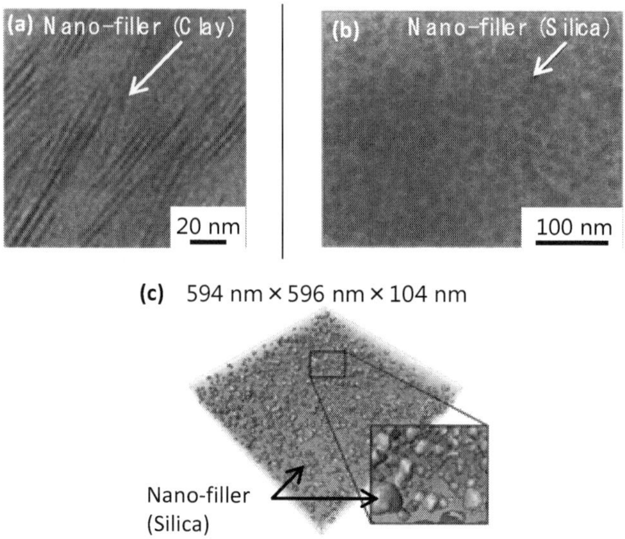

Figure 3.18 TEM images of nano-filler dispersion state in nanocomposites. (a) Epoxy/clay nanocomposite [43]. (b) Polyester-imide/silica nanocomposite [44]. (c) Epoxy/silica nanocomposite [45].

3.5.3 Energy Dispersive X-ray Spectroscopy

Energy dispersive X-ray spectroscopy (EDX) detects the characteristic X-rays generated by electron irradiation, and energy dispersive spectroscopy enables elemental analysis. EDX provides information regarding elements within the excited volumes within the specimen and can be used in conjunction with both SEM or TEM.

Figure 3.19 shows elemental maps obtained from a nano-micro composite containing a few wt% of TiO_2 nano-filler plus 65 wt% of SiO_2 micro-filler [46]. These elemental maps reveal the configuration of the TiO_2 nano-fillers and SiO_2 micro-fillers; the area adjacent to the SiO_2 micro-silica fillers is filled with TiO_2 nano-fillers.

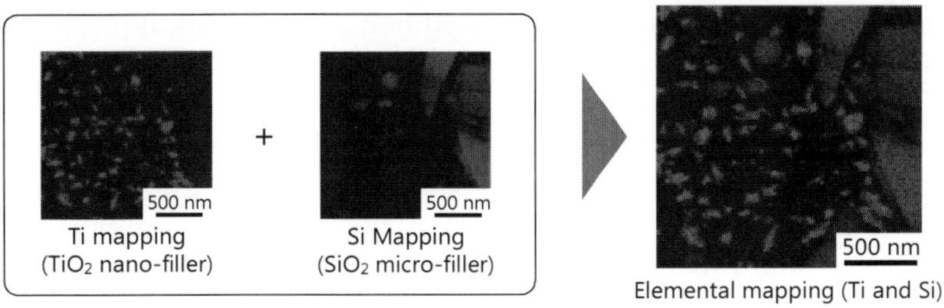

Figure 3.19 Elemental mapping using EDX in nano-micro composite containing TiO_2 nano-filler and SiO_2 micro-filler [46].

3.5.4 Wide Angle X-ray Diffraction

Wide angle X-ray diffraction (WAXD) can provide information regarding the exfoliation of clay nano-fillers within clay nanocomposites.

Figure 3.20a shows the influence of hardener (curing agent) reactivity on the structure of nanocomposites [47]. In nanocomposites including organically modified clays containing octadecylammonium ions (H_2N^+-$(CH_2)_{17}$-CH_3), the basal spacing (d_{001} spacing) is dependent on the kind of hardener. Clay nanocomposite cured with (i) PACM, has a diffraction peak indicative of a 3.7 nm of basal spacing while the basal spacing of the clay nanocomposite cured with (ii) DDDHM expands to 4.0 nm.

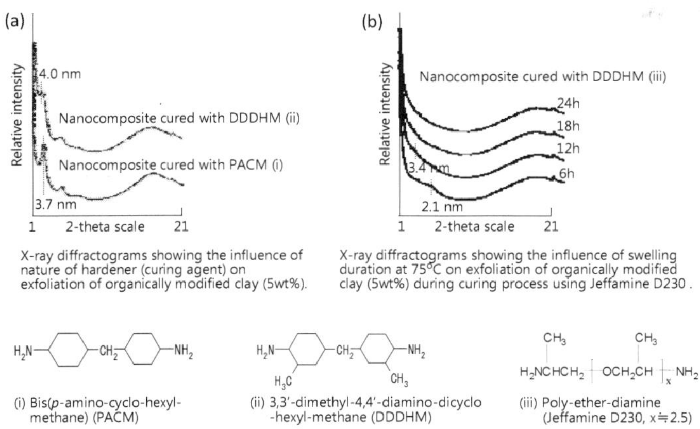

Figure 3.20 Evaluation of clay exfoliation using WAXD in epoxy/clay nanocomposites [47]. (a) Influence of nature of amine hardeners (curing agents). (b) Influence of mixing time of epoxy resin and clays.

Figure 3.20b also shows the influence of mixing time on the structure of the resulting epoxy resin/clay nanocomposites. The epoxy resin and the clays modified by octadecylammonium ions were mixed for 6, 12, 18, and 24 h before each mixture with (iii) poly-ether-diamine (Jeffamine D230). The WAXD spectra of the nanocomposites show that the basal spacing of the clay expands from 2.1 nm to 3.4 nm with increase in mixing time. The diffraction peak, which is attributed to the laminated structure of the clay, disappears in the nanocomposite after 24 h of mixing, indicating a loss of the initial periodic structure of the clay.

3.5.5 Small Angle X-ray Scattering

Small angle X-ray scattering (SAXS) can provide information on filler diameter, filler shape, distance between filler particles and the filler size distribution in nanocomposites. For example, when inter-particle scattering is observed in SAXS analysis of filler-dispersed materials, the filler size distribution can be calculated. The SAXS intensity depends on the difference in electron density between the particles and surrounding polymer and is represented by the following formula:

$$I(q) = |\Omega(q)|^2 \quad (q = 4\pi \sin\theta/\lambda),$$

where λ is the incident wave length.

Moreover, when the scattering particles are spherical with radius R, the SAXS spectrum is represented by the following formula:

$$\Omega(q,R) = \frac{4\pi R^3}{(qR)^3}(\sin(qR) - qR\cos(qR))$$

Fittings of these formulae to the observed SAXS pattern therefore lead to an estimate of the average dimensions of the disperse phase.

Figure 3.21a shows a SAXS patterns obtained from epoxy/silica nanocomposite containing colloidal silica nano-fillers with an average diameter of 110 nm [48]. Peaks attributed to inter-particle scattering appear in the small angle range (q [1/nm] = 0.02–0.06), which shift to higher q with increasing nano-filler content (vol%). Moreover, the separation between nanoparticles is estimated based on the nano-filler content and the shift in peak position, as shown in Fig. 3.21b. These distances are in accordance with direct observation of the nanocomposites using SEM, as shown in Fig. 3.21c.

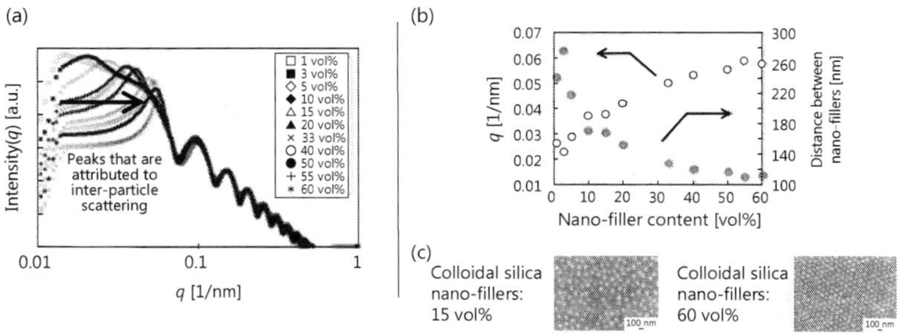

Figure 3.21 Evaluation of nano-fillers dispersion state using SAXS in epoxy/colloidal silica nanocomposites [48]. (a) SAXS patterns. (b) Distance between the colloidal silica nanoparticles. (c) SEM images of nanocomposites containing colloidal silica nano-fillers.

3.6 Conclusions

The main objective of this chapter is to illustrate the range of fabrication methods available for the production of polymer nanocomposite dielectrics. The properties of nanocomposites depend on the fabrication method and, in particular, the nano-filler dispersion state has a major impact. The first half of this chapter describes the selection of combinations of nano-fillers and polymers and fabrication methods, such as sol-gel, direct dispersion and clay dispersion methods. The latter half of the chapter then focuses on nano-micro composites as an effective approach to the practical use of nanocomposites and considers preparation methods and the selection of combinations of nano-fillers and micro-fillers.

A lot of research and development of nanocomposites have been conducted, and improvements in fabrication methods continue to this day. However, most work concerning nanocomposite fabrication has involved the laboratory scale, ranging from several hundreds of grams to several kilograms. Fabrications of industrial-scale (more than several tons) are needed for the practical use of nanocomposites; some nanocomposites have nevertheless been commercialized. Generally, scale-up from the laboratory- to the industrial-scale seems to have technical challenges. However, nanocomposites offer many industrial advantages and, consequently, they are certain to become leading candidates for use as near-future high performance materials.

References

1. Henk, P. O., Kortsen, T. W., and Kvarts, T. (1999). Increasing the electrical discharge endurance of acid anhydride cured DGEBA epoxy resin by dispersion of nanoparticle silica, *High Perform. Polym.*, **11**, pp. 281–296.
2. Woodcock, L. V. (1987). Proceedings of a workshop held at Zentrum fur Interdisziplinare Forschung University Bielefield, November 11–13, edited by Dorfmiiller, T., and Williams, G. ((1987) *Lect. Notes Phys.*, **277**, pp. 113–124).
3. Imai, T., Sawa, F., Yoshimitsu, T., Ozaki, T., and Shimizu, T. (2004). Preparation and insulation properties of epoxy-layered silicate nanocomposite, *Annual Report of IEEE Conference on Electrical Insulation and Dielectric Phenomena (CEIDP)*, pp. 402–405.
4. Chen, Y., Imai, T., Ohki, Y., and Tanaka, T. (2010). Tree initiation phenomena in nanostructured epoxy composites, *IEEE Trans. Dielectr. Electr. Insul.*, **17**, pp. 1509–1515.
5. Tanaka, T., Ohki, Y., Ochi, M., Harada, M., and Imai, T. (2008). Enhanced partial discharge resistance of epoxy/clay nanocomposite prepared by newly developed organic modification and solubilization methods, *IEEE Trans. Dielectr. Electr. Insul.*, **15**, pp. 81–89.
6. Tanaka, T., and Iizuka, T. (2010). Generic PD resistance characteristics of polymer (epoxy) nanocomposite, *Annual Report of IEEE Conference on Electrical Insulation and Dielectric Phenomena (CEIDP)*, pp. 518–521.
7. Krivda, A., Tanaka, T., Fréchette, M., Gorur, R., Morshuis, P., Gubanski, S., Kindersberger, K., Vaugahn, A., Pelissou, S., Tanaka, Y., Schmidt, L. E., Iyer, G., Andritsch, T., Seiler, J., and Anglhuber, M. (2012). Characterization of epoxy microcomposite and nanocomposite materials for power engineering applications, *IEEE Electr. Insul. Magazine*, **28**, No. 21, pp. 38–51.
8. Imai, T., Sawa, F., Ozaki, T., Inoue, Y., Shimizu, T., and Tanaka, T. (2006). Comparison of insulation breakdown properties of epoxy nanocomposites under homogeneous and divergent electric fields,

Annual Report of IEEE Conference on Electrical Insulation and Dielectric Phenomena (CEIDP), pp. 306–309.

9. Kozako, M., Kuge, S., Imai, T., Ozaki, T., Shimizu, T., and Tanaka, T. (2005). Surface erosion due to partial discharges on several kinds of epoxy nanocomposites, *Annual Report of IEEE Conference on Electrical Insulation and Dielectric Phenomena (CEIDP)*, pp. 162–165.
10. Nelson, J. K., and Hu, Y. (2004). The impact of nanocomposite formulations on electrical voltage endurance, *Proceedings of the IEEE International Conference on Solid Dielectrics (ICSD)*, pp. 832–835.
11. Nelson, J. K., Utracki, L. A., Maccrone, R. K., and Reed, C. W. (2004). Role of the interface in determining the dielectric properties of nanocomposites base resin, *Annual Report of IEEE Conference on Electrical Insulation and Dielectric Phenomena (CEIDP)*, pp. 314–317.
12. Hayakawa, N., Kurimoto, M., Fujii, Y., Shimomura, J., Hanai, M., Hoshina, Y., Takei, M., and Okubo, H. (2010). Influence of dispersed nanoparticle content on dielectric property in epoxy/alumina nanocomposites, *Annual Report of IEEE Conference on Electrical Insulation and Dielectric Phenomena (CEIDP)*, pp. 572–575.
13. Kochetov, R., Andritsch, T., Morshuis, P. H. F., and Smit, J. J. (2012). Anomalous behaviour of the dielectric spectroscopy response of nanocomposite, *IEEE Trans. Dielectr. Electr. Insul.*, **19**, pp. 107–117.
14. Huang, X., Li, Y., Liu, F., Jiang, P., Iizuka, T., Tatsumi, K., and Tanaka, T. (2014). Electrical properties of epoxy/POSS composites with homogeneous nanostructure, *IEEE Trans. Dielectr. Electr. Insul.*, **21**, pp. 1516–1528.
15. El-Hag, A. H., Simon, L. C., Jayaram, S. H., and Cherney, E. A. (2006). Erosion resistance of nano-filled silicone rubber, *IEEE Trans. Dielectr. Electr. Insul.*, **13**, pp. 122–128.
16. Kozako, M., Fuse, N., Ohki, Y., Okamoto, T., and Tanaka, T. (2004). Surface degradation of polyamide nanocomposites caused by partial discharges using IEC (b) electrodes, *IEEE Trans. Dielectr. Electr. Insul.*, **11**, pp. 833–839.
17. Zhang, M. Y., Niu, Y., Chen, A. Y., Chang, W., and Hao G. W. (2007). Study on mechanical and corona-resistance property of polyimide/silica hybrid films, *Proceedings of the IEEE International Conference on Solid Dielectrics (ICSD)*, pp. 353–356.
18. Tanaka, T., Bulinski, A., Castellon, J., Fréchette, M., Gubanski, S., Kindersberger, J., Montanari, G. C., Nagao, M., Morshuis, P., Tanaka, Y., Pélissou, S., Vaughan, A., Ohki, Y., Reed, C. W., Sutton, S., and Han, S. J.

(2011). Dielectric properties of XLPE/SiO_2 nanocomposites based on CIGRE WG D1.24 cooperative test results, *IEEE Trans. Dielectr. Electr. Insul.*, **18**, pp. 1484–1517.

19. Hui, L., Smith, R., Nelson, J. K., and Schadle, L. S. (2009). Electrochemical treeing in XLPE/silica nanocomposites, *Annual Report of IEEE Conference on Electrical Insulation and Dielectric Phenomena (CEIDP)*, pp. 111–114.
20. Hayase, Y., Aoyama, H., Matsui, K., Tanaka, Y., Takada, T., and Murata. Y. (2006). Space charge formation in LDPE/MgO nano-composite film under ultra-high DC electric stress, *IEEJ Trans. Fundamentals and Materials*, **126**, pp. 1084–1089.
21. Nagao, M., Watanabe, S., Murakami, Y., Murata, Y., Sekiguchi, Y., and Goshowaki, M. (2008). Water tree retardation of MgO/LDPE and MgO/XLPE nanocomposites, *Proc. IEEJ International Symposium on Electrical Insulating Materials (ISEIM)*, pp. 483–486.
22. Kikuchi, H., and Hanawa, H. (2012). Inverter surge resistant enameled wire with nanocomposite insulating material, *IEEE Trans. Dielectr. Electr. Insul.*, **19**, pp. 99–106.
23. Montanari, G. C., Fabiani, D., Palmieri, F., Kaempfer, D., Thomann, R., and Mülhaupt, R. (2004). Modification of electrical properties and performance of EVA and PP insulation through nanostructure by organophilic silicates, *IEEE Trans. Dielectr. Electr. Insul.*, **11**, pp. 754–762.
24. Evonik Industries AG, *Silica Nanocomposites–NANOCRYL-NANOPOX-NANOPAL-*, pp. 148–155.
25. Kozako, M., Higashikoji, M., Tominaga, T., Hikita, M., Ueta, G., Okabe, S., and Tanaka, T. (2012). Fabrication of silicone rubber nanocomposites and quantitative evaluation of dispersion state of nanofillers, *IEEE Trans. Dielectr. Electr. Insul.*, **19**, pp. 1760–1767.
26. Kurimoto, M., Okubo, H., Kato, K., Hanai, M., Hoshina, Y., Takei, M., and Hayakawa, N. (2010). permittivity characteristics of epoxy/alumina nanocomposite with high particle dispersibility by combining ultrasonic wave and centrifugal force, *IEEE Trans. Dielectr. Electr. Insul.*, **17**, pp. 662–670.
27. Bian, S., Jayaram, S., and Cherney, E. A. (2012). Electrospinning as a new method of preparing nanofilled silicone rubber composites, *IEEE Trans. Dielectr. Electr. Insul.*, **19**, pp. 777–785.
28. Imai, T., Sawa, F., Ozaki, T., and Shimizu, T. (2004). Partial discharge resistance enhancement of organic insulating materials by nano-

size particle dispersion technique, *IEEJ Paper of Technical Meeting on Dielectrics and Electrical Insulation*, No. DEI-04-78, pp. 35–38 (in Japanese).

29. Chujo, K., and Abe, K. (2001). *Technological Trend of Polymer Nanocomposites*, CMC Publishing Co., Ltd. (in Japanese).
30. Fukushima, Y., and Inagaki, S. (1987). Synthesis of an intercalated compound of montmorillonite and 6-polyamide, *J. Inclusion Phenom.*, **5**, pp. 473–482.
31. Kato, M. (2009). *Eng. Mater.*, **57**, No. 7, pp. 31–35 (in Japanese).
32. Harada, M., Miyamoto, T., and Ochi, M. (2007). Dispersibility and mechanical properties of epoxy/clay nanocomposites prepared by different clay treatment methods, *Proceedings of the 10th Pacific Polymer Conference (PPC10)*, p. 240.
33. Park, J., and Lee, J. (2010). A new dispersion method for the preparation of polymer/organoclay nanocomposite in the electric fields, *IEEE Trans. Dielectr. Electr. Insul.*, **17**, pp. 1516–1522.
34. Matsumoto, A., Otsuka, K., Kimura, H., and Yamaoka, K. (2009). *Proceedings of the 59th the Network Polymer Symposium Japan*, pp. 233–234 (in Japanese).
35. Hasegawa, N., and Usuki, A. (2001). Development of polyamide-clay nanocomposites, *Proceedings of the Annual Meeting of the Japan Society of Polymer Processing (JSPP)*, pp. 153–154 (in Japanese).
36. Imai, T., Sawa, F., Nakano, T., Ozaki, T., Shimizu, T., Kuge, S., Kozako, M., and Tanaka, T. (2005). Insulation properties of nano- and micro-filler mixture composite, *Annual Report of IEEE Conference on Electrical Insulation and Dielectric Phenomena (CEIDP)*, pp. 171–174.
37. Fréchette, M. F., Larocque, R. Y., Trudeau, M. L., Veillette, R., Cole, K. C., and That, M.-T. T. (2005). Nanodielectric surface performance when submitted to partial discharges in compressed air, *Annual Report of IEEE Conference on Electrical Insulation and Dielectric Phenomena (CEIDP)*, pp. 727–731.
38. Park, J., Lee, C., Lee, J., and Kim, H. (2011). Preparation of epoxy/micro- and nano-composites by electric field dispersion process and its mechanical and electrical properties, *IEEE Trans. Dielectr. Electr. Insul.*, **18**, pp. 667–674.
39. Nguyen, V., Vaughan, A. S., Lewin, P. L., and Krivda, A. (2011). Stoichiometry and effects of nano-sized and micro-sized fillers on an epoxy based system, *Annual Report of IEEE Conference on Electrical Insulation and Dielectric Phenomena (CEIDP)*, pp. 302–305.
40. Imai, T., Komiya, G., Murayama, K., Ozaki, T., Sawa, F., Shimizu, T., Harada, M., Ochi, M., Ohki, Y., and Tanaka, T. (2008). Improving

epoxy-based insulating materials with nano-fillers toward practical application, *Proceedings of the IEEE International Symposium on Electrical Insulation (ISEI)*, pp. 201–204.

41. Tanaka, T., Wang, Z., Iizuka, T., Kozako, M., and Ohki, Y. (2011). High thermal conductivity epoxy/bn composites with sufficient dielectric breakdown strength, *Annual Report of IEEE Conference on Electrical Insulation and Dielectric Phenomena (CEIDP)*, pp. 691–694.
42. Li, Z., Okamoto, K., Ohki, Y., and Tanaka, T. (2011). The role of nano and micro particles on partial discharge and breakdown strength in epoxy composites, *IEEE Trans. Dielectr. Electr. Insul.*, **18**, pp. 675–681.
43. Imai, T., Sawa, F., Ozaki, T., Nakano, T., Shimizu, T., and Yoshimitsu, T. (2004). Preparation and insulation properties of epoxy-layered silicate nanocomposite, *IEEJ Trans. Fundamentals and Materials*, **124**, pp. 1065–1072 (in Japanese).
44. Kikuchi, H., and Asano, K. (2006). Development of organic/inorganic nano-composite enameled wire, *IEEJ Trans. Power and Energy*, **126**, pp. 460–465 (in Japanese).
45. Meichsner, C., Clark, T., Groeppel, P., Winter, B., Butz, B., and Spiecker, E. (2012). Formation of a protective layer during IEC(b) test of epoxy resin loaded with silica nanoparticles, *IEEE Trans. Dielectr. Electr. Insul.*, **19**, pp. 786–792.
46. Imai, T., Sawa, F., Ozaki, T., Inoue, Y., Shimizu, T., and Tanaka, T. (2007). Roles of fillers on properties of nano-TiO_2 and micro-SiO_2 filler mixed composites, *Proceedings of the IEEE International Conference on Solid Dielectrics (ICSD)*, pp. 407–410.
47. Kornmann, X., Lindberg, H., and Berglund, L. A. (2001). Synthesis of epoxy–clay nanocomposites. Influence of the nature of the curing agent on structure, *Polymer*, **42**, pp. 4493–4499.
48. Senoo, M., Takeuchi, K., Oka, A., Shimonabe, Y., Kuwamoto, S., Urushibara, Y., Matsui, S., and Nakamae, M. (2009). *Proceedings of the 59th the Network Polymer Symposium Japan*, pp. 29–32 (in Japanese).

Chapter 4

Nanoparticle Surface Modification for Dielectric Polymer Nanocomposites

Xingyi Huang and Pingkai Jiang

Shanghai Key Laboratory of Electrical Insulation and Thermal Ageing,
Department of Polymer Science and Engineering,
Research Center of Dielectrics and Electrical Insulation,
Shanghai Jiao Tong University, 800 Dongchuan Road, Shanghai, China

xyhuang@sjtu.edu.cn

Interfaces are considered as one the crucial factors determining the properties of the polymer/particulate nanocomposites. Often, a prerequisite for obtaining the desired properties in the nanocomposites is the proper surface modification of the nanoparticles. This chapter presents the state of the art in methods and strategies for engineering the nanoparticles surfaces, including the utilization of silane coupling agents, titanate coupling agents, phosphonic acid/phosphonates, dopamine, polymer coating and inorganic coating. For each strategy, case studies are provided for illustrating the importance of the nanoparticle surface modification in determining the electrical properties of the polymer nanocomposites.

Tailoring of Nanocomposite Dielectrics: From Fundamentals to Devices and Applications
Edited by Toshikatsu Tanaka and Alun S. Vaughan

ISBN 978-981-4669-80-1 (Hardcover), 978-981-4669-81-8 (eBook)
www.panstanford.com

4.1 Introduction

Nanoparticles have a large specific surface area and unique physical properties, offering us many chances for designing dielectric polymer nanocomposites with desirable performance. However, there are also many challenges in the realization of high electrical performance, one of which is the homogeneous dispersion of the nanoparticles in a polymer matrix. Owing to having the large specific surface area and, consequently, many surface functional groups, nanoparticles are characterized by a high surface energy and strong particle/particle interaction. In this case, they commonly form large aggregates and agglomerates rather than single nanoparticles, when dispersed in a polymer matrix, which results in poor electrical properties. Another challenge is the ability to tailor the nanoparticle/polymer interface. This is because, apart from nanoparticle dispersion, the physicochemical properties of the interfacial region also play a critical role in determining the electrical properties of the nanocomposites. Taking these challenges into consideration, it is believed that the key route to achieving desirable electrical properties in polymer nanocomposites is to engineer the nanoparticle surface appropriately. This chapter summarizes the main strategies for engineering the nanoparticles surface, including silanization, grafting of titanate coupling agents, phosphorylation, modification by dopamine, polymer coating and inorganic coating. The effect of nanoparticle surface modification on the electrical properties of the polymer nanocomposites will also be briefly reviewed.

4.2 Nanoparticle Surface Chemistry

Depending on the preparation methods, nanoparticles will carry different chemical groups on their surfaces, but most of them are oxygen-containing terminations such as –OH, –COOH, –C=O, and –O–; these groups can be exploited to introduce the required functionality during surface modification. Some nanoparticles, such as carbon nanotubes (CNTs) and boron nitride nanotubes (BNNTs), have no reactive groups for further surface modification, in which case, pre-treatment is necessary. For example, the treatment of carbon nanotubes is usually executed by strong acid (e.g., 3:1 mixture of concentrated H_2SO_4 and HNO_3), which can introduce oxygenated groups (mainly carboxyl and hydroxyl) onto the graphitic surface (Kim and Sigmund, 2004). Treatment of BNNTs with H_2O_2 has been performed, leading to the introduction of –OH (Zhi et al., 2009). Some nanoparticles, such as $BaTiO_3$, carry a small number of reactive groups (i.e., –OH), which may not result in effective surface modification of the nanoparticles. In this case, the introduction of more –OH is necessary and it has been shown that treatment of $BaTiO_3$ by H_2O_2 is effective means of achieving this (Xie et al., 2011).

Inorganic nanoparticles such as SiO_2, TiO_2, $BaTiO_3$, and ZnO usually carry –OH on their surface. Apart from –OH, other oxygen-containing groups such as –COOH, –C=O and –O– groups have been found as the main surface groups of manmade diamonds, acid-treated CNTs and graphene oxide (GO) (Dreyer et al., 2010; Krueger and Lang, 2012). Some non-oxygen-containing groups such as NH_2 may be also found on nanoparticle (e.g., BN) surfaces (Zhi et al., 2008).

4.3 General Strategies for Nanoparticle Surface Modification

Nanoparticle surface modification can be performed through covalent or non-covalent routes, or through a combination of both. Non-covalent surface modification differs from covalent surface modification in that it does not involve the formation of covalent bonds between the nanoparticle and the modifiers. Non-covalent approaches can be divided into three main categories: electrostatic (i.e., hydrogen bonding); π-effects (i.e., π-π interactions); van der Waals forces (i.e., dipole–dipole interaction). Although non-covalent surface modification is effective in improving nanoparticle dispersion, covalent surface modification is more attractive in designing and tailoring the nanoparticle/polymer interface. In the following, six types of covalent surface modification methods will be introduced.

4.4 Nanoparticle Surface Modification by Silane Coupling Agents

4.4.1 The Chemistry of Silanes

Silane coupling agents are generally characterized by the formulae of $YCH_2CH_2CH_2Si(X)_3$ and $YCH_2CH_2Si(CH_3)(X)_2$, as shown in Fig. 4.1. Here, X represents some hydrolysable group, which can react with -OH groups on the nanoparticle surfaces to form covalent bonds between nanoparticles and the silanes. X is an usually alkoxy group (e.g., OCH_3 and OCH_2CH_3) or a halogen (e.g., Cl). Y can be, for example, a long alkyl chains, vinyl, epoxy, amino, methacryloxy, mercapto, or isocyanate group. As such, Y can be a non-reactive group, as in the case of alkyl chains, or may be reactive, such that reactions between the silane and the polymer result in the formation of covalent interactions between the nanoparticles and the matrix.

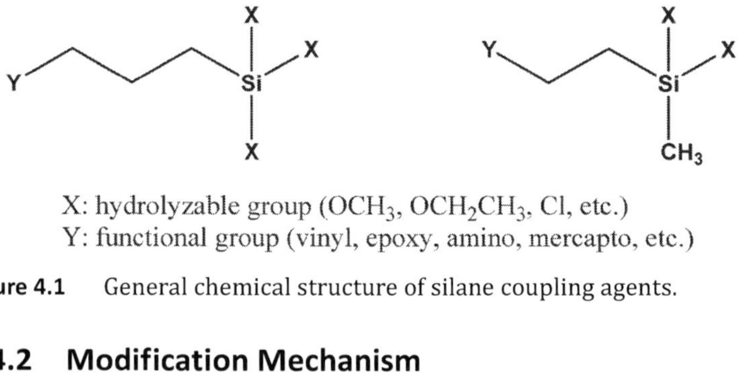

X: hydrolyzable group (OCH_3, OCH_2CH_3, Cl, etc.)
Y: functional group (vinyl, epoxy, amino, mercapto, etc.)

Figure 4.1 General chemical structure of silane coupling agents.

4.4.2 Modification Mechanism

In 1977, Arkles proposed a reaction process between silanes and the -OH groups of a substrate surface, which is shown in Fig. 4.2 (Arkles, 1977), which can be considered in terms of four steps. First, the alkoxy groups of the silane are hydrolyzed to form silanol functional groups. Then, the silanol groups form hydrogen bonds with -OH groups on the substrate surface. Finally, a covalent bond between the substrate and the hydrolyzed silane is formed. It should be noted that condensation, hydrogen bonding and the formation of covalent bonds may occur simultaneously.

Generally, not all silanol groups can react with -OH groups on the substrate surface. Taking (3-aminopropyl)triethoxysilane

(APTES) as an typical example, the possible reaction routes involved in nanoparticle silanization by APTES are shown in Fig. 4.3. The remaining silanol groups can be free or may form hydrogen bonds with the amino groups of APTES. Alternatively,

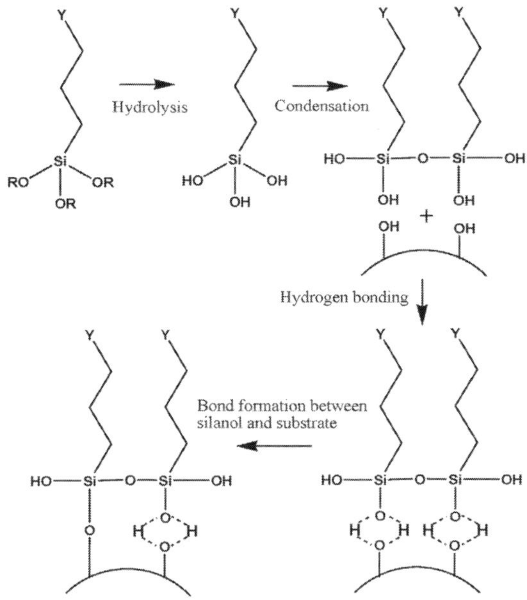

Figure 4.2 The process of silanization reaction.

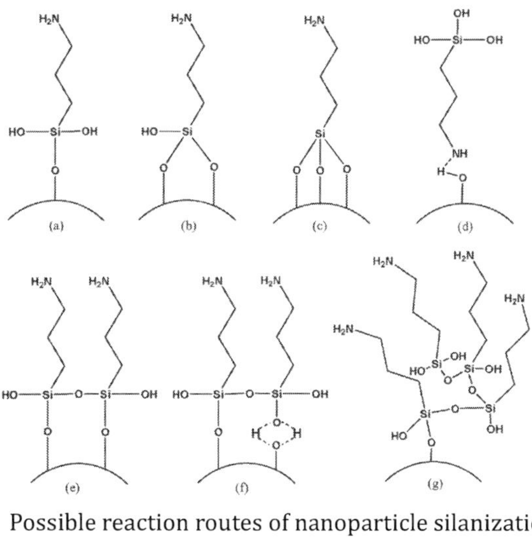

Figure 4.3 Possible reaction routes of nanoparticle silanization.

they may react with the silanol groups of other hydrolyzed silanes, resulting in a multilayer of silane on the nanoparticle surfaces. In this case, the amino groups may be buried in the inner silane layer, making some of them not available for further reaction with other chemicals.

4.4.3 Nanoparticle Modification Methods

Two methods for surface silanization are commonly used in the laboratory: the anhydrous (dry) method and hydrous (wet) method. In the dry method, the silanes are first resolved in an anhydrous solvent (e.g., toluene) and then slowly added into the solvent/nanoparticle mixture, whereupon the reaction can be easily performed at extended reaction time (e.g., several to 10 h) and elevated temperatures (e.g., 50 °C to the boiling point of the solvent). Increasing the reaction temperature reduces the reaction time, as does adding a catalyst (e.g., p-toluenesulfonic acid, PTSA). Indeed, it has been shown that the modification of the silica nanoparticles is complete within 2 h when using methoxysilanes in refluxing toluene (i.e., ~110 °C), with PTSA as catalyst (Garcia et al., 2007). It has also been reported that the successful modification of $BaTiO_3$ by APTES can be carried out in toluene within 24 h at 80 °C (Xie et al., 2011). The advantage of this dry method is that the silanes tend to form a single layer on the nanoparticle surface.

In the wet method, the silanization of nanoparticles is carried out in the presence of water, either in water or in a water/alcohol solution. In this case, the pH significantly affects the reaction between silanes and the –OH groups on substrate surface, because the reaction starts with the hydrolysis of the silanes. At low pH values, the silanes have a high rate of hydrolysis and the silanols so formed are stable. However, the reactivity of silanols increases at pH < 1.5 or pH > 4.5. Combining these two aspects, surface modification is usually carried out at a pH in the range 3–4 (Pujari et al., 2014). When the pH is about 7, the hydrolysis rate of silanes is very low and the silanols are not stable. At pH > 8, hydrolysis of silanes occurs rapidly the resulting silanols are unstable and condense quickly to from Si–O–Si networks. Under such conditions, the formation of Si–O–Si cannot be controlled, resulting in multilayers of silanes on the substrate surfaces (Fig. 4.3g).

It should be noted that both the alkoxy functional groups (X in Fig. 4.1) and the organic substituents (Y in Fig. 4.1) in the silane affect the hydrolysis rate. As for the alkoxy functional groups, the order of hydrolysis rate varies according to methoxy > ethoxy >> propoxy. Also, the hydrolysis rate of silanes can be increased by increasing the polarity of the organic substituents, such that amino > epoxy > mercapto > methacryloxy ≈ vinyl > alkyl.

Industrially, silane-modified nanoparticles have been produced on a large scale. For this, the silanes (pure or solution) are usually uniformly sprayed onto the nanoparticle surfaces in the presence of high rate of shear, before being heat treated. Alternatively, the silane can be directly added into the matrix polymers during processing, either prior to or together with the addition of nanoparticles.

4.4.4 Choosing Proper Silanes for Nanoparticles Surface Modification

The surface modification of nanoparticles by silane coupling agents can improve the electrical properties of the dielectric polymer nanocomposites by modifying the surface chemistry (e.g., displacing adsorbed water molecules), improving the nanoparticle dispersion and enhancing nanoparticle/polymer interactions. In some cases, the silane functional groups also play an important role in determining the electrical performance of the nanocomposites. It is also expected that the electrical parameters such as polarizability and electronegativity of the silane functional groups may have an important influence on the electrical properties of the resulting polymer nanocomposites. In addition to enhancing electrical properties, silane surface modification can also be aimed at improving the mechanical, processing and barrier properties of polymer nanocomposites.

A list of silane functional groups, their compatible polymers and the corresponding methoxysilanes are given in the appendix. Generally, the following principles can be roughly used as a guide to choose a proper silane coupling agent for a specific polymer matrix:

(i) The functional groups have a good compatibility with the polymer matrix. For example, alkylsilanes can be used to modify the nanoparticles when polyolefins are used as the matrix.

(ii) The functional groups can react with the polymer matrix or have a strong interaction with the polymer matrix by non-covalent bonds (e.g., hydrogen bonding). For example, when epoxy resins are used, silanes with epoxy or amino functional groups which can take part in the curing process are suitable.

4.4.5 Effect of Nanoparticle Surface Modification on the Electrical Properties of Polymer Nanocomposites

When nanoparticles surfaces are modified with silane coupling agents, the inherent electrical characteristics of the nanoparticles do not usually change. However, the physicochemical properties of the nanoparticle surfaces may be significantly changed which, ultimately, influences the electrical properties of the polymer nanocomposite. In the following, influence of silane surface modification on the dielectric constant, dielectric loss tangent, electrical conductivity (absorption current), voltage endurance time and breakdown strength of polymer nanocomposites will be discussed.

4.4.5.1 Dielectric constant and dielectric loss tangent

Surface modification by silanes can remove impurities (e.g., water molecules) adsorbed on nanoparticle surfaces. The impurities usually not only have large dielectric constant (e.g., the water has a dielectric constant of 80) and high electrical conductivity, but also are able to migrate when an electric field is applied, which can result in strong interfacial polarization and high leakage currents. The nanoparticle dispersion also plays an important role in determining the dielectric constant of polymer nanocomposites, particularly when the nanoparticles are electrically conductive and/or have an intrinsic high dielectric constant.

Figure 4.4 shows the frequency dependent of the dielectric constant of a range of epoxy nanocomposites containing $BaTiO_3$ nanoparticles, with and without silane treatment. After surface modification of the nanoparticles by silanes, the nanocomposites exhibit a significant decrease in dielectric constant and slightly lower dielectric loss tangent values. This decrease in both the dielectric constant and the dielectric loss tangent should be attributed to a reduction in–OH groups on the nanoparticle surfaces as a result of their modification.

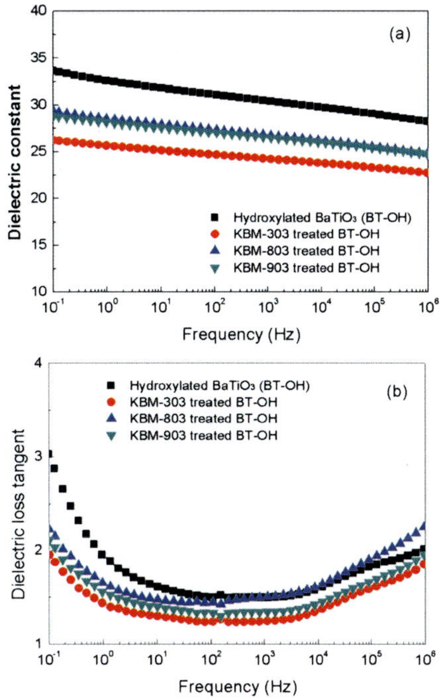

Figure 4.4 Dielectric constant and dielectric loss tangent of epoxy nanocomposites with silane surface-treated and untreated $BaTiO_3$. KBM-303: 2-(3,4 epoxycyclohexyl) ethyltrimethoxysilane; KBM-803: Mercaptopropyl-trimethoxysilane; KBM-903: 3-Aminopropyltrimethoxysilane (Huang et al., 2014).

4.4.5.2 Electrical conductivity or absorption current

Absorption current is defined as the polarizing current originating from the dipole alignment within the dielectrics under the applied electric field. The current initially shows a high value but then gradually decreases because of the increasing polarization of dipoles in the material. While silane surface modification of nanoparticles may directly result in a significant decrease in the number of charge carriers in the system, surface modification can also improve nanoparticle dispersion, which increases the average particle-particle distance, resulting in low interparticle tunneling currents and/or hopping currents. Figure 4.5 compares the time dependence of the charging current in linear low-density polyethylene (LLDPE) nanocomposites containing untreated and octyltrimethoxysilane-treated Al nanoparticles (nano-Al).

Compared with the nanocomposites with untreated nano-Al, those containing silane-treated nano-Al show much lower electrical current.

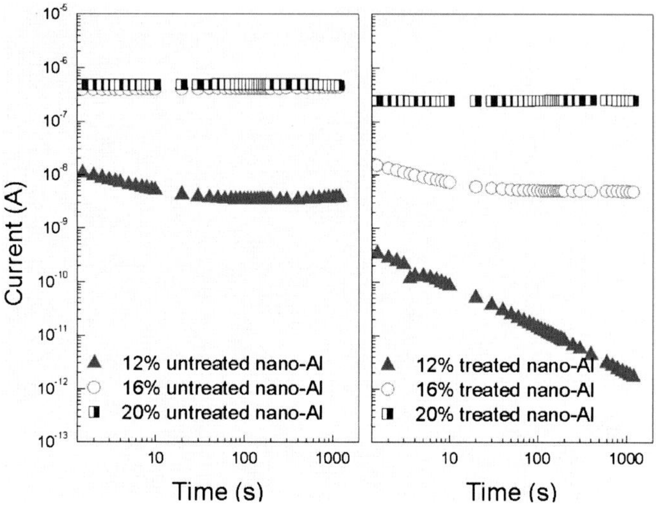

Figure 4.5 Time dependence of charging current for LLDPE with untreated and octyltrimethoxysilane-treated nano-Al (electric field: 1 kV/mm, 303 K) (Huang et al., 2009a).

4.4.5.3 Space charge

Space charge in insulating materials originates mainly from ionization of impurities and the injection of charges at electrodes. The former results in heterocharge, whereas the later produces homocharge. The surface modification of nanoparticles by silanes can remove impurities associated with nanoparticle surfaces, resulting in a significant decrease of heterocharges in polymer nanocomposites. However, the surface modification of nanoparticles by silanes can also change the trap density, trap depth distribution, and charge carrier mobility, which exert a significant influence on both charge injection from the electrodes and the charge dynamics within the material. Figure 4.6 shows the space charge distribution during depolarization in LLDPE and LLDPE nanocomposites containing 2.5 wt% of untreated and octyltrimethoxysilane-treated SiO_2 nanoparticles after applying a dc electric field of –50 kV/mm for 90 min. It can be seen from Fig. 4.6 that (a) homocharge injection appears to dominate in the pure LLDPE; (b) significant heterocharge is evident in the LLDPE nanocomposite containing the untreated SiO_2 nanoparticles; (c) only a small amount of homocharge is observed in the LLDPE nanocomposite where the SiO_2 nanoparticles have been silane treated. It is the silane surface modification that has resulted in the suppression of homocharge injection and the disappearance of heterocharge in the nanocomposites.

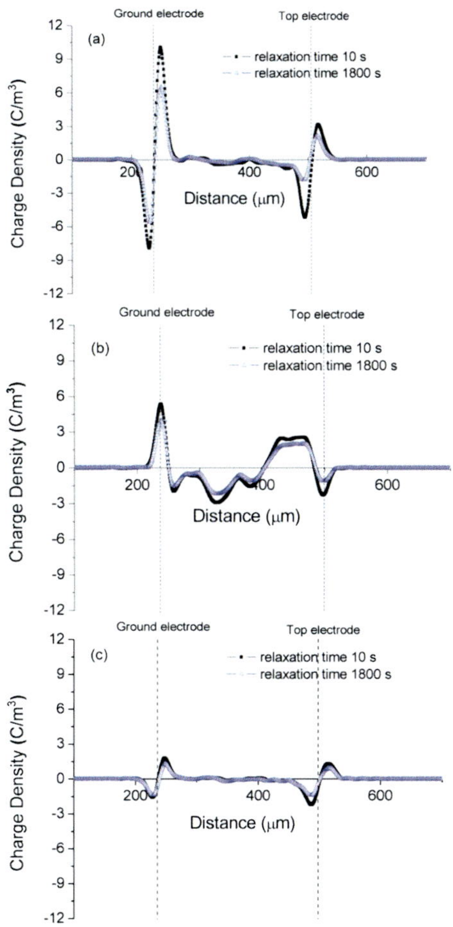

Figure 4.6 Space charge distribution during depolarization in (a) LLDPE, (b) LLDPE filled with 2.5 wt% untreated SiO_2 nanoparticles (c) and LLDPE filled with 2.5 wt% octyltrimethoxysilane-treated SiO_2 nanoparticles after applying a dc electric field of −50 kV/mm for 90 min (Huang et al., 2009b).

4.4.5.4 Breakdown strength

At given filler content, the breakdown strength of composites is mainly determined by the following factors: electrical mismatch between the filler and the polymer matrix (e.g., dielectric constant, electrical conductivity); filler dispersion; defect features (e.g., voids). Figure 4.7 shows Weibull plots of breakdown strength obtained from LLDPE nanocomposites containing untreated and octyltrimethoxysilane-treated Al nanoparticles. From this, it can be seen that the nanocomposite with the silane-treated nano-Al is characterized by a much higher breakdown strength. The silane surface modification improves the nano-Al dispersion in LLDPE and results in fewer defects by enhancing the interface adhesion, which is considered as the main reason for the improved breakdown strength.

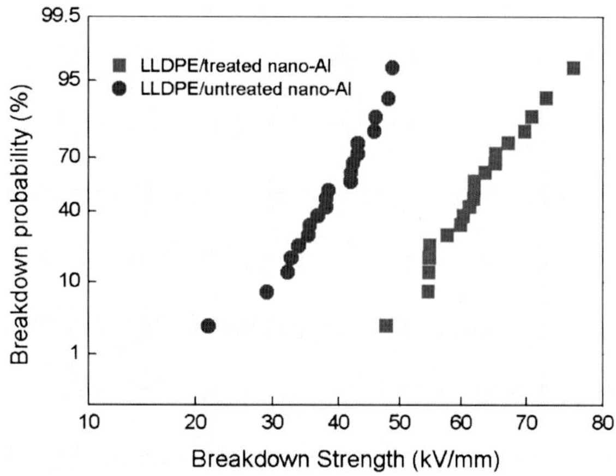

Figure 4.7 Weibull plots of breakdown strength of LLDPE nanocomposites with 8 vol% untreated and octyltrimethoxysilane-treated nano-Al (Huang et al., 2009c).

4.5 Nanoparticle Surface Modification by Titanate Coupling Agents (Monte, 2005)

4.5.1 Chemistry of Titanate Coupling Agents

The general chemical structure of titanate coupling agents is shown in Fig. 4.8. There are six functional sites for a specific titanate coupling agent and their functions are presented below:

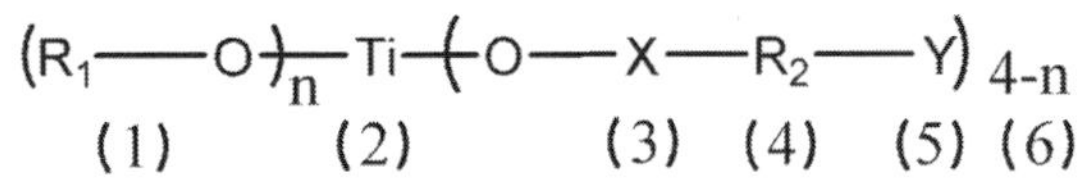

Figure 4.8 General chemical structure of titanate coupling agents.

Table 4.1 Chemical structure and applications of typical titanates

Functional group	Typical titanates and chemical structure	Proper matrix
Monoalkoxy	Titanium IV 2-propanolato, tris isooctadecanoato-O 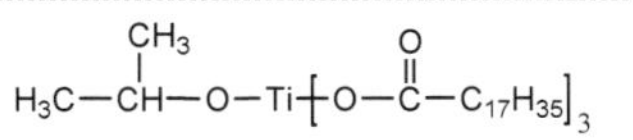	Polyolefin, EPDM, EPR, SBS, SBR, polybutadiene rubber
	Titanium IV 2-propanolato, tris(dioctyl) pyrophosphato-O 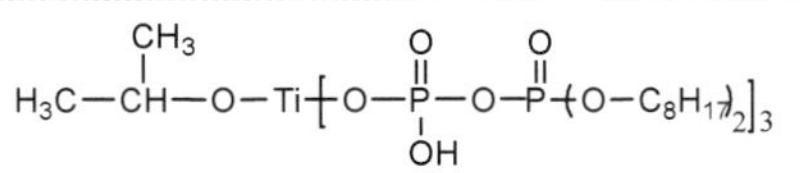	PS, nylon, alkyd resin, EPDM, EPR, polybutadiene rubber, natural rubber
Coordinate	Titanium IV tetrakis (bis-2-propenolatomethyl)-1-butanolato adduct 2 mol (ditridecyl)hydrogen phosphite 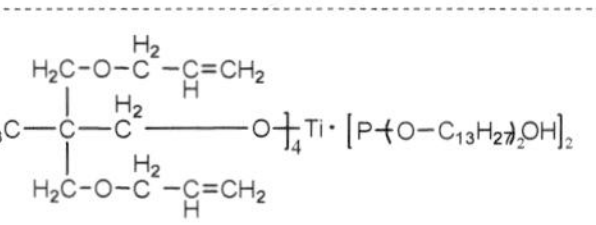	Epoxy resin, unsaturated polyester resin, PVC, PU,
Chelate	Titanium IV bis(dioctyl)pyrophosphato-O, oxoethylenediolato, (adduct), (dioctyl) (hydrogen)phosphite-O 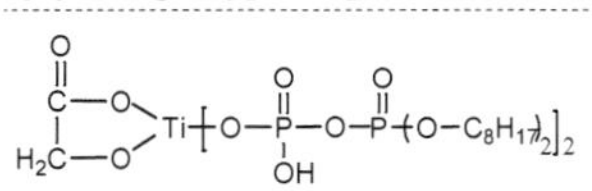	Polyolefin, PVC, epoxy resin, phenolic resin

Functional groups (1) are hydrolysable. The hydrolyzed titanates can react with hydroxyl groups on filler surfaces, leading to the formation of a covalently bridged interface.

Functional groups (2) are Ti-O bonds, which can be disassociated, resulting in transesterification and transalkylation.

Functional groups (3) include phosphato, pyrophosphato, sulfonyl and carboxyl, etc., which are binder functional groups and affect the composite performance.

Functional groups (4) are thermoplastic functional groups, including aliphatic and non-polar groups (isopropyl, butyl, octyl, isostearoyl), slightly polar dodecylbenzyl groups or aromatic groups such as benzyl, cumyl and phenyl, which determine the particle/polymer compatibility by van der Waals interaction.

Functional groups (5) are those that can react with the polymer matrix, including acryl, methacryl, mercapto, and amino moieties.

Functional groups (6) are equal to 1, 2 or 3, indicating mono, di or tri-organofunctional hybrid titanates, respectively.

The chemical structures of typical titanate coupling agents are shown in Table 4.1.

4.5.2 Effect of Nanoparticle Surface Modification on Electrical Properties of Polymer Nanocomposites

Compared with silane coupling agents, the titanates have not been extensively used as nanoparticle modifiers for improving the electrical properties of the polymer nanocomposites. However, it is believed that the surface modification of nanoparticles by titanate coupling agents can also improve the electrical properties of the polymer nanocomposites. This is because that the titanates can play a similar role to that of silanes in improving the nanoparticle dispersion, tailoring the nanoparticle surface properties and/or enhancing nanoparticle/polymer interfacial interactions. A typical example is the surface modification of hydroxylated $BaTiO_3$ nanoparticles by isopropyl dioleic(dio ctylphosphate) titanate (Gao et al., 2014). In preparing high-dielectric-constant polymer nanocomposites for energy storage applications, the hydroxylated $BaTiO_3$ rather than untreated $BaTiO_3$ nanoparticles have been frequently used, because of their high potential of dielectric enhancement. However, the introduction of hydroxylated $BaTiO_3$ often also results in high viscosity and high dielectric loss in the nanocomposites. Nevertheless, in the case of isopropyl dioleic(dioctylphosphate) titanate-modified $BaTiO_3$, it was found that the use of the titanate not only enhanced the dielectric constant but also suppressed the dielectric loss in PVDF nanocomposites (Gao et al., 2014). In another study, the surface modification of carbon black by titanate coupling agents led to a low percolation threshold in PVDF/PP(polyproplene) blends (Xu et al., 2007).

4.6 Nanoparticle Surface Modification by Phosphonic Acid or Phosphonates

4.6.1 Chemistry of Phosphonic Acid and Phosphonates

Both phosphonic acid and phosphonate ester derivatives can modify hydroxylated nanoparticles by forming strong covalent bridges between –P–O–OH and the hydroxyl groups on the nanoparticle surfaces. The chemical structure of phosphonic acid and phosphonate ester derivatives is shown in Fig. 4.9.

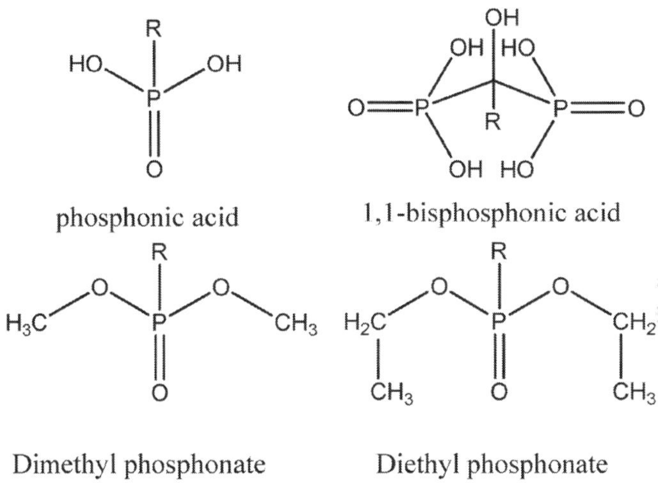

Figure 4.9 Phosphonic acids and typical phosphonates ester derivatives, R is alkyl or aryl group.

4.6.2 Modification Mechanism and Methods

It has been reported that chemical interactions between phosphonic acids and inorganic nanoparticle surfaces are significantly affected by the reaction conditions, including temperature, solvent, pH value, concentration and oxide types (Pujari et al., 2014). Usually there are two mechanisms depending on the acidity of the nanoparticles, which are shown in Fig. 4.10. In the case of nanoparticles with strong Lewis acidity (**mechanism I**), the phosphoryl oxygen of P=O first coordinates with Lewis sites on the nanoparticle surfaces, which makes the P atom more electrophilic and induces a heterocondensation reaction between P–OH and the H–O– groups on the nanoparticle surface, resulting in covalent bonding between the phosphonic acid and the nanoparticle. For nanoparticles with weak Lewis acidity (**mechanism II**), it is hydrogen bonding that promotes the heterocondensation reaction. Regarding phosphonates ester derivatives, the hydrolysis of P–O–C bonds takes place prior to the heterocondensation reaction. It should also be noted that the surface density of hydroxyl groups is a key factor in influencing phosphonate based surface modification. Phosphonic acids can be easily introduced onto nanoparticle surfaces with a high hydroxyl density.

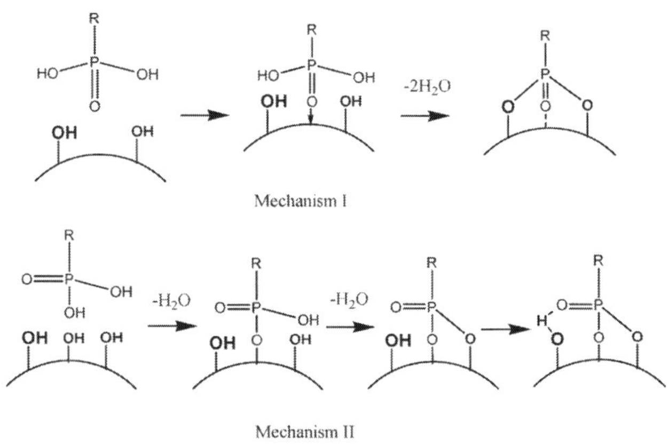

Figure 4.10 Two reaction mechanisms for nanoparticle surface modification by phosphonic acids.

Phosphonic acid and phosphonate based surface modification is usually performed in solution, which has several advantages: (i) phosphonic acids and typical phosphonates ester derivatives tend to form monolayers on the nanoparticle surfaces; (ii) phosphonic acids are soluble in water; (iii) phosphonate esters are highly soluble in many organic solvents and can be converted into phosphonic acids; (iv) phosphonate esters are less susceptible to self-condensation reactions, making surface modification easily controllable. For example, the removal of water in the solvent is not necessary and, indeed, nanoparticle surface modification by phosphonic acids can be performed in water. Taking the modification of $BaTiO_3$ nanoparticles by pentafluorobenzyl phosphonic acid (PFBPA) as an example, the nanoparticles were first dispersed in 95% ethanol solution under ultrasonic radiation; then the PFBPA was added to the $BaTiO_3$ dispersion before, finally, surface modification was completed with a further ultrasonic treatment (10 min) and heating the mixture at 80 °C for 1 h with stirring (Kim et al., 2009).

4.6.3 Choosing Proper Phosphonates for Nanoparticles Surface Modification

When choosing a phosphonic acid or a phosphonate ester as a modifier, one should consider the compatibility and/or reactivity between the relevant R group and the polymer matrix. The electronic nature (e.g., electron donating or electronic withdrawing) of the R groups may also be utilized to realize the desired property in the ultimate nanocomposites.

4.6.4 Examples of Improving Electrical Properties of Polymer Nanocomposites by Phosphonate Surface Modification

Several types of phosphonic acids (see in Fig. 4.11) have been used as the nanoparticle modifiers to improve the electrical properties of polymer nanocomposites. Following surface modification of the nanoparticles by phosphonic acids, property enhancement of the resulting nanocomposites primarily occurs as a consequence of the following mechanisms. Surface modification improves nanoparticle dispersion and consequently reduces the number of large aggregates present in the system. Surface modification enhances nanoparticle/polymer interfacial adhesion and thereby reduces interfacial defects such as voids and/or pores. The introduction of phosphonic acids on the nanoparticle surfaces reduces the density of ionizable species in the system (i.e., hydroxyl groups), resulting in a decrease in the concentration and mobility of charge carriers. In addition, the electronic nature of the R groups of phosphonic acid modifiers may directly affect the electrical properties of polymer nanocomposites. For example, phosphonic acids with electron-rich or electron-poor R groups have been used to modify TiO_2 nanoparticles. This work demonstrated that nanoparticles modified with a phosphonic acid with electropositive phenyl, electron-withdrawing functional groups significantly reduced the dielectric loss and leakage currents in the epoxy/TiO_2 nanocomposites (Siddabattuni et al., 2013). In another study, it was found that surface modification of TiO_2 by 2-ethylhexyl phosphoric acid ester (EHP) or 2-aminoethyl dihydrogen phosphate (AEP) significantly enhanced the breakdown strength of epoxy/TiO_2 nanocomposites (Schuman et al., 2010), as shown in Fig. 4.12. PC and PVDF-HFP nanocomposites, respectively, containing PFBPA (Fig. 4.11e) and 2-ethylhexyl phosphoric acid ester (Fig. 4.11f)-modified $BaTiO_3$ nanoparticles, have been shown to exhibit high breakdown strength and high energy storage density (Kim et al., 2007).

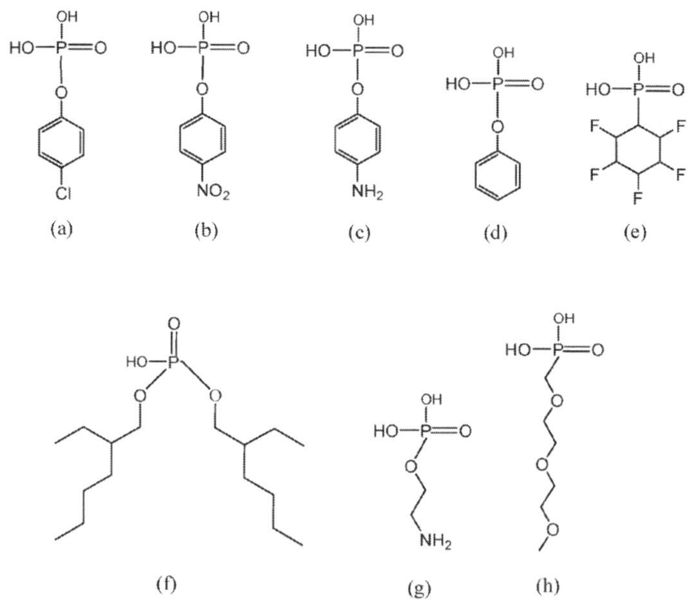

Figure 4.11 Some phosphonic acids used for nanoparticle modification: (a) chlorophenyl phosphate, (b) nitrophenyl phosphate, (c) aminophenyl phosphate, (d) phenyl phosphate, (e) pentafluorobenzyl phosphonic acid, (f), 2-ethylhexyl phosphoric acid ester, (g) 2-aminoethyl dihydrogen phosphate and (h) and [2-[2-(2-methoxyethoxy)ethoxy]ethyl] phosphonic acid.

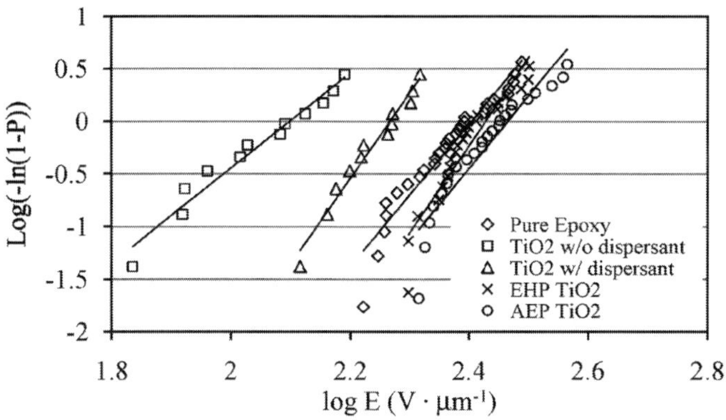

Figure 4.12 Weibull plot of breakdown strength of epoxy nanocomposites with 5 vol% TiO_2 nanoparticles. EHP: 2-Ethylhexyl phosphoric acid ester; AEP: 2-aminoethyl dihydrogen phosphate (Schuman et al., 2010).

4.7 Nanoparticle Surface Modification by Dopamine

4.7.1 Chemistry of Dopamine

Dopamine, or 3,4-dihydroxy-L-phenylalanine, is a type of catechol. It has been found at high concentrations in the adhesive-pad proteins of marine mussels, which can cause extremely strong binding between the adhesive-pad and substrates. The chemical structure of dopamine is shown in Fig. 4.13.

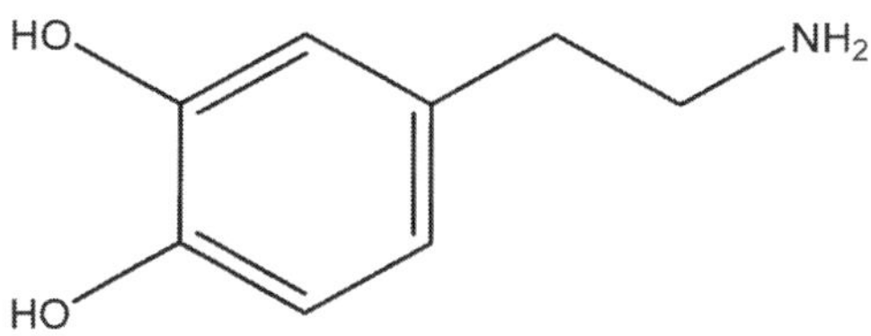

Figure 4.13 Chemical structure of dopamine.

4.7.2 Modification Mechanism and Methods

It has been shown that dopamine can provide strong adhesion with particles or a substrate via the formation of coordinative chelate structures, hydrogen bonds, metal-ligand complexes and quinhydrone charge-transfer complexes (Yang et al., 2011). Dopamine can be attached on the surfaces of various types of materials, including metals, oxides, ceramic, semiconductors and polymers. The processes of surface attachment of dopamine to a surface are complex, particularly in the presence of excess dopamine, but usually involve surface anchoring by the catechol moiety followed by its oxidation, which finally results in the formation of highly cross-linked polydopamine. The suggested possible chemical structures of polydopamine are shown in Fig. 4.14.

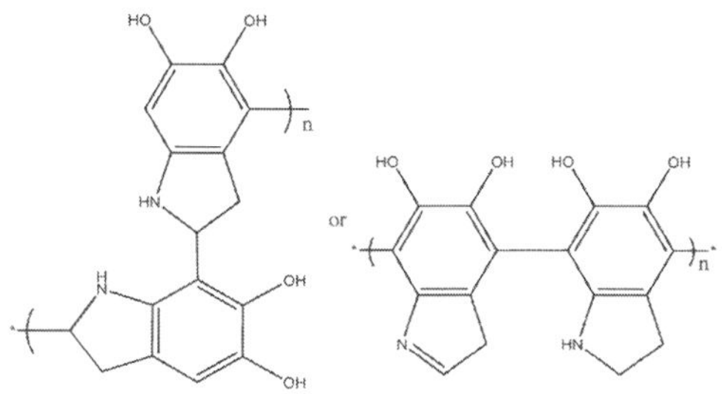

Figure 4.14 Suggest chemical structure of polydopamine.

In addition to its wide applicability, the surface modification of nanoparticles by dopamine has other merits.

First, the modification process is easily carried out; nanoparticle surface modification can be performed in a tris(hydroxymethyl)-aminomethane-HCl $[(HOCH_2)_3CNH_2]$-HCl buffer solution (e.g., 10 mM, pH = 8.5) at room temperature or above.

Second, functional groups exist on the polydopamine surfaces (i.e., –OH) for further functionalization.

Third, many dopamine derivatives can be selected for optimization of adhesion strength and tailoring the electrical properties of the modified nanoparticle surfaces.

4.7.3 Improved Electrical Properties of Polymer Nanocomposites Resulting from Dopamine Surface Modification

Dopamine has been used to modify $BaTiO_3$ nanoparticles and $BaTiO_3$ nanofibers in order to enhance the dielectric properties of polymer nanocomposites (Song et al., 2012). $BaTiO_3$ nanoparticles exhibit a high dielectric constant of several hundred, which usually causes a large decrease in the breakdown strength of their polymer nanocomposites because of the large dielectric constant mismatch and weak interfacial strength between the nanoparticles and the polymer matrix. However, it has been found that dopamine-modified $BaTiO_3$ nanofibers can significantly enhance the breakdown strength of poly(vinylidene fluoride-trifluoroethylene) (PVDF-TrFE) nanocomposites; as shown in Fig. 4.15, adding 3.0 vol% of dopamine-modified $BaTiO_3$ nanofibers increased the breakdown strength of PVDF-TrFE from 170.9 ± 9.1 kV/mm to 225.8 ± 11.6 kV/mm. The authors attributed this breakdown strength enhancement to the dopamine-modified $BaTiO_3$ nanofibers strongly limiting macromolecular chain movement in the interfacial region, which may reduce the probability of charge carriers transferring through the polymer matrix.

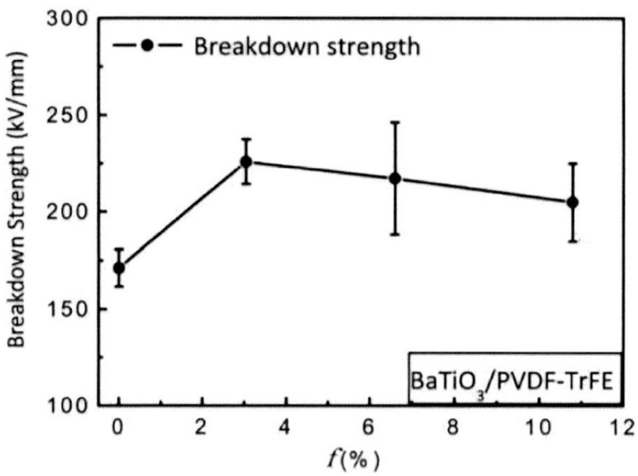

Figure 4.15 The breakdown strength of $BaTiO_3$/PVDF-TrFE nanocomposites as a function of volume fractions of dopamine-modified $BaTiO_3$ nanofibers (Song et al., 2012).

It was also found that dopamine-modified $BaTiO_3$ nanoparticles can not only suppress dielectric loss but also enhance the dielectric constant of PVDF/$BaTiO_3$ nanocomposites, particularly at high $BaTiO_3$ loading levels (Lin et al., 2011). The suppression of dielectric loss is also attributed to the strong limitation effect of dopamine-modified $BaTiO_3$ nanoparticles on macromolecular chain movement in the interfacial region. However, the increase in the dielectric constant has been ascribed to reduced defects (voids and pores) and enhanced nanoparticle dispersion. In addition, nanocomposites containing dopamine-modified $BaTiO_3$ nanoparticles show weak frequency dependent dielectric properties in comparison with equivalent nanocomposites formulated with unmodified $BaTiO_3$.

4.8 Nanoparticle Surface Modification by Polymer Coating

4.8.1 Coating Strategies

Two strategies have been developed to graft polymers onto the nanoparticle surfaces: "grafting from" and "grafting to."

The "grafting from" strategy corresponds to the formation of a polymer coating layer on the nanoparticle surfaces by surface initiated in situ polymerization. Some powerful techniques that have been well used to modify nanoparticles include controlled/living radical polymerization such as atom transfer radical polymerization (ATRP) (Xie et al., 2011), and reversible addition-fragmentation chain transfer (RAFT) polymerization (Yang et al., 2012). The preparation process and TEM image of poly(methyl methacrylate) (PMMA)-coated $BaTiO_3$ by ATRP are shown in Figs. 4.16 and 4.17, respectively.

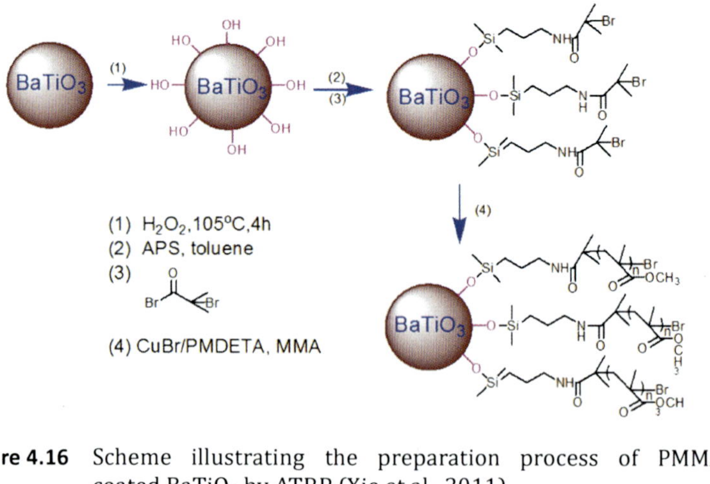

Figure 4.16 Scheme illustrating the preparation process of PMMA-coated $BaTiO_3$ by ATRP (Xie et al., 2011).

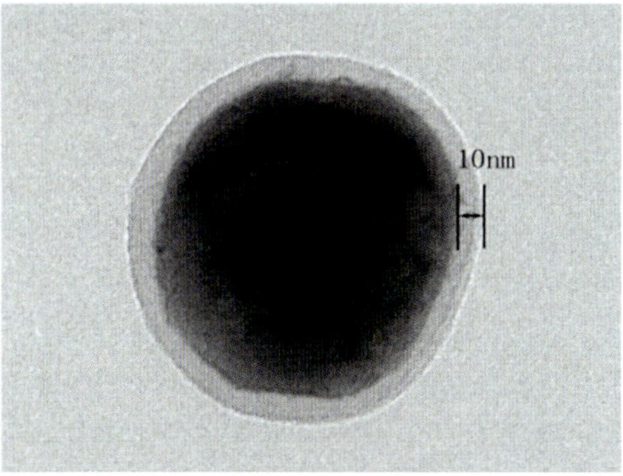

Figure 4.17 TEM image showing the morphology of a PMMA-coated $BaTiO_3$ nanoparticle (Xie et al., 2011).

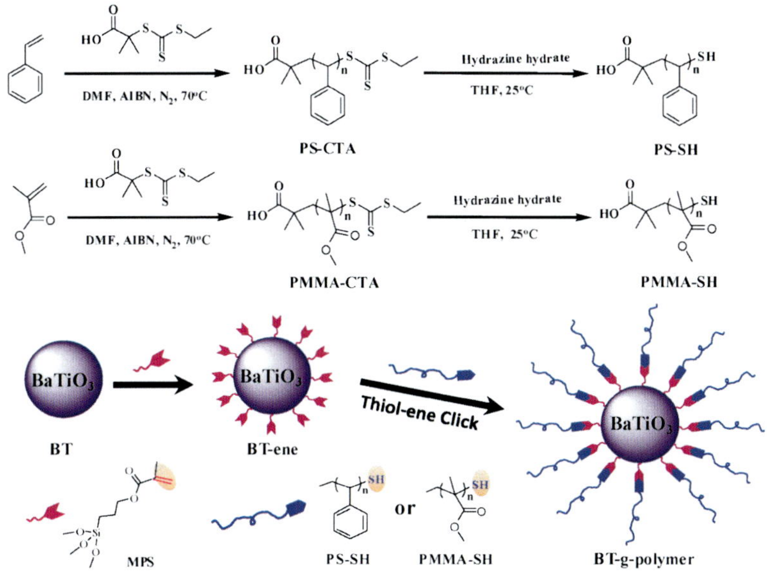

Figure 4.18 Scheme showing the preparation of thiol-terminated polystyrene (PS-SH) and thiol-terminated PMMA (PMMA-SH) chains via RAFT polymerization and the preparation of PS- and PMMA-coated $BaTiO_3$ nanoparticles by thiol-ene click reaction (Yang et al., 2014).

The "grafting to" strategy corresponds to the formation of a polymer coating layer by grafting the pre-prepared polymer chains onto the nanoparticle surfaces, which can be carried out via a large range of reactions between the end-groups of polymer chains and the groups on the nanoparticle surfaces. The "grafting to" strategy makes the control of the molecular composition and molecular weight of the polymer chains easier compared with the "grafting from" strategy.

Preparation processes for PMMA- and PS-coated $BaTiO_3$ by a "grafting to" strategy are shown in Fig. 4.18 (Yang et al., 2014).

4.8.2 Electrical Properties of Polymer Nanocomposites Containing Polymer-Coated Nanoparticles

Compared with nanoparticle modification by small molecules such as silane and titanate coupling agents, surface modification by polymer coating has many merits (Huang and Jiang, 2015). First, chemically, the nanoparticles become encapsulated by one of more shells of other materials. This not only changes the surface chemistry but also effectively isolates each nanoparticles and thus inhibits nanoparticle agglomeration leading to a homogeneous dispersion of nanoparticles in the polymer matrix. Second, the addition of a shell enables constitutes a versatile means of tuning overall nanocomposites properties, by appropriately tailoring the physical properties of the shells. For example, as shown in Fig. 4.19, computational analysis reveals that, compared with the particles without a core–shell structure, core–shell structured particles can significantly reduce the local electric field enhancement in a composite, resulting in increased breakdown strength (Wang et al., 2011). In addition, the electric field enhancement in the composites can be tuned by appropriately tailoring the ratio of the shell dielectric constant to that of the matrix. This work indicates that one can choose a shell materials with a specific dielectric constant for a targeted application. It has been shown experimentally that the dielectric loss and breakdown strength of nanocomposites containing polymer-coated nanoparticles can be tuned by tailoring the electrical properties of the coating layer (Zhu et al., 2014), the thickness of the coating layer and the interaction between the coated polymer layer and the polymer matrix (Yang et al., 2013).

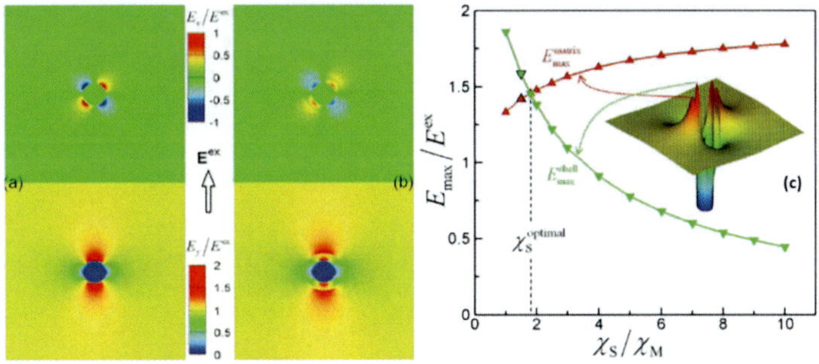

Figure 4.19 Electric field distribution in dielectric polymer nanocomposites filled by particles (a) without and (b) with a core–shell structure. E^{ex} indicates the direction of the applied electric field, and E_x and E_y correspond to the x-axis and y-axis components of the total electric field. The color scale indicates the magnitude of the electric field enhancement. (c) Effect of the dielectric constant of the shell on the local electric field enhancement in a dielectric polymer composites containing core–shell structured particles with $(k_{matrix} - 1)/(k_{core} - 1) = 1 : 100$ and a ratio of shell thickness to core diameter of 2 : 3. The Inset in (c) shows the electric field distribution in a dielectric polymer composite containing core–shell structured particles where $(k_{shell} - 1)/(k_{matrix} - 1) = 1.5 : 1$. E_y is shown both by a color scale and height in 3D visualization. Here, $k = 1 + \chi$ (Wang et al., 2011).

4.9 Nanoparticle Surface Modification by Inorganic Coating

As shown in the preceding section, the electrical properties of polymer nanocomposites can be optimized via tailoring the electrical properties of the core–shell nanofillers. In addition to the polymeric case considered above, inorganic compounds or ultra-small nanoparticles have also been used to coat nanoparticles and Fig. 4.20 includes examples of (a) a dielectric nanoparticles coated with a dielectric layer, (b, c) dielectric layer-coated metallic nanoparticles, and (d) dielectric nanoparticles decorated with ultra-small metallic nanoparticles. Such nanofillers may be prepared in order to realize electrical properties that cannot be achieved by other methods.

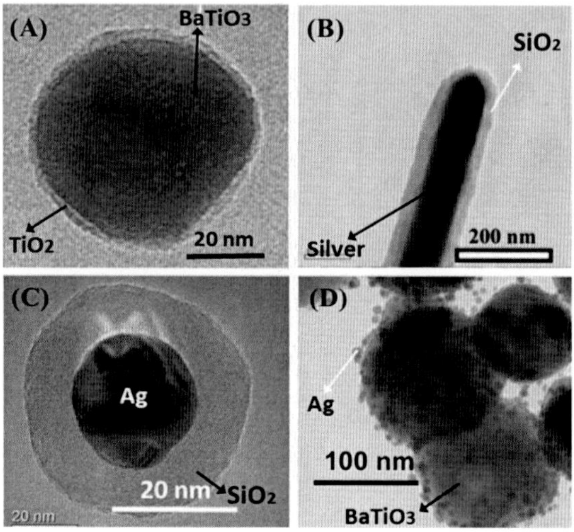

Figure 4.20 TEM images showing (a) TiO_2-coated $BaTiO_3$ nanoparticles (Rahimabady et al., 2013), (b, c) a silver nanowire (Chen et al., 2014) and a silver nanoparticle (Quinsaat et al., 2013) both coated with SiO_2 and (d) ultra-small-Ag decorated $BaTiO_3$ nanoparticles (Xie et al., 2013).

For example, $BaTiO_3$ nanoparticles have been widely used to prepare high-dielectric-constant polymer nanocomposites for capacitor dielectric and electric stress control applications. For this purpose, the optimal case is that the polymer nanocomposites simultaneously have high dielectric constant and high breakdown strength. However, conventional surface modification methods usually increase the dielectric constant at the expense of reduced breakdown strength. By forming inorganic layer(s) with low dielectric constant (i.e., TiO_2) on the surface of high dielectric constant nanoparticles (e.g., $BaTiO_3$), as shown in Fig. 4.20a, this problem has been well solved.

Polymer nanocomposites containing TiO_2-coated $BaTiO_3$ can simultaneously exhibit enhanced dielectric constant and breakdown strength in comparison with the $BaTiO_3$ nanocomposites (Rahimabady et al., 2013), as illustrated in Fig. 4.21. Composites prepared by dispersing an electrically conducting filler into an insulating matrix usually exhibit very high dielectric constant when the filler concentration is near the percolation threshold. However, such percolative composites also show very high dielectric loss and extremely low breakdown strength, making such systems unattractive for use in many dielectric applications. However, coating the electrically conductive nanoparticles (e.g., silver nanoparticles or nanowires) with an electrically insulating layer (e.g., silica) may prevent the nanoparticle from forming electrically percolating structures, so resulting simultaneously in a high dielectric constant and low dielectric loss (Chen et al., 2014).

Ultra-small metallic or semi-conductive nanoparticles may exhibit quantum confinement and Coulomb blockade effects, which may result in novel electrical phenomena that cannot be observed in traditional dielectric polymer nanocomposites. It has been found that nanocomposites filled with ultra-small-Ag decorated $BaTiO_3$ nanoparticles show decreased dielectric loss and enhanced breakdown strength compared with $BaTiO_3$ based nanocomposites (Xie et al., 2013).

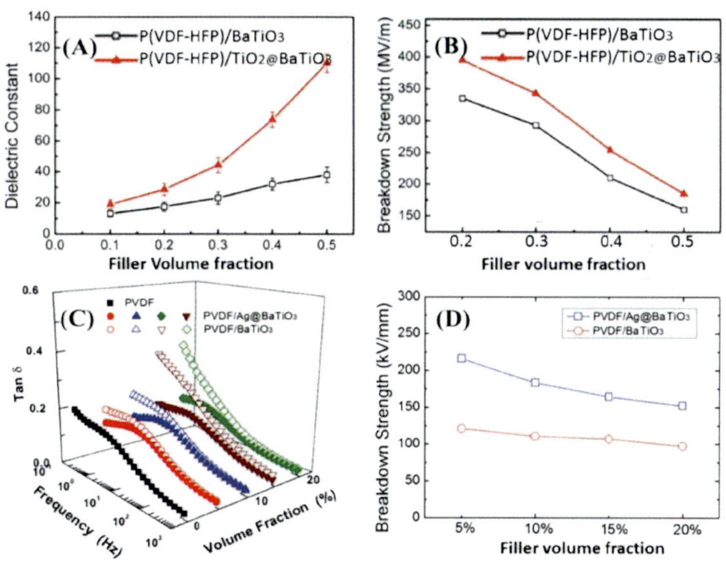

Figure 4.21 Filler loading dependence of (a) the dielectric constant at 1 kHz and (b) breakdown strength of P(VDF-HFP) nanocomposites containing $BaTiO_3$ and TiO_2-coated $BaTiO_3$ nanoparticles (Rahimabady et al., 2013); (c) the frequency dependent of the dielectric loss tangent and (d) DC breakdown strength of PVDF nanocomposites with $BaTiO_3$ and ultra-small-Ag decorated $BaTiO_3$ (Xie et al., 2013).

4.10 Conclusions

Nanoparticle surface modification allows control over nanoparticle surface chemistry and physics, thereby providing opportunities to tailor and optimize the microstructure, processing and properties of dielectric polymer nanocomposites. This chapter has summarized the state-of-the-art in strategies for nanoparticle surface modification, which can be categorized into the following different methods:

- modification by silane or titanate coupling agents
- modification by phosphonates
- modification by dopamine
- modification by polymer coating
- modification by inorganic coating

While the material used to form the nanoparticle usually plays a central role in many applications, because of its inherent physical properties such as high/low dielectric constant, nanoparticle surface modification can further improve the properties of nanocomposites or even realize unexpected properties, for one or more of the following reasons:

(i) Modification can improve the nanoparticle dispersion in the polymer matrix.
(ii) Modification can enhance the nanoparticle/polymer interaction.
(iii) Modification can change the physicochemical properties of the nanoparticle surfaces.
(iv) The modifiers may themselves have unique physical properties and therefore directly play an important role.

Compared with other modification methods, nanoparticle surface modification by polymer or inorganic coating is often more powerful, because the introduced polymer or inorganic layer(s) may itself play a direct role in determining the final properties of the resulting nanocomposites. Nevertheless, the selection of modification method should depend on the specific conditions and, consequently, a deep understanding of the modification mechanism and the role of interface is beneficial in choosing the most appropriate modification method.

Acknowledgments

X.Y.H. thanks the SMC Excellent Young Faculty Award of Shanghai Jiao Tong University, Shanghai Pujiang Program under Grant PJ14D018, the National Natural Science Foundation of China (Nos. 51107081, 51477096 and 51522703). P.K.J. and X.Y.H. thank the National Natural Science Foundation of China (No. 51277117) and the Special Fund of the National Priority Basic Research of China under Grant 2014CB239503 for financial support. The authors thank Prof. Alun S. Vaughan for the English polishing of the manuscript.

References

Arkles B (1977). Tailoring Surfaces with Silanes. *Chemtech*, **7**: 766–778.

Chen C, Tang YJ, Ye YS, Xue ZG, Xue Y, Xie XL and Mai YW (2014). High-performance epoxy/silica coated silver nanowire composites as underfill material for electronic packaging. *Compos Sci Technol*, **105**(0): 80–85.

Dreyer DR, Park S, Bielawski CW and Ruoff RS (2010). The chemistry of graphene oxide. *Chem Soc Rev*, **39**(1): 228–240.

Gao L, He JL, Hu J and Li Y (2014). Large enhancement in polarization response and energy storage properties of poly(vinylidene fluoride) by improving the interface effect in nanocomposites. *J Phys Chem C*, **118**(2): 831–838.

Garcia N, Benito E, Guzman J and Tiemblo P (2007). Use of p-toluenesulfonic acid for the controlled grafting of alkoxysilanes onto silanol containing surfaces: preparation of tunable hydrophilic, hydrophobic, and super-hydrophobic silica. *J Am Chem Soc*, **129**(16): 5052–5060.

Huang XY and Jiang PK (2015). Core–shell structured high-k polymer nanocomposites for energy storage and dielectric applications. *Adv Mater*, **27**(3): 546–554.

Huang XY, Jiang PK, and Yin Y (2009c). Nanoparticle surface modification induced space charge suppression in linear low density polyethylene. *Appl Phys Lett*, **95**(24): 242905.

Huang XY, Kim C, Jiang PK, Yin Y and Li Z (2009a). Influence of aluminum nanoparticle surface treatment on the electrical properties of polyethylene composites. *J Appl Phys*, **105**(1): 014105.

Huang XY, Ma ZS, Wang YQ, Jiang PK, Yin Y and Li Z (2009b). Polyethylene/aluminum nanocomposites: improvement of dielectric strength by nanoparticle surface modification. *J Appl Polym Sci*, **113**(6): 3577–3584.

Huang XY, Xie LY, Yang K, Wu C, Jiang PK, Li ST, Wu S, Tatsumi K and Tanaka T (2014). Role of interface in highly filled epoxy/$BaTiO_3$ nanocomposites. Part I-correlation between nanoparticle surface chemistry and nanocomposite dielectric property. *IEEE Trans Dielectr Electr Insul*, **21**(2): 467–479.

Kim P, Doss NM, Tillotson JP, Hotchkiss PJ, Pan MJ, Marder SR, Li JY, Calame JP and Perry JW (2009). High energy density nanocomposites based on surface-modified $BaTiO_3$ and a ferroelectric polymer. *ACS Nano*, **3**(9): 2581–2592.

Kim P, Jones SC, Hotchkiss PJ, Haddock JN, Kippelen B, Marder SR and Perry JW (2007). Phosphonic acid-modified barium titanate polymer nanocomposites with high permittivity and dielectric strength. *Adv Mater*, **19**(7): 1001.

Kim B and Sigmund WM (2004). Functionalized multiwall carbon nanotube/gold nanoparticle composites. *Langmuir*, **20**(19): 8239–8242.

Krueger A and Lang D (2012). Functionality is key: recent progress in the surface modification of nanodiamond. *Adv Funct Mater*, **22**(5): 890–906.

Lin MF, Thakur VK, Tan EJ and Lee PS (2011). Surface functionalization of $BaTiO_3$ nanoparticles and improved electrical properties of $BaTiO_3$/polyvinylidene fluoride composite. *RSC Adv*, **1**(4): 576–578.

Pujari SP, Scheres L, Marcelis ATM and Zuilhof H (2014). Covalent surface modification of oxide surfaces. *Angew Chem Int Ed*, **53**(25): 6322–6356.

Quinsaat JEQ, Nuesch FA, Hofmann H and Opris DM (2013). Dielectric properties of silver nanoparticles coated with silica shells of different thicknesses. *RSC Adv*, **3**(19): 6964–6971.

Rahimabady M, Mirshekarloo MS, Yao K and Lu L (2013). Dielectric behaviors and high energy storage density of nanocomposites with core-shell $BaTiO_3$@TiO_2 in poly(vinylidene fluoride-hexafluoro-propylene). *Phys Chem Chem Phys*, **15**(38): 16242–16248.

Schuman TP, Siddabattuni S, Cox O and Dogan F (2010). Improved dielectric breakdown strength of covalently-bonded interface polymer-particle nanocomposites. *Composite Interfaces*, **17**(8): 719–731.

Siddabattuni S, Schuman TP and Dogan F (2013). Dielectric properties of polymer-particle nanocomposites influenced by electronic nature of filler surfaces. *ACS Appl Mater Interfaces*, **5**(6): 1917–1927.

Song Y, Shen Y, Liu HY, Lin YH, Li M and Nan CW (2012). Enhanced dielectric and ferroelectric properties induced by dopamine-modified $BaTiO_3$ nanofibers in flexible poly(vinylidene fluoride-trifluoroethylene) nanocomposites. *J Mater Chem*, **22**(16): 8063–8068.

Wang YU, Tan DQ and Krahn J (2011). Computational study of dielectric composites with core-shell filler particles. *J Appl Phys*, **110**(4): 044103.

Xanthos M (2005). *Functional Fillers for Plastics*, ed. Monte SJ, Chapter 5, Titanate Coupling Agent in Functional Fillers for Plastics. Weinheim, Germany, Wiley-VCH.

Xie LY, Huang XY, Li BW, Zhi CY, Tanaka T and Jiang PK (2013). Core-satellite Ag@$BaTiO_3$ nanoassemblies for fabrication of polymer nanocomposites with high discharged energy density, high breakdown strength and low dielectric loss. *Phys Chem Chem Phys*, **15**(40): 17560–17569.

Xie LY, Huang XY, Wu C and Jiang PK (2011). Core-shell structured poly(methyl methacrylate)/$BaTiO_3$ nanocomposites prepared by in situ atom transfer radical polymerization: a route to high dielectric constant materials with the inherent low loss of the base polymer. *J Mater Chem*, **21**(16): 5897–5906.

Xu HP, Dang ZM, Yao SH, Jiang MJ and Wang DY (2007). Exploration of unusual electrical properties in carbon black/binary-polymer nanocomposites. *Appl Phys Lett*, **90**(15): 152912.

Yang K, Huang XY, Huang YH, Xie LY and Jiang PK (2013). Fluoro-polymer@ $BaTiO_3$ hybrid nanoparticles prepared via RAFT polymerization: toward ferroelectric polymer nanocomposites with high dielectric constant and low dielectric loss for energy storage application. *Chem Mater*, **25**(11): 2327–2338.

Yang K, Huang XY, Xie LY, Wu C, Jiang PK and Tanaka T (2012). Core-shell structured polystyrene/$BaTiO_3$ hybrid nanodielectrics prepared by in situ RAFT polymerization: a route to high dielectric constant and low loss materials with weak frequency dependence. *Macromol Rapid Commun*, **33**(22): 1921–1926.

Yang K, Huang XY, Zhu M, Xie LY, Tanaka T and Jiang PK (2014). Combining RAFT polymerization and thiol-ene click reaction for core-shell structured polymer@$BaTiO_3$ nanodielectrics with high dielectric constant, low dielectric loss, and high energy storage capability. *ACS Appl Mater Interfaces*, **6**(3): 1812–1822.

Yang LP, Phua SL, Teo JKH, Toh CL, Lau SK, Ma J and Lu XH (2011). A biomimetic approach to enhancing interfacial interactions: polydopamine-coated clay as reinforcement for epoxy resin. *ACS Appl Mater Interfaces*, **3**(8): 3026–3032.

Yu K, Niu YJ, Zhou YC, Bai YY and Wang H (2013). Nanocomposites of surface-modified $BaTiO_3$ nanoparticles filled ferroelectric polymer with enhanced energy density. *J Am Ceram Soc*, **96**(8): 2519–2524.

Zhi CY, Bando Y, Tang CC, Huang Q and Golberg D (2008). Boron nitride nanotubes: functionalization and composites. *J Mater Chem*, **18**(33): 3900–3908.

Zhi CY, Bando Y, Terao T, Tang CC, Kuwahara H and Golberg D (2009). Chemically activated boron nitride nanotubes. *Chem Asian J*, **4**(10): 1536–1540.

Zhu M, Huang XY, Yang K, Zhai X, Zhang J, He JL and Jiang PK (2014). Energy storage in ferroelectric polymer nanocomposites filled with core–shell structured polymer@$BaTiO_3$ nanoparticles: understanding the role of polymer shells in the interfacial regions. *ACS Appl Mater Interfaces*, **6**(22): 19644–19654.

Appendix

Silane functional groups, their compatible matrix and the corresponding methoxysilanes

Functional groups	Proper matrix	Corresponding methoxysilane
Alkyl	Polyolefin	Propyltrimethoxysilane
		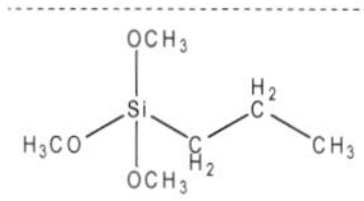
		Octyltrimethoxysilane
		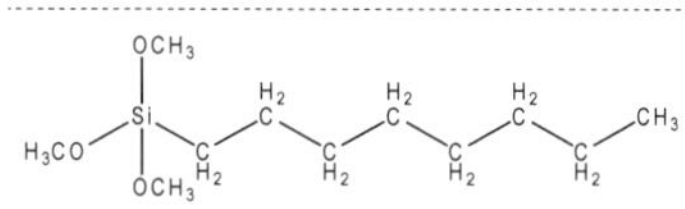
		Hexadecyltrimethoxysilane
		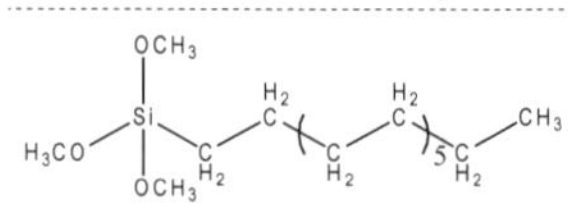

Functional groups	Proper matrix	Corresponding methoxysilane
Acryloxy	Unsaturated polyester, EPDM, EPR, acrylic resin ABS, PC, PS, polyolefin, urethane resin	Acryloxypropyltrimethoxysilane 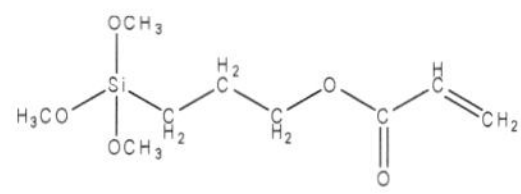
Methacryoxy		Methacryloxypropyltrimethoxysilane 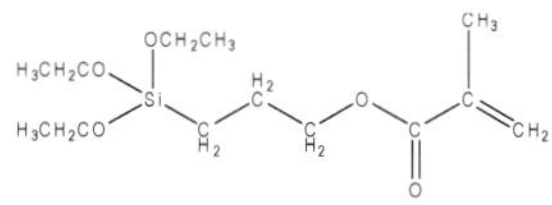
Amino	Most of thermoplastic and thermoset resins, EPDM, EPR, nitrile rubber, neoprene rubber, butyl rubber, urethane rubber, polysulfide	3-Aminopropyltrimethoxysilane 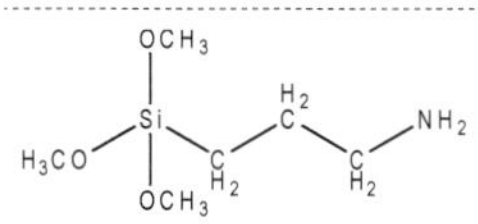
		N-2(Aminoethyl)-3-aminopropyltrimethoxysilane 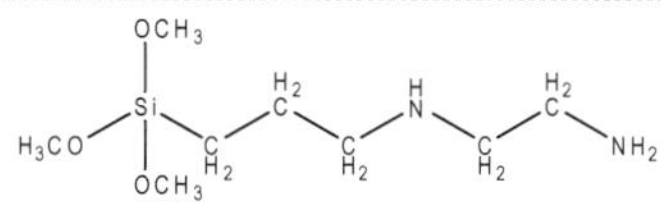
		N-2(Aminoethyl)-3-aminopropylmethyldimethoxysilane 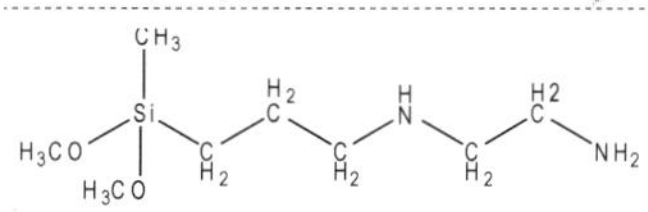
Epoxy	Most of thermoplastic and thermoset resins, SBR, nitrile rubber, butyl rubber, urethane rubber, polysulfide	3-Glycidoxypropyltrimethoxysilane 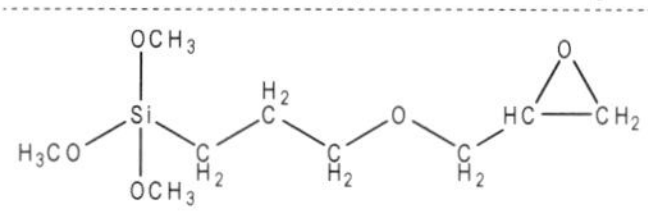
		2-(3,4 Epoxycyclohexyl) ethyltrimethoxysilane 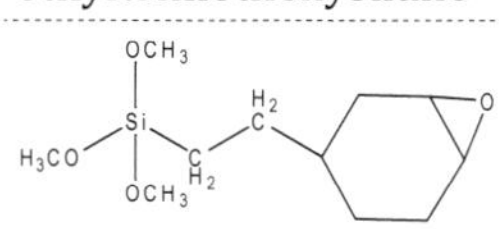

Functional groups	Proper matrix	Corresponding methoxysilane
Isocyanate	Urethane rubber, PC, polyimides, epoxy resin, phenolic resin, melamine resin, ABS, nylon,	3-Isocyanatepropyltrimethoxysilane 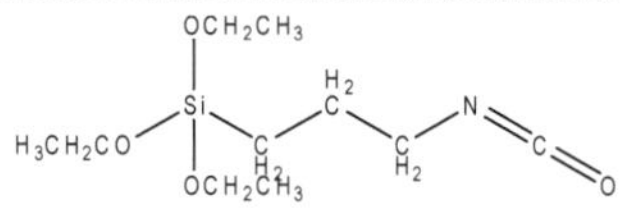
Mercapto	Most of rubbers, ABS, PVC, PS Epoxy resin, phenolic resin, melamine resin, urethane	Mercaptopropyltrimethoxysilane 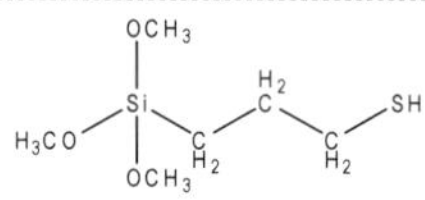
Styryl	Acrylic resin, PS	p-Styryltrimethoxysilane 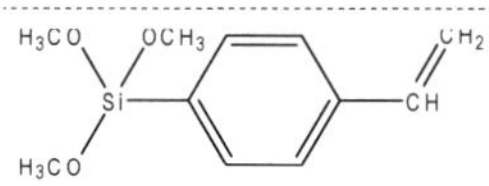
Sulfide	EPR, EPDM, SBR, butyl rubber	Bis(triethyoxysilylpropyl)tetrasulfide 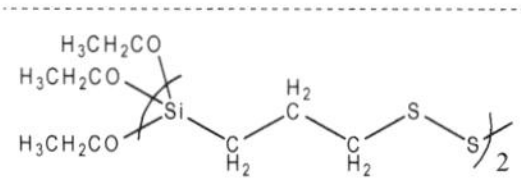
Vinyl	Polyolefin, EPR, EPDM, EVA, silicone acrylic resin	Vinyltrimethoxysilane 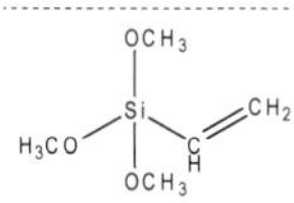

Part 2
Characterization and Computer Simulation

Chapter 5

Characterization of Nanocomposites

Thomas Andritsch and Alun S. Vaughan

ESC, Faculty of Physical Sciences and Engineering, University of Southampton, Southampton SO17 1BJ, UK

t.andritsch@soton.ac.uk

Many workers in the field of nanodielectrics, or more generally nanocomposites, have suggested that the key to understanding these systems lies at the interface between nanoparticle and matrix. The argument is that local interactions will define both the structure and molecular dynamics in consequent interphase regions. Along with the processing methodology, this will then determine the distribution of the nanoparticles throughout the bulk. Generic terms like "polyethylene" or "epoxy resin" cover a vast array of different molecular forms and microstructures, resulting in different material properties, depending on various factors like, for example, the processing temperature. As an example, there are more than 50 compounds that fit the definition of epoxy resin, which can be combined with several hundred types of hardener [1, 2]. Polyethylene is also not limited to varying degrees of cross-linking, but can exhibit various crystalline structures spanning the size range of nanometers to micrometers [3]. When surface modified, nanometric particles of varying aspect ratios are added to the mix, the terminology "nanodielectric" gets

Tailoring of Nanocomposite Dielectrics: From Fundamentals to Devices and Applications
Edited by Toshikatsu Tanaka and Alun S. Vaughan

ISBN 978-981-4669-80-1 (Hardcover), 978-981-4669-81-8 (eBook)
www.panstanford.com

even fuzzier. However, in the literature it is often assumed that nanocomposites must contain inorganic filler particles in order to be classified as such. The definition of "nanoparticle" is equally vague. It is generally accepted, that engineered materials with at least one dimension smaller than 100 nm are considered nanofillers. This includes such diverse structures as spherical nanosilica, graphene sheets, nanoscale rods or urchin-like structures that are created by RAFT methods. A series of questions needs to be answered, in order clearly to define such a composite material built from a combination of such diverse components. Questions like: what is the size distribution of the particles? How well dispersed are the particles in the polymer matrix? What is the inter-particle distance, i.e., the average distance between neighboring particles? How large are the agglomerates and how many are there? A good particle dispersion does not guarantee a good particle distribution (see Fig. 5.1). What is the surface chemistry of the particles and how does it affect the polymer? Are there by-products left in the polymer from the manufacturing process? All these and many other factors affect the dielectric performance of nanostructured materials, albeit it is not always clear to what extent [4]. A wide range of different techniques are therefore relevant to characterize these systems.

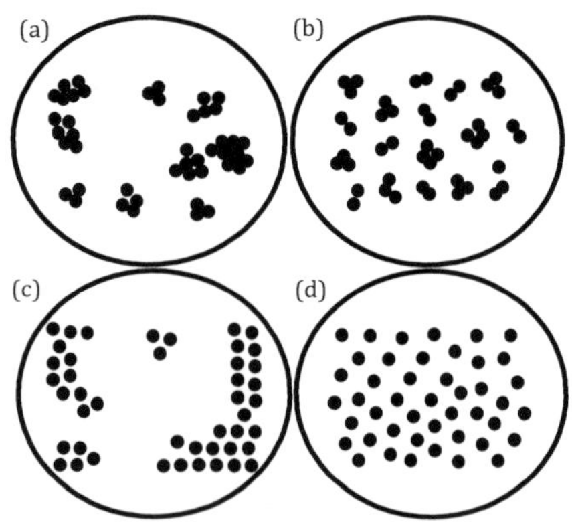

Figure 5.1 Illustration of distribution and dispersion: (a) bad distribution and dispersion, (b) bad dispersion but good distribution, (c) bad distribution but good dispersion, (d) good distribution and dispersion. Good local dispersion of particles does not guarantee good distribution throughout the bulk.

5.1 Molecules, Interfaces, and Interphases

Interphase regions are defined as those parts of a nanocomposite that exist between the filler and unperturbed matrix and which exhibit properties that are typical of neither material in the continuous bulk [5, 6]. One way to describe such interphases is the multi-core model, which includes a number of different interphase regions [7]. However, there are a number of other models that have been proposed, for example, to explain the change of dielectric permittivity of nanocomposites [8, 9]. Recently, molecular dynamics simulations have been employed to probe material adjacent to a nanoparticle in detail [10]. Indications of a rigid layer around small filler particles were found decades ago in a number of polymers, as indicated by multiple glass transitions, which suggest distinct and dissimilar polymer structures within the same material [11, 12]. In nanocomposites, the interphase should constitute a significant volume fraction of the complete system and, therefore, a comprehensive understanding of this structural element constitutes a key pre-requisite to understanding the system as a whole. It is now understood that these interphases are the key to the performance of some of the oldest mass produced nanocomposites: rubber tires. However, in many polymers it is very difficult to discern a distinct interphase. In amorphous polymers like polystyrene (PS) there is still no indication of a significant structural change due to the introduction of filler particles [13].

Identifying and analyzing the change in polymer structure at these interfaces requires the study of the environments within which the constituent molecules exist and, consequently, spectroscopic techniques that directly probe molecular and atomic factors are of considerable utility. For example, Raman and infrared (IR) spectroscopies. These techniques relate to molecular vibrational states, which are determined by both symmetry factors and the molecules' local environment. Consequently, chemical interactions between polymer molecules and nanoparticles would be expected to modify the vibrational modes of the polymer, while the limited extent of a nanoparticle may induce specific surface and interface vibrational contributions into the observed spectrum. Nuclear magnetic resonance (NMR) spectroscopy generally concerns the fine structure in the resonance of one nuclear species, which is dependent upon both spin-spin coupling and chemical shift effects related to the local environment around the nucleus of interest. Electron spin resonance (ESR) spectroscopy concerns unpaired electronic spins and is therefore the electronic analogue of NMR. Interactions between unpaired electrons and their local environments serve to modify the shape of the observed ESR spectral lines.

5.2 Dispersion in Real Space

In evaluating the dispersion of a nanofiller within a matrix, it is necessary to consider not only the typical local aggregation state, but also its variability from place to place throughout the bulk (as illustrated in Fig. 5.1). As an example, for the case of dielectric breakdown, the practical acceptance of the Weibull distribution [14] as a means of describing experimental data stems from the fact that they relate to the failure of some localized "weak link." Thus, in the case of a nanodielectric, it is conceivable that the existence of just a few large aggregates may be more important for long-term performance than the average dispersion state. Following this logic, scattering techniques that provide average measures may be less useful for characterizing such composites than high-resolution real space imaging techniques.

Imaging theory indicates that the wavelength of the illuminating radiation sets a fundamental limit upon the spatial resolution of an ideal far-field optical system. Consequently, electron optics is most relevant for spatially imaging nanocomposite systems. Indeed, even a cursory examination of the literature will reveal that the use of transmission electron microscopy (TEM) is almost de rigueur in any nanocomposite study. This is both because of the high resolution attainable and because one of the contrast mechanisms that operate in the TEM is ideally suited to the imaging of relatively electron dense nanoparticles embedded within a matrix composed predominantly of carbon and hydrogen. Especially in epoxy resin based nanocomposites, TEM provides clear images enabling identification of particle distribution in real space, while scanning electron microscopy (SEM) often delivers ambiguous results [15].

However, the preparation of suitably thin and representative specimens can be problematical. Hence, the examination of a statistically significant sample volume is extremely labor intensive. For these reasons, SEM has much to recommend it, provided a suitable contrast mechanism can be employed [16]. In SEM, most users exploit electrons emitted from the sample surface to generate an image. Secondary electrons (SE) typically have an energy of ~50 eV and, consequently, are only able to escape from the specimen when generated close to the surface and are readily swept into the detector by the application of a relatively small positive potential. As a consequence, SE images are related to local surface topography and roughness. They are characterized by a high resolution (comparable to the diameter of the exciting electron beam) and contain little information related to the elemental composition of the specimen. Conversely, backscattered electrons (BSE) have an energy comparable to that of the exciting electron beam. They are generated and can escape from throughout the excitation volume within the specimen, which is typically of the order of 1 μm in each dimension. Because of their high energy, BSE have linear trajectories between the sample and the detector and, consequently, BSE images often appear to contain sharp shadows, as if the surface were illuminated with a highly directional light source. In the case of flat surfaces, BSE image contrast is related to variation in atomic number (Z-contrast) and, as such, BSE images contain compositional information. However, image resolution is generally comparable to that of an optical microscope, because of the extent of the excitation volume.

An alternative means of building up an image of a sample, point by point, involves any one of the plethora of scanning probe microscopies (SPM) that have evolved from the initial concept of scanning tunneling microscopy (STM). By effectively working in the near-field regime, many different contrast mechanisms can be exploited, although scanning large areas to provide comprehensive data sets is more problematical than for TEM, even though sample preparation is easier. One inherent problem of these techniques is that they generally provide a two-dimensional snapshot of a three-dimensional spatial distribution. In short, the selection of an appropriate imaging technique is usually a compromise between resolution, contrast mechanism and the need to examine a statistically significant quantity of the specimen.

5.3 Scattering and Reciprocal Space

Scattering techniques average structural measures throughout the inspected volume and therefore complement investigations by real space techniques. However, there are serious issues associated with the conversion of reciprocal space data into meaningful real space parameters. The use of the Bragg equation, for example, may be an attractive (i.e., convenient) means of processing X-ray data. However, its application is often inappropriate in nanocomposites. When the system contains appreciable disorder, as it does in the case of a nanocomposite, it is necessary to consider the specimen to be an assembly of scattering volume elements. The observed scattering can be considered to arise from the local crystal lattice, combined with superimposed larger scale structural features, which are termed motifs. By way of simple illustration, consider the practically important example of a semicrystalline polymer, where the lamellar nature of the material can be represented as a long-range superlattice fluctuation (motif), superimposed upon the local crystalline lattice. The measured scattered intensity is the Fourier transform (FT) of the electron density distribution which, for a well-ordered lamellar stack, would be a convolution of the FT of the functions representing the two structural factors. In practice, the superlattice will be disordered in such a way that information relating to it is only preserved in the zeroth order peak. The analysis of such functions requires the application of some general, assumed structural model. However, it is rarely possible to transform intensity data alone into a unique structure. Crucial phase information is generally lost during recording said data. In short, when analyzing scattering data it is essential to consider that ordering will generally occur at a range of dimensional levels, all which are likely to exert some influence on the diffraction pattern. Aforementioned Bragg's law merely represents the most idealized case and, consequently, must be used with caution, particularly in connection with highly disordered systems.

5.4 Spectroscopy

5.4.1 UV/Visible Spectroscopy

Since ultra violet/visible (UV/vis) spectroscopy involves electronic transitions, its utilization with respect to nanocomposites is associated with systems that readily undergo such processes and, predominantly, those containing metallic components. Surface plasmon resonance effects are commonly studied using this technique [17]. Colloidal systems composed of metallic nanoparticles have attracted interest in connection with many applications, including as novel sensor systems that exhibit macroscopic changes as a result of changes in aggregation state in response to an external stimulus. To this end, Shiraishi et al. [18] examined the effect of surface chemistry on pH-dependent color changes in colloidal dispersions of gold nanoclusters. This work demonstrated that significant red-shifts and peak broadening occur in the surface plasmon band of gold nanoclusters as the pH of the system is reduced. Similar effects have also been associated with catalysis effects involving charge transfer [19] and dipole-dipole interactions leading to aggregation [20]. Li et al. [21] synthesized cadmium sulfide (CdS) nanoparticles on nanofibers of bacterial cellulose. Subsequent characterization with UV/vis spectroscopy revealed a small absorption peak at 426 nm. As a result of the finite size of the nanoparticles, the absorption was found to be blue shifted by 90 nm, compared to bulk CdS. In contrast, in polyaniline (PANi) systems containing both titania (TiO_2) nanoparticles and a magnetic Fe_3O_4 micro-filler, the UV-vis spectrum was found to be independent of composition [22]. This implies that the presence of the filler affects neither the structure of the PANi nor the electronic transitions that occur within it. The π -π^* transition, for example, was reported to occur at an invariant wavelength of 371 nm in all samples. An alternative interpretation of this is that if any interactions do occur between matrix and nanofiller, the volume that is affected is too small to have a detectable effect on the spectrum. Similarly, Koh et al. [23] synthesized silver loaded TiO_2 systems templated in a graft copolymer film. UV/vis spectroscopy of the final system exhibited the characteristic plasmon peak of silver nanoparticles at 410 nm, together with another feature at 294 nm. The authors assigned this "to the typical absorption edge of TiO_2." An alternative use of UV/vis spectroscopy is as a means of simply evaluating optical clarity at different wavelengths. It has been suggested that improved dispersion results in smaller aggregates, resulting in increased clarity [24]. While this philosophy might be valid for amorphous polymers, it cannot be applied simplistically to semicrystalline systems. Elsewhere, the same approach has been used to determine the degree of "yellowing" in clay/PET system, an effect attributed to decomposition of the organic compatibilizers [25]. Information about structure and stability of polypyrrole-based nanocomposites with graphene oxide filler have been gathered using UV/vis spectroscopy by Bora and Dolui [26]. Shi and Zhitomirsky had similar findings for polypyrrole/CNT-based electrodes [27], while Yu et al. used UV/vis in order to investigate nanocomposites consisting of Pt-ZnO metal/semiconductors [28]. Kumar et al. used UV/vis as a complementary characterization method along with Fourier transform infra-red (FT-IR) spectroscopy, Raman spectroscopy, X-ray diffraction (XRD) and thermogravimetric analysis (TGA), in order to investigate covalently crafted high-K graphene-polyaniline nanocomposites [29].

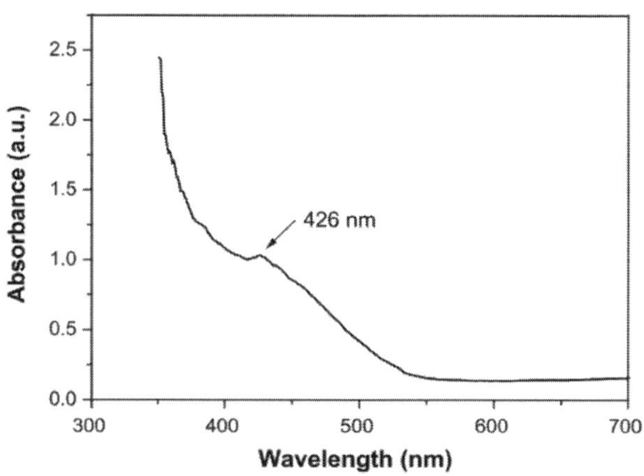

Figure 5.2 UV/vis absorption spectrum for CdS/BC composites, showing an absorption peak at 426 nm. Reprinted from *Carbohydrate Polymers*, vol. 76, X. Li, S. Chen, W. Hu, S. Shi, W. Shen, X. Zhang, et al., In situ synthesis of CdS nanoparticles on bacterial cellulose nanofibers, 509–512, Copyright (2009), with permission from Elsevier.

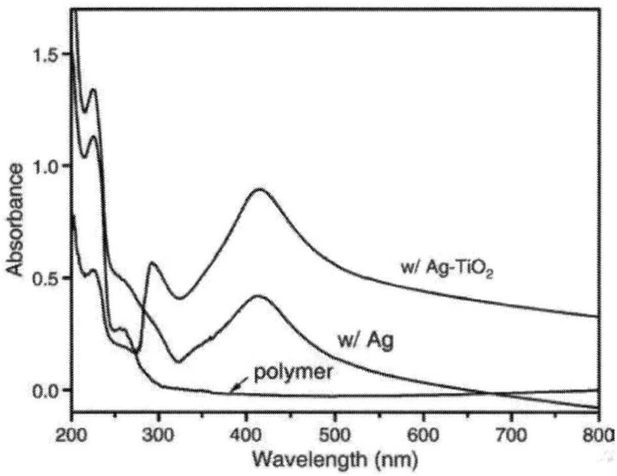

Figure 5.3 UV/vis absorption spectrum, showing characteristic plasmon peak of silver at 410 nm and another feature at 294 nm attributed to TiO_2. Reprinted from *Materials Letters*, vol. 63, J. K. Koh, J. A. Seo, J. H. Koh, and J. H. Kim, Templated synthesis of Ag loaded TiO_2 nanostructures using amphiphilic polyelectrolyte, 1360–1362, Copyright (2009), with permission from Elsevier.

5.4.2 Molecular Vibration Spectroscopies

In their excellent review, "Raman spectroscopy of nanomaterials: how spectra relate to disorder, particle size and mechanical properties," Gouadec and Colomban [30] describe two models by which particle sizes can be derived from Raman spectral data: the phonon confinement model (PCM) and the elastic sphere model (ESM). Such models contain parameters that have the dimensions of distance, but the precise interpretation of such parameters is far from straightforward. For example, the PCM of Richter et al. [31] is concerned with phonons confined within nanospheres of diameter L. Comparison of values of L derived from Raman scattering studies of CeO_2 with TEM data derived from the same systems revealed an order of magnitude discrepancy [32]. In this case, the key factor in determining L was not the size of the CeO_2 particles, but the distance between defects in the oxygen lattice within the ceria. In germanium particles, Carles et al. [33] showed that the correct interpretation of L was itself size dependent: for larger grains, L does indeed correspond to the actual size of the grains. The Raman behavior of disordered graphitic carbon has been studied extensively, both experimentally and theoretically. Vaughan et al. [34] used Raman spectroscopy to study disordered graphitic carbon within the channels of electrical trees. For such systems, the ratio of the intensity of the so-called G and D bands has been suggested as a means of determining an average graphitic domain size [35, 36]. Although this approach has been widely used [37–39], theoretical calculations indicate that size is not the only factor [40–42]; it has been suggested that resonance effects are also important, such that the graphitic domains that are sampled in practice will simply be those that are of the correct size to be excited by the chosen laser [43, 44]. In short, although it is theoretically possible to use confinement effects to estimate phase size, this is not always practical and may be potentially misleading. In the case of nanocomposites containing multi-walled carbon nanotubes (MWCNTs), shifts in the position of the G and D bands have been associated with interactions between the polymer and the nanotubes. Proposed explanations include charge exchange between the polymer and the CNTs and chemical interactions between the matrix and the nanotube, which serve to affect the vibrational modes of the latter [45–47]. Jeon et al. [48] used Raman microprobe spectroscopy as part of a study of nanocomposites based upon high density polyethylene (HDPE) and single walled carbon nanotubes (SWCNT). A key feature of the Raman response of CNTs is the radial breathing mode (RBM), which is located in the Raman spectrum between 100 and 400 cm^{-1}. The RBM is affected by factors like the nanotube dimensions, chirality, the surrounding matrix and the aggregation state of the nanotubes. In this work, a Raman feature around 266 cm^{-1} was ascribed to individual SWCNTs, while another feature close to 232 cm^{-1} was related to aggregated forms of SWCNT. Additionally, these peaks were found to shift to higher frequencies in the presence of the polymer matrix. It is proposed that this is related to "some compressive effect on the nanotubes." Raman was also used to characterize thermoresponsive graphene-containing hydrogels [49]. Jiang et al. used Raman to show the successful population of NiO particles onto graphene sheets [50]. The resulting highly porous material with large relative surface area has potential for the next generation of capacitors. Monti et al. used Raman for the comparison of graphene-epoxy composites obtained by different techniques and was successfully in identifying them [51]. Srivastava et al. used Raman in combination with other techniques to show a layer-by-layer assembly and dispersion of $NiCo_2O_4$ particles atop graphene oxide, quoting improvement of electrochemical performance [52].

Infrared (IR) spectroscopy has been very widely used in the study of nanocomposite systems, where a key issue is related to nanofiller-matrix interactions. Therefore, it would seem pertinent to quantify the surface chemistry of the nanofiller prior to its introduction into the matrix, as exemplified by the work of Huang et al. [53] and Chen et al. [54]. There are a number of examples of the successful use of FTIR to reveal effects related to interactions between a polymer matrix and a nanofiller. For example, in polyamide-10,6/organoclay samples, the intensity of the organoclay hydroxyl bands was found to diminish in the nanocomposite [55]. Elsewhere, in systems based upon gelatin, a range of chemical changes within the hydroxyapatite galleries has been reported [56]. Additional spectral features have been observed in the 960–1140 cm^{-1} region in montmorillonite (MMT) filled polyethylene systems prepared by melt mixing, with a dimethyl-dialkyl-ammonium salt used as a compatibilizer [57]. These changes are associated with interactions between the MMT and the alkyl chains of the compatibilizer or the polymer itself. Although the precise origin of the reported effects is unclear, evidence such as this does indicate that interactions do occur at the nanofiller-polymer interface and that such effects can be detected spectroscopically [58]. While most IR studies have involved the conventional wavenumber range, some attempts have also been made to use near infrared (NIR) spectroscopy. Moghaddam et al. [59] report on the effects of processing in thermoplastic polyurethane (TPU)-clay nanocomposites, as observed by NIR spectroscopy, with evidence of variations in peak intensities as processing progressed. Although Moghaddam et al. ascribed some of these effects to specific processes, the origin of others remained unclear. To conclude, the spectral changes seen due to the introduction of nanoparticles into a polymer are often subtle and interpretation is difficult, due to the inherent complexity of these systems.

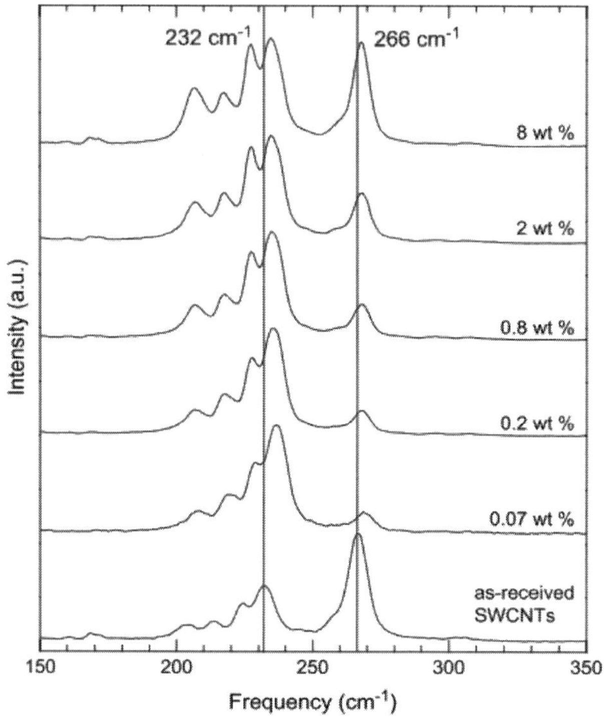

Figure 5.4 Raman spectra in RBM region, showing feature at 266 cm^{-1} corresponding to individual SWCNT and 232 cm^{-1} corresponding to their aggregate forms. Reprinted from Polymer, vol. 48, K. Jeon, L. Lumata, T. Tokumoto, E. Steven, J. Brooks, and R. G. Alamo, Low electrical conductivity threshold and crystalline morphology of single-walled carbon nanotubes—high density polyethylene nanocomposites characterized by SEM, Raman spectroscopy and AFM, 4751–4764, Copyright (2007), with permission from Elsevier.

5.4.3 NMR and ESR Spectroscopies

A particularly attractive feature of nuclear magnetic resonance (NMR) and electron spin resonance (ESR) techniques is that they hold the potential for rapid quantification of the dispersion state of nanofillers within a matrix—a critical factor in terms of manufacturing and quality control. In 2003 Gilman et al. [60] published a paper entitled "High throughput methods for polymer nanocomposites research: Extrusion, NMR characterization and flammability property screening," which discussed the potential of NMR as an on-line tool for determining nanofiller dispersion in polystyrene systems containing MMT. This approach relied upon the presence of paramagnetic Fe^{3+} ions in the MMT, which affects the proton longitudinal relaxation time in the polymer (T_1^H). T_1^H will therefore be affected by both the concentration of Fe^{3+} ions and their proximity to the polymer molecules. Hence, for a fixed composition it can be concluded that the greater the dispersion, the greater the effect on T_1^H. Similar processes have been used in connection with nylon-6/MMT systems, where ^{27}Al was used as a probe to reveal surface modification of the clay as a consequence of polymer/matrix interactions [61]. Elsewhere, a number of different NMR-effects have been associated with the interaction of nanofiller surfaces with their immediate environment. A ^{13}C NMR study of styrene-butadiene rubber (SBR) and titania revealed significant peak shifts, indicating nanoparticle/matrix interactions. However, spin lattice relaxation experiments suggested that the occurring interactions did not significantly affect the polymer mobility [62]. With the introduction of MWCNTs into poly(3-hexylthiophene) (P3HT), π-π stacking between benzene rings in the polymer and the graphitic benzene rings of the nanotubes was detected by NMR, along with evidence of CH-π interactions [63]. Pardal et al. [64] used proton and ^{13}C techniques in concert to study the grafting of short chain oligomers onto the surface of nanosilica, while ^{29}Si NMR has been used in a similar way and as a means of studying the sol-gel reactions within epoxy matrices [65, 66].

ESR spectroscopy has been used in a number of studies to examine polymer/nanoparticle interfaces, interphases and interactions, in an analogous way to NMR. In systems based on a pyrrole derivative and layered double hydroxide (LDH) nanoparticles, hyperfine structure in the ESR spectrum was interpreted as interactions between the aluminum nuclei within the interlayer galleries and the neighboring carboxylate radicals [67]. While the chemical and physical nature of interphase layers are important, the electrical nature of such regions is also relevant to understanding the behavior of nanodielectrics. Ai et al. [68] studied nanocomposite films of poly(vinyl alcohol) (PVA) including polyoxometalates (POM) (i.e., nano-sized metal-oxygen cluster anions) and reported evidence of charge transfer between the PVA and the POM. Earlier in this section, the use of ions as probes of local structures was described in connection with NMR—similar processes can also be exploited in ESR. By introducing a so-called spin label (i.e., a stable free radical—commonly nitroxide), materials that do not have an intrinsic paramagnetic response can be studied by ESR. In suitably modified poly(methyl acrylate) (PMA) systems containing fluoromica, this approach showed that significant interactions occur between the polymer and the nanofiller when the latter is suitably exfoliated. As a consequence, the mobility of PMA chains is reduced within a rigid interphase region, that was estimated to be approximately 5–15 nm in thickness [69]. Miwa et al. also used nitroxy radicals attached to PEO chains to study molecular dynamics within with the galleries of a fluoromica clay [70]. These studies, which involved a combination of experiment and simulation, revealed a range of dynamic effects associated with both rapid and slow chain dynamics, depending on the proximity of the relevant chain segment to the gallery walls.

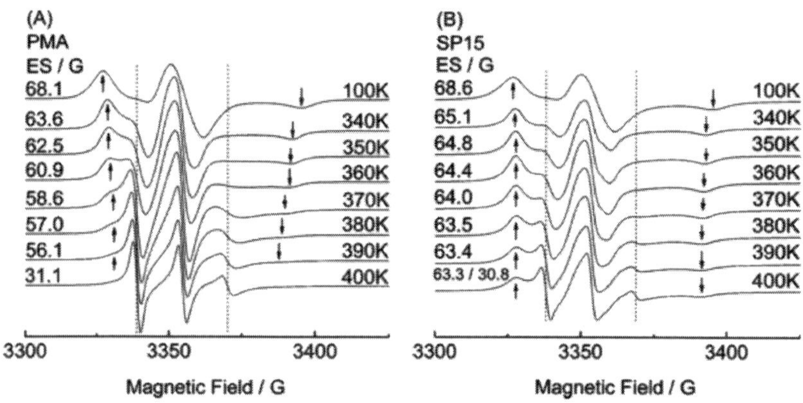

Figure 5.5 ESR studies on PMA and PMA/clay composites; at 400 K the PMA is in the motionally narrowed regime, but the spectrum of SP15 consists of two components, indicating restriction of PMA chains in a rigid interphase region of about 5 to 15 nm. Reprinted with permission from Y. Miwa, A. R. Drews, and S. Schlick, Detection of the direct effect of clay on polymer dynamics: the case of spin-labeled poly(methyl acrylate)/clay nanocomposites studied by ESR, XRD, and DSC, *Macromolecules*, vol. 39, pp. 3304–3311, 2006. Copyright (2006) American Chemical Society.

5.4.4 Further Related Techniques

Broadband dielectric spectroscopy (BDS) covers a frequency range from 10^{-6} to 10^{12} Hz and facilitates similar investigations to some of the spectroscopy techniques presented above. Due to its importance for the analysis of nanodielectrics, BDS is discussed extensively later in this book.

Dynamic mechanical analysis (DMA) and dynamic mechanical thermal analysis (DMTA) work in a similar manner to BDS, albeit that instead of an external electric field of adjustable magnitude and frequency, a mechanical stimulus of adjustable magnitude and frequency is used. DMTA (as a function of frequency and temperature) delivers comparable spectra to BDS. The storage modulus, E', is thereby equivalent the real part of the complex relative permittivity, ε', while the loss modulus, E'', corresponds to the imaginary part of the complex relative permittivity, ε''. This method is particularly useful for investigating changes in the glass transition temperature T_g.

DMTA responds to relaxation peaks at the same frequencies but with different magnitudes to BDS and, therefore, the two techniques are particularly useful in combination [71]. Combining the spectra enables the easier identification of relaxation peaks. Chua investigated the dynamic mechanical properties of unidirectional glass-fiber-reinforced polyester composites, showing the effect of volume fraction and amount of aminosilane treatment on the interfacial shear strength and tan δ. In combination with the shift in the glass transition the conclusion was drawn that DMA can be used to characterize the quality of the interphase [72].

Calorimetry is another straightforward way of characterizing nanodielectrics; such measurements have been carried out since the middle of the 18th century. Today, thermogravimetric analysis (TGA) and differential scanning calorimetry (DSC) are essential tools that are used widely for material characterization, with automatic and recording balances widely available since the 1950s. While TGA measures the change in sample weight as a function of increasing temperature, DSC measures changes in the heat capacity as a function of increasing temperature. These two methods provide information relating to certain physical and chemical phenomena, including, but not limited to: absorption;

adsorption; chemisorption; crystalline transitions; decomposition; desolvation; desorption; fusion; oxidation; reduction; sublimation; vaporization; solid-gas, solid-state or second-order transitions. The main use of TGA in polymer science is for the investigation of the degradation temperature and the change thereof, the level of oligomer in the polymer and quantification of moisture content [73]. It is a useful tool to confirm the reliability of the nanodielectric synthesis procedure, by measuring the content of filler particles in polymer nanocomposites [74]. A vital application of TGA is the characterization of the surface groups on nanoparticles, which is a fast and simple way to assess changes to the particle surface [75, 76].

5.5 Microscopy

5.5.1 Transmission Electron Microscopy

Transmission electron microscopy is probably the most widely used technique for studying polymer-based nanocomposites. This is due to the combination of high resolution and good image contrast. It is easy to image the distribution of nanoparticles using bright field techniques, provided suitably thin, representative specimens can be prepared. This simple and direct approach has been used to study a myriad of systems, including the first recorded CNT [77]. More recent examples include the study of the dispersion of an LDH nanofiller in a range of ethylene/vinyl acetate systems of varying polarity, which considered different processing effects [78]. Solution intercalation was found to be superior to melt processing in this case. Hess et al. [79] considered the dispersion of ZnO nanoparticles within PMMA and reported fractal structuring of the ZnO nanoparticles. This work showed uniform dispersion on the micrometer scale, but localized clustering, and thereby highlights a key issue that must be considered in connection with any microscopy system, namely the degree to which the examination of extremely small sample volumes can be considered typical of the bulk. Another issue concerns projection effects, which are particularly pertinent for layered nanofillers such as graphite oxide, nanoclays or mica [80–82]. Images are generally chosen to reveal edge-on projections, where the so-called Z-contrast between the filler and matrix is high. While it is difficult to image aggregates of a few layers when oriented orthogonal to the electron beam, useful images of suitably small tactoids can be produced [83]. More questionable is the assertion that it is possible directly to detect changes in surface chemistry by conventional bright field imaging [84], since the interpretation of such images is not straightforward. Ngo et al. [85] have reported on the structure of TiO_2 nanoparticles, using high-resolution lattice imaging. In this case, the crystal lattice of the TiO_2 could be seen throughout the nanoparticles prior to surface grafting, indicating no evidence of disordered surfaces. Conversely, after surface grafting, disordered surfaces layers of about 1–5 nm in thickness could be identified. Elsewhere, specialist high-resolution imaging techniques have been used to examine cerium-based nanoparticles within PMMA. Both lattice images and selected area electron diffraction (SAED) patterns were obtained [86].

A key element in the production of nanocomposites from systems with filler materials such as clays, graphite and hexagonal born nitride concerns the exfoliation of the constituent layer. The structure of such fillers is lamellar and they can be expanded into nanosheets for dispersal into a matrix through the application of appropriate processing techniques. TEM has been applied to analyze such systems, as for example in [87]. While bright field imaging techniques are most commonly used, SAED is also applicable, as shown in various studies, such as, polyethylene containing an LDH [88], nanocomposites based upon graphite oxide within a styrene-butyl acrylate copolymer [89] and polystyrene containing graphite nanosheets [90]. While it is well appreciated that the addition of nanoparticles can markedly affect the structure of the matrix system, Chen et al. [90] demonstrated with SAED that their graphite nanosheets were no longer single crystals after processing. Houdayer et al. [91] employed dark field imaging to study the structure of antimony/PANi nanocomposites. Ultramicrotomy was used to prepare thin sections of a nanocomposite composed of a biodegradable elastomer, containing needle-like nano-hydroxyapatite (n-HAp) [92]. From this, the authors concluded that strong interfacial interactions exist between the two components.

High-resolution TEM (HRTEM) and high-resolution scanning TEM (HRSTEM) investigations can help to understand interfaces on the molecular scale, as exemplified by Garrett et al., who examined cubic born nitride (BN) and found hexagonal BN grains at the surface, which were caused by slow solution-precipitation processes [93]. An increase in the sintering temperature resulted in increased hBN growth, which in turn resulted in an increased interface layer and deteriorated mechanical properties. A layered shell of boron oxide was found on cBN particles by Andritsch et al., who attributed high amounts of space charge to the B_2O_3 shells covering cBN [94]. The shells were identified to be about 3 nm in thickness.

5.5.2 Scanning Electron Microscopy

Although images can be generated in the SEM using a number of different approaches, the technique is primarily associated with image formation using either SE or BSE emissions from the surface of a specimen. As such, the technique can be applied to examine external or internal surfaces, although the latter is generally preferable if it is the bulk structure of a material that is of interest. Examples of situations where the direct examination of nanocomposites via SEM is valid including the study of wear behavior [95], mechanical failure processes [96], the characterization of nanoparticles before incorporation into a host matrix [97] and the effect of nanoparticle addition on degradation processes, including flammability [98], laser ablation [99] and electrical discharge erosion [100].

Clearly, to examine the internal structure of a specimen it is necessary to expose an internal surface and to ensure that some contrast mechanism exists that will lead to useful structural contrast in the image. While a range of techniques exists that vary somewhat in complexity, a commonly used and straightforward technique is cryo-fracture. Although the inclusion of the term "cryo" merely indicates that fracturing occurred well below the glass transition temperature (T_g) of the specimen, even for polymers with a relatively high T_g, it would seem to be commonplace to cool the specimen prior to inducing brittle fracture [101–103]. Examples of nanocomposite systems in which this very conventional approach has been used include a polyurethane/titania nanohybrid membrane [104] and an ethylene/(vinyl acetate) copolymer containing nano-titania [105]. Conversely, Filippi et al. [106] used cryo-fracture in conjunction with an environmental scanning electron microscope (ESEM) as a means of examining internal surfaces. Their study revolved around the preparation procedure and morphology of nanocomposites of HDPE grafted with maleic anhydride (g-MA) and an organoclay. By using an ESEM, the fracture surfaces could be examined without the need for prior gold coating. It was claimed that this procedure resulted in images with enhanced contrast. Elsewhere, the Z-contrast inherent in BSE imaging was exploited in the study of polystyrene systems containing copper nanowires [107]. An accelerating voltage of 20 kV was used, which would generate an extensive excitation volume for BSE production. Despite this, the authors were able to resolve structures as small as a few nanowires. By varying the acceleration voltage, Loos et al. could improve the contrast in SWNT polystyrene composites and, by correlating the secondary electron yield, obtain quasi-three dimensional information about the nanotube-dispersion in the host polymer [108]. The investigated SWNT network showed that the conductive SWNT network consisted of both straight and bent tubes.

Microtomy, or some comparable approach, constitutes an alternative way of exposing an internal surface within a sample prior to examination in the SEM. As in the case of fracture, this is generally best done at a temperature below T_g, to minimize the generation of artifacts within the surface that are related to the cutting process. Frechette et al. [109] employed this approach to study epoxy cross-sections through resin-based systems containing both micro-silica and a layered silicate. The objective was to explore the effect of the nano-additive on electrical surface erosion performance. Hence, BSE images were sufficient to reveal both the depth of erosion and the distribution of the micro-silica within the material (Z-contrast). Where some comparable inherent contrast mechanism is not available, or where high-resolution SE images are required, it is necessary to induce some variations in surface topography after cutting, which are related to the underlying structure of the material; e.g., the distribution of the nano-additive or the effect of the nanofiller on the matrix microstructure. Although this is potentially fraught with problems, for many commonly used polymeric systems appropriate true etching techniques have been developed and proven, since the concept was first developed in the late 1970s. The initial motivation for etching polymers was to reveal the lamellar texture of melt crystallized polyethylene and, as such, relied upon the different etching rates of crystalline and amorphous regions of the same polymer. Such techniques can be highly discriminating. Despite this, relatively few attempts seem to have been made to apply this approach to the study of nanocomposites, even where proven polymeric etchants exist. Vaughan and co-workers have applied conventional permanganic reagents based upon an acidic solution of potassium permanganate to a range of nanocomposites based upon PE and PS [13, 110, 111]. These composites contained a wide selection of nanofillers, such as MMT, lead zirconate titanate (PZT), silica (SiO_2) and BN. The authors have shown that this approach is able simultaneously to reveal both the nanofiller distribution and the effect of this on the morphology of the matrix polymer. Elsewhere, Wang et al. [112] used the rather less discriminating approach of solvent etching in order to examine the phase structure of a tertiary system, consisting of a rubber, poly(trimethylene terephthalate) and an organoclay.

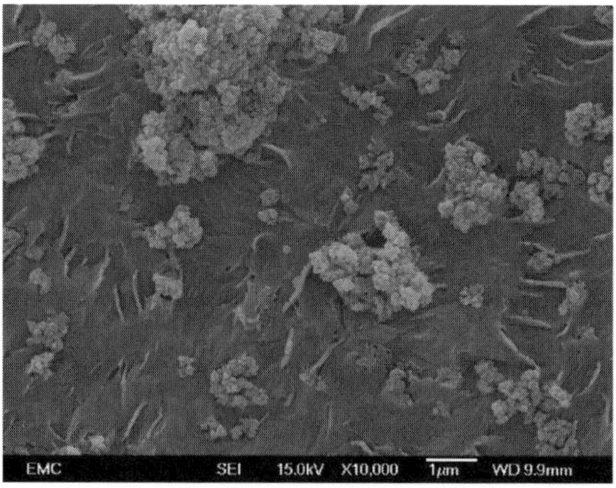

Figure 5.6 SEM micrograph, showing an etched surface in polyethylene/ nanosilica system. The micrograph simultaneously reveals the primary nanosilica particle size, the range of agglomeration states and the background morphology of the matrix polymer.

5.5.3 Spectroscopy in the Electron Microscope

When electrons interact with a specimen, a certain percentage of them will undergo inelastic scattering; in case of an SEM, it is multiple scattering that gives rise to the excitation volume introduced above. Consequently, it is not only BSEs that are produced within this region of space, but also X-rays. The X-ray spectrum produced as a result of interactions between electrons and sample contains two components. The first is a continuous background radiation, termed Bremsstrahlung, onto which, characteristic peaks that originate from electronic transitions within the atoms are superimposed. These characteristic X-ray emissions, which are generally produced below the sample surface, contain information concerning the elemental composition of the specimen, but may undergo further interactions with the sample before exiting the surface. The resulting absorption and fluorescence makes the quantification of such X-ray spectra not entirely straightforward. Also, as a result of X-ray production throughout the excitation volume, the available resolution of X-ray spectroscopy within the SEM is insufficient to enable individual nanoparticles to be detected. Nevertheless, the technique can provide a means of probing gross dispersion and agglomeration effects, as in Owpradit and Jongsomjit's [113] study of polyethylene systems filled with titania. This methodology has also been used to evaluate the relative uniformity of dispersion of titania nanoribbons and nanotubes within polystyrene [114], where it indicated that the latter tend to form clusters, presumably as a result of increased entanglement. Similarly, Yeh et al. used X-ray spectroscopy to study dispersion in poly(methyl methacrylate) (PMMA)–silica hybrid sol–gel systems [115]. While the production route used did not give rise to the formation of large silica aggregates, the images presented in the paper do suggest some spatial variations in composition on the scale of micrometers for low silica loading levels. At higher levels it would appear that the nanosilica forms elongated features, although this is not discussed. Most studies that have exploited X-ray spectroscopy in the SEM to study nanocomposites have used this approach to generate X-ray maps, in which the intensity of some characteristic X-ray peak is plotted as a function of 2-D position. An alternative is to show representative spectra. Considering the case of an ideal flat specimen, the spatial variation in X-ray intensity should correlate well with elemental composition. Strictly speaking, it would be better to consider the relative intensities of two lines—one representing the polymer and one representing the nanofiller. One example of this approach is shown by Casciola et al. [116], who examined zirconium phosphate filled poly(vinylidene fluoride) (PVDF) nanocomposites; the Zr and P lines were taken to represent the nanoparticles and the F line to represent the polymer. As would be anticipated, the ratio of the intensity of the Zr and P lines was found to be spatially invariant, whereas the Zr/F ratio varied significantly from location to location.

For TEM, it is an intrinsic requirement that the specimen is sufficiently thin for electrons to be transmitted without multiple scattering. Therefore, the potential resolution of X-ray spectroscopy is much greater in this case than when comparable instrumentation is used in conjunction with SEM. However, as discussed above, the electron density in nanoparticles is generally much greater than in the surrounding polymer matrix and provides a simple means of imaging dispersion directly. As such, X-ray spectroscopy appears to be most commonly used in TEM as a means of characterizing nanoparticles [117–119]. X-ray generation in the TEM occurs as a result of inelastic scattering with the specimen. Consequently, for energy to be conserved and assuming that the kinematical approximation applies, the transmitted electron beam must contain complementary information to the X-ray spectrum. The resulting technique is termed electron energy loss spectroscopy (EELS). Although this approach does not seem to have been used extensively for the study of nanostructured materials thus far, the approach has been used to study mesoporous carbon [120], titanium carbide/amorphous carbon nanocomposites [121], surface functionalized carbon nanotubes [122] and surface functionalized alumina nanoparticles [123]. In the case of polymer-based nanocomposites, the used of EELS-related techniques appears even more limited. However, the technique does appear to be of some value, as exemplified by Li et al. [124], who studied systems composed of a vinyl ester resin containing polyhedral oligomeric silsesquioxane (POSS). While conventional bright field TEM was unable to reveal the POSS, energy filtering was shown to produce clear images.

5.5.4 Scanning Probe Microscopies

Scanning tunneling microscopy (STM) was developed in 1981 and led to its inventors, Binnig and Rohrer, receiving the Nobel prize in Physics in 1986. Since then, a multitude of different scanning probe microscopy (SPM) techniques have been developed, that exploit a range of properties to generate images of a sample surface. Nevertheless, perhaps the most commonly used instrument of this type is the atomic force microscope (AFM) and, consequently, it has been used to examine many different nanocomposite systems.

AFM may be used in a number of modes to produce images that reflect the local variation in, for example, mechanical properties or surface topography. In a study of gold nanoparticles in polyacrylamide [125], tapping mode was used to produce topographical images of the distribution of gold nanoparticles within the material surface. The gold particles were found to stand out quite distinctly from the matrix, resulting in high contrast images. However, this is not always the case. Rao and Chen [126] used AFM as part of a study of titania-based nanocomposites, in which a low surface roughness (0.45 nm) was observed. They associated this with a high degree of nanoparticle dispersion. However, examination of the TEM images presented in the paper indicate a degree of granularity. The ability of AFM to produce topographical images has been exploited not just in terms of material structure, but also in terms of the resistance of nanocomposites to surface erosion. In one study, the effect of nanoalumina was investigated and the resulting images revealed increasing roughness with increasing degradation [127]. This work indicated the accumulation of nanoparticles in the degraded surface layer, which served to decrease the rate of erosion. While it is convenient to produce a specimen and then examine the external surface directly using a surface imaging technique such as AFM, this does require that the external surface is representative of the bulk. This is frequently not the case and, hence, it is generally preferable to examine an internal surface prepared using microtomy, etching, freeze-fracture, etc., as in the case of SEM. For polypropylene (PP) nanocomposites filled with MWNT, Ganss et al. [128] used AFM after microtomy to study the structure of the system, which was found to be composed of loosely packed assemblies of MWNTs. These assemblies were approximately 1 μm in size and dispersed relatively homogeneous within the polymer matrix. Other examples of investigations of nanoparticle distributions through topographical images with AFM include: poly(ethylene terephthalate) (PET) containing MMT [129] and polyurethane with nanosilica [130]. In contrast, Zubitur et al. [131] employed AFM tapping mode phase imaging to reveal tactoids of a functionalized organoclay in poly(p-dioxanone). There, the image contrast is presumably related to the difference in mechanical properties between the filler and the matrix. Seiler and Kinderberger used AFM and electrostatic force microscopy (EFM) to investigate the dispersion of SiO_2 nanoparticles in silicone rubber. Using the intermittent contact AFM mode, they could not only distinguish between nanoparticles and host, but also determine the extent of what they interpreted as bound interphase layers around the particles [132]. These interphase volumes were identified as reasons for increased HV arcing resistance in the investigated polymer nanocomposites.

Other scanning probe techniques rely on spatial variations in the electrical properties of the system of interest; e.g., the tunneling current that flows from the probe tip to the sample to generate images in STM. This was used by Tamburri et al. [133] to form images of SWCNT in conducting polymer nanocomposites, based upon local variations in electrical properties. However, while the tunneling current will vary with substrate conductivity, it is also strongly dependent on the separation between probe and sample. Consequently, interpreting images of anything other than atomically flat specimens can be problematical. In block copolymers, nanoparticles can become preferentially located within one of the phases, leading to local variations in electrical properties. EFM uses local variations in the electric field gradient to produce image contrast; this has been applied in studies of systems containing functionalized graphene [134] and silver nanoparticles [135]. Although quantitative interpretation of EFM data is not straightforward, a method for determining the local dielectric constant by EFM has been proposed by Gramse et al. [136]. Many variants of EFM can be devised to exploit the sample of interest in such a way as to generate useful image contrast. For example, in dynamic contact-electrostatic force microscopy (DC-EFM), a DC voltage is first applied to the probe tip and used locally to pole the specimen, before the sample is subsequently imaged. Ramasundaram et al. [137] applied this to nanocomposites films of PVDF containing organically modified silicate, whereupon the final images were based upon local variations in the ferroelectric behavior of the specimen. Other techniques that have been used in the study of nanocomposites include Kelvin probe microscopy [138] and scanning near field optical microscopy (SNOM) [139].

5.6 Scattering

5.6.1 X-Ray Scattering

Microscopy provides qualitative data concerning a small volume of space being imaged. Imaging is therefore particularly well suited to the identification of anomalous features within materials. Many physical processes are driven by structural anomalies, such as defects. Scattering techniques, on the other hand, involve relatively large sampling volumes and can therefore provide data that are typical of the bulk. Also, since scattering is related to reciprocal space, it is generally necessary to apply some theory to convert from reciprocal to real space, such that quantitative parameters are extracted. As mentioned earlier, while simple approaches (e.g., Bragg's law, the Scherrer equation, etc.) may prove to be adequate in many circumstances, a critical element that must always be remembered is that these are convenient simplifications. Understanding when these are and are not appropriate is key to deriving meaningful data.

One area where X-ray scattering is widely applied in relation to nanocomposites is in connection with the dispersion of nanoclays. Key issues here concern the extent of aggregation of the clay and the effect of the intercalant that is introduced into the material to vary the inter-layer spacing and to aid processing.

Many studies have adopted a rather simplistic approach to the problem of nanoclay dispersion and claims such as "the absence of diffraction peaks, indicating the silicate layer are completely exfoliated in the PU matrix" [140], should be treated with caution. Complete exfoliation would indeed result in no periodicity, and therefore no Bragg peak. However, the existence of aggregated but disordered structures that are not characterized by any periodicity would also mean that Bragg's law would not apply. Conversely, "the results suggest that the single sheets of LDH have been exfoliated in the polymer matrix" was made despite the existence of clear X-ray scattering peaks, which would indicate periodicity [141]. The value of combining imaging and scattering techniques is that, while the former defines the nature of the structure, the latter facilities its quantification. Therefore, studies that apply both in concert are of particular value. For example, wide angle X-ray scattering (WAXS) and TEM were used by Gârea et al. [142] to reveal the structure of nanocomposites based upon polybenzoxazine with MMT. The TEM images presented reveal the presence of tactoids of different sizes. However, the WAXS patterns contained broad features that are indicative of a disordered structure. In a study of PP with MMT and calcium carbonate [143], WAXS and TEM were used together to characterize the resulting nanocomposites. The WAXS data contained a weak Bragg peak that indicated the presence of a structure with a periodicity of 1.3 nm, and high contrast structures were seen in the TEM. Despite this, the authors stated that the "TEM image for sample PP1 [...] shows individual silicate layers which can be seen to be well-dispersed (exfoliated) in the matrix" [143]. While this statement may well be strictly true, when all the data are combined and analyzed, it may also be unrepresentative. An interesting use of X-ray scattering was reported by Hernandez et al. [144], who attempted to derive the size distribution of tactoids present in their samples from X-ray results. Although the TEM images reveal the presence of significant aggregates, the X-ray-based tactoid size distribution indicated that this is dominated by small structures. The significance of the different filler fractions within this distribution is likely to be dependent upon the topic or material property of interest. Where barrier properties are concerned, the presence of a small number of large aggregates may not be an issue. In the case of dielectric breakdown, a few large "defects" may be dominant. In short, while X-ray scattering is an extremely powerful technique, interpretation of the resulting data is not straightforward, particularly when derived from disordered systems. Simply using the Bragg equation in such cases can be fraught with danger.

A major issue concerning the introduction of nanoclays into polymers concerns compatibilization. Nanoclays are made up of primary layers, which are separated from one another by gallery regions. As a consequence of impurities within the layered structures themselves, the clay layers are generally electrically charged (anionic or cationic clays) and this charge is then balanced by the presence of oppositely charged counterions that reside within the gallery regions. Replacing these ions with amphiphilic moieties therefore constitutes a commonly used means of introducing functional groups that act to compatibilize the clay and the chosen polymer matrix. This aims to assist the dispersion and/or exfoliation of the clay. Replacing small Na^+ ions with bulky surfactant-like molecules also has a significant effect on the interlayer spacing. These molecules are made up of a polar head, which associates with the clay, and an organic tail. They are designed both to reduce cohesion between platelets and to lead to favorable interactions with the matrix polymer. Indeed, it has been shown that with the polar head tethered to polar clay surfaces, alkyl chains tend to adopt layered trans/gauche conformations [145–147]. In the case of nylon-6 and MMT, Fornes et al. [148] showed that compatibilizers containing a single long alkyl tail led to better dispersion than those with alternative structures. The rationale being that the system based upon the single alkyl chain resulted in a more optimal reduction in inter-platelet cohesion. This would mean easier exfoliation, followed by increased polar interactions between the clay and the nylon, resulting in improved stabilization. An alternative strategy towards dispersion is based on introducing monomers into the galleries and then polymerizing them in situ. WAXS was used to study the exchange of chlorine anions within LDHs with sulfonate monomers. The observed increase in the basal spacing was interpreted in terms of the inclusion of two monomer layers [149].

The preceding paragraphs have exclusively considered the wide angle scattering regime, but small angle X-ray scattering (SAXS) is also relevant for the characterization of nanocomposites, although "this technique has not been widely used except in a few laboratories probably because most laboratories do not have SAXS facilities or experience in interpreting the results" [150]. While it is possible to employ the Bragg equation in connection with WAXS patterns, the interpretation of SAXS data generally requires more complex approaches. For example, Sedláková et al. [151] used Guinier analysis, which involves plotting ln $I(q)q^2$ versus q^2, where q is the scattering vector and $I(q)$ is the observed variation in the scattering intensity. This approach indicated that, in their poly(butyl methacrylate)/MMT nanocomposite latexes, the MMT existed in the form of sheet-like entities of ~1.3 nm in thickness. Kim and Macosko [152] used SAXS as part of a study of polyester/exfoliated graphite nanocomposites. In this case, plotting the SAXS data in the form log $I(q)$ against log q led these workers to deduce that the graphite layers were present in their system in the form of a fractal structure with dimensionality 2.0–2.3. This interpretation is consistent with the reported WAXS data, which lacked any strong diffraction peaks, and TEM images indicating a high degree of exfoliation.

5.6.2 Dynamic Light Scattering

The wavelength of visible light is of the order of 500 nm and light scattering techniques are therefore directly relevant to the study of structural features of comparable dimensions. In the case of a system of well dispersed nanoparticles of the order of tens of nanometers in size, extensive scattering would not be anticipated and, consequently, optical clarity can be used as an indirect measure of dispersion [153]. Dynamic light scattering (DLS) is a technique that relies on Rayleigh scattering of a coherent light source and the analysis of the resulting speckle pattern. This makes it immediately relevant to the study of suspensions of scatterers, including nanoparticles. Consequently, it has been applied to many different systems, including solutions of fluorinated block copolymer and SWCNT [154], nanocomposites of poly(styrene-co-hexylacrylate) and cellulose whiskers [155], and C60/polymer colloid nanocomposites [156]. In a study of poly *N*-isopropylacrylamide (PNIPA)-clay nanocomposite hydrogels [157], DLS was applied to the study of the nanoparticles and the polymer initially in water, then separately and finally mixed together. For the water-clay system, the hydrodynamic radius of the clay was found to be 16 nm and did not vary with temperature. Conversely, the hydrodynamic radius of the polymer in isolation increased dramatically, from 5 nm to 260 nm, as the temperature of the system passed through the lower critical solution temperature (LCST). However, in the presence of the clay, the increase in molecular size on passing through the LCST was found to be markedly reduced, an effect that was associated with strong interactions between the clay and the polymer leading to reduced PNIPA aggregation above the LCST. Foster et al. [158] used DLS to explore processing effects in polypropylene and carbon nanofiber (CNF) samples. CNFs were extracted from each nanocomposite, their dimensions determined in suspension by DLS and the resulting data used to evaluate a theoretical Young's modulus for each system, which was ultimately compared with experiment.

5.6.3 Optical Coherence Tomography

Optical coherence tomography (OCT) is a well-established medical technique for the production of three-dimensional images within tissue specimens; that is, within scattering media. Reference to the literature indicates a great deal of interest in methodologies that can enhance image contrast and the use nanoparticles for this purpose has received significant attention. For example, Al Rawashdeh et al. [159] compared the effectiveness of two core/shell nanoparticle systems as potential OCT contrast enhancer in both a model scattering system and in ex vivo chicken muscle tissue. These workers concluded that the PS spheres had significant potential for use in this application. Such results suggest that OCT may also provide a useful means of evaluating the dispersion of nanoparticles within nanocomposites. To this end, Broughton et al. [160] have recently published an assessment of this approach using a range of nanocomposite systems based upon epoxy resins and polypropylene, containing a wide range of different nanofillers (e.g., a nanoclay, MWCNT, zinc oxide, etc.). The effectiveness of the approach is dependent upon a number of material factors, including particle size, the refractive index contrast within the system, nanoparticle loading level and optical absorption. However, this work demonstrates that, in appropriate systems, OCT can provide useful data relating to the nanoparticle dispersion state in nanocomposites.

5.7 Conclusions

A very wide range of different techniques has been applied successfully to the study of nanocomposite systems. While each has value in its own right, interpretation of the resulting data is not always straight forward and cases exist where limited data may have been over interpreted. Since different techniques can provide different perspectives of the same problem, the use of complementary techniques in concert has much to recommend it. Physical interpretations that are consistent with a number of "orthogonal" viewpoints would seem to have much greater general validity than conclusions drawn from more limited approaches.

References

1. J. D. Lemay and F. N. Kelley, Structure and ultimate properties of epoxy-resins, *Advances in Polymer Science,* vol. 78, pp. 115–148, 1986.
2. N. M. Augustsson, *Epoxy Handbook.* Sweden: Nils Malmgren AB, 2004.
3. S. N. Kolesov, The influence of morphology on the electric strength of polymer insulation, *Electrical Insulation, IEEE Transactions on,* vol. EI-15, pp. 382–388, 1980.
4. T. Andritsch, D. Fabiani, and I. Ramirez Vazquez, Nanodielectrics: examples of preparation and microstructure, *IEEE Electrical Insulation Magazine,* vol. 29, pp. 21–28, 2013.
5. M. R. Piggott, The effect of the interface interphase on fiber composite properties, *Polymer Composites,* vol. 8, pp. 291–297, 1987.
6. S. Raetzke and J. Kindersberger, Role of interphase on the resistance to high-voltage arcing, on tracking and erosion of silicone/SiO_2 nanocomposites, *Dielectrics and Electrical Insulation, IEEE Transactions on,* vol. 17, pp. 607–614, 2010.
7. T. Tanaka, Dielectric nanocomposites with insulating properties, *IEEE Transactions on Dielectrics and Electrical Insulation,* vol. 12, pp. 914–928, 2005.

8. M. G. Todd and F. G. Shi, Complex permittivity of composite systems: a comprehensive interphase approach, *Dielectrics and Electrical Insulation, IEEE Transactions on,* vol. 12, pp. 601–611, 2005.

9. P. Maity, N. Gupta, V. Parameswaran, and S. Basu, On the size and dielectric properties of the interphase in epoxy-alumina nanocomposite, *Dielectrics and Electrical Insulation, IEEE Transactions on,* vol. 17, pp. 1665–1675, 2010.

10. D. Brown, V. Marcadon, P. Mélé, and N. D. Albérola, Effect of filler particle size on the properties of model nanocomposites, *Macromolecules,* vol. 41, pp. 1499–1511, 2008.

11. J. Y. Park and G. B. McKenna, Size and confinement effects on the glass transition behavior of polystyrene/o-terphenyl polymer solutions, *Physical Review B,* vol. 61, pp. 6667–6676, 2000.

12. J. O'Brien, E. Cashell, G. E. Wardell, and V. J. McBrierty, NMR investigation of interaction between carbon-black and cis-polybutadiene, *Macromolecules,* vol. 9, pp. 653–660, 1976.

13. J. Topham, O. Boorman, I. L. Hosier, M. Praeger, R. Torah, A. S. Vaughan, et al., Dielectric studies of polystyrene-based, high-permittivity composite systems, in *Electrical Insulation and Dielectric Phenomena (CEIDP), 2014 IEEE Conference on,* pp. 711–714, 2014.

14. W. Weibull, A statistical distribution function of wide applicability, *Journal of Applied Mathematics and Mechanics,* vol. 18, pp. 293–297, 1951.

15. T. Andritsch, Epoxy based nanodielectrics for high voltage DC applications: synthesis, dielectric properties and space charge dynamics, Ph.D., Electrical Sustainable Energy, University of Delft, Delft, 2010.

16. Y. Liu, X. Zhou, J. Carr, C. Butler, S. L. Mills, J. O'Connor, et al., Visualisation of conductive filler distributions in polymer composites using voltage and energy contrast imaging in SEM, *Polymer,* vol. 54, pp. 330–340, 2013.

17. L. Bois, F. Chassagneux, S. Parola, F. Bessueille, Y. Battie, N. Destouches, et al., Growth of ordered silver nanoparticles in silica film mesostructured with a triblock copolymer PEO-PPO-PEO, *Journal of Solid State Chemistry,* vol. 182, pp. 1700–1707, 2009.

18. Y. Shiraishi, D. Arakawa, and N. Toshima, pH-dependent color change of colloidal dispersions of gold nanoclusters: effect of stabilizer, *Eur Phys J E Soft Matter,* vol. 8, pp. 377–83, 2002.

19. C. Novo, A. M. Funston, and P. Mulvaney, Direct observation of chemical reactions on single gold nanocrystals using surface plasmon spectroscopy, *Nat Nanotechnol,* vol. 3, pp. 598–602, 2008.

20. N. Satoh, H. Hasegawa, K. Tsujii, and K. Kimura, Photoinduced coagulation of Au nanocolloids, *Journal of Physical Chemistry,* vol. 98, pp. 2143–2147, 1994.

21. X. Li, S. Chen, W. Hu, S. Shi, W. Shen, X. Zhang, et al., In situ synthesis of CdS nanoparticles on bacterial cellulose nanofibers, *Carbohydrate Polymers,* vol. 76, pp. 509–512, 2009.

22. S. W. Phang, M. Tadokoro, J. Watanabe, and N. Kuramoto, Effect of Fe_3O_4 and TiO_2 addition on the microwave absorption property of polyaniline micro/nanocomposites, *Polymers for Advanced Technologies,* vol. 20, pp. 550–557, 2009.

23. J. K. Koh, J. A. Seo, J. H. Koh, and J. H. Kim, Templated synthesis of Ag loaded TiO_2 nanostructures using amphiphilic polyelectrolyte, *Materials Letters,* vol. 63, pp. 1360–1362, 2009.

24. M.-C. Lai, K.-C. Chang, W.-C. Huang, S.-C. Hsu, and J.-M. Yeh, Effect of swelling agent on the physical properties of PET–clay nanocomposite materials prepared from melt intercalation approach, *Journal of Physics and Chemistry of Solids,* vol. 69, pp. 1371–1374, 2008.

25. J. W. Chung, S.-B. Son, S.-W. Chun, T. J. Kang, and S.-Y. Kwak, Thermally stable exfoliated poly(ethylene terephthalate) (PET) nanocomposites as prepared by selective removal of organic modifiers of layered silicate, *Polymer Degradation and Stability,* vol. 93, pp. 252–259, 2008.

26. C. Bora and S. K. Dolui, Fabrication of polypyrrole/graphene oxide nanocomposites by liquid/liquid interfacial polymerization and evaluation of their optical, electrical and electrochemical properties, *Polymer,* vol. 53, pp. 923–932, 2012.

27. K. Shi and I. Zhitomirsky, Polypyrrole nanofiber-carbon nanotube electrodes for supercapacitors with high mass loading obtained using an organic dye as a co-dispersant, *Journal of Materials Chemistry A,* vol. 1, pp. 11614–11622, 2013.

28. C. Yu, K. Yang, Y. Xie, Q. Fan, J. C. Yu, Q. Shu, et al., Novel hollow Pt-ZnO nanocomposite microspheres with hierarchical structure and enhanced photocatalytic activity and stability, *Nanoscale,* vol. 5, pp. 2142–2151, 2013.

29. M. Kumar, K. Singh, S. K. Dhawan, K. Tharanikkarasu, J. S. Chung, B.-S. Kong, et al., Synthesis and characterization of covalently-

grafted graphene-polyaniline nanocomposites and its use in a supercapacitor, *Chemical Engineering Journal,* vol. 231, pp. 397–405, 2013.

30. G. Gouadec and P. Colomban, Raman Spectroscopy of nanomaterials: how spectra relate to disorder, particle size and mechanical properties, *Progress in Crystal Growth and Characterization of Materials,* vol. 53, pp. 1–56, 2007.
31. H. Richter, Z. P. Wang, and L. Ley, The one phonon Raman-spectrum in microcrystalline silicon, *Solid State Communications,* vol. 39, pp. 625–629, 1981.
32. W. H. Weber, K. C. Hass, and J. R. McBride, Raman-study of CeO_2–2nd-order scattering, lattice-dynamics, and particle-size effects, *Physical Review B,* vol. 48, pp. 178–185, 1993.
33. R. Carles, A. Mlayah, M. Amjoud, A. Reynes, and R. Morancho, Structural characterization of Ge microcrystals in Ge_xC_{1-x} films, *Japanese Journal of Applied Physics Part 1-Regular Papers Short Notes & Review Papers,* vol. 31, pp. 3511–3514, 1992.
34. A. S. Vaughan, I. L. Hosier, S. J. Dodd, and S. J. Sutton, On the structure and chemistry of electrical trees in polyethylene, *Journal of Physics D-Applied Physics,* vol. 39, pp. 962–978, 2006.
35. D. S. Knight and W. B. White, Characterization of diamond films by Raman-spectroscopy, *Journal of Materials Research,* vol. 4, pp. 385–393, 1989.
36. M. J. Matthews, M. A. Pimenta, G. Dresselhaus, M. S. Dresselhaus, and M. Endo, Origin of dispersive effects of the Raman D band in carbon materials, *Physical Review B,* vol. 59, pp. R6585–R6588, 1999.
37. H. Darmstadt, D. Pantea, L. Summchen, U. Roland, S. Kaliaguine, and C. Roy, Surface and bulk chemistry of charcoal obtained by vacuum pyrolysis of bark: influence of feedstock moisture content, *Journal of Analytical and Applied Pyrolysis,* vol. 53, pp. 1–17, 2000.
38. S. Prawer, K. W. Nugent, Y. Lifshitz, G. D. Lempert, E. Grossman, J. Kulik, et al., Systematic variation of the Raman spectra of DLC films as a function of sp(2):sp(3) composition, *Diamond and Related Materials,* vol. 5, pp. 433–438, 1996.
39. D. G. McCulloch, S. Prawer, and A. Hoffman, Structural investigation of xenon-ion-beam-irradiated glassy-carbon, *Physical Review B,* vol. 50, pp. 5905–5917, 1994.
40. M. Rigolio, C. Castiglioni, G. Zerbi, and F. Negri, Density functional theory prediction of the vibrational spectra of polycyclic aromatic

hydrocarbons: effect of molecular symmetry and size on Raman intensities, *Journal of Molecular Structure,* vol. 563, pp. 79–87, 2001.

41. F. Negri, C. Castiglioni, M. Tommasini, and G. Zerbi, A computational study of the Raman spectra of large polycyclic aromatic hydrocarbons: toward molecularly defined subunits of graphite, *Journal of Physical Chemistry A,* vol. 106, pp. 3306–3317, 2002.
42. C. Castiglioni, E. Di Donato, M. Tommasini, F. Negri, and G. Zerbi, Multi-wavelength Raman response of disordered graphitic materials: models and simulations, *Synthetic Metals,* vol. 139, pp. 885–888, 2003.
43. C. Castiglioni, C. Mapelli, F. Negri, and G. Zerbi, Origin of the D line in the Raman spectrum of graphite: a study based on Raman frequencies and intensities of polycyclic aromatic hydrocarbon molecules, *Journal of Chemical Physics,* vol. 114, pp. 963–974, 2001.
44. A. C. Ferrari, Determination of bonding in diamond-like carbon by Raman spectroscopy, *Diamond and Related Materials,* vol. 11, pp. 1053–1061, 2002.
45. D. Baskaran, J. W. Mays, and M. S. Bratcher, Noncovalent and nonspecific molecular interactions of polymers with multiwalled carbon nanotubes, *Chemistry of Materials,* vol. 17, pp. 3389–3397, 2005.
46. K. E. Wise, C. Park, E. J. Siochi, and J. S. Harrison, Stable dispersion of single wall carbon nanotubes in polyimide: the role of noncovalent interactions, *Chemical Physics Letters,* vol. 391, pp. 207–211, 2004.
47. G. G. Silva, A. W. Musumeci, L. Jiang-Wen, W. N. Martens, and E. R. Waclawik, Structure and conductivity of multi-walled carbon nanotube/poly(3-hexylthiophene) composite films, *Polymer,* vol. 48, pp. 1667–1678, 2007.
48. K. Jeon, L. Lumata, T. Tokumoto, E. Steven, J. Brooks, and R. G. Alamo, Low electrical conductivity threshold and crystalline morphology of single-walled carbon nanotubes—high density polyethylene nanocomposites characterized by SEM, Raman spectroscopy and AFM, *Polymer,* vol. 48, pp. 4751–4764, 2007.
49. R. Sanna, D. Sanna, V. Alzari, D. Nuvoli, S. Scognamillo, M. Piccinini, et al., Synthesis and characterization of graphene-containing thermoresponsive nanocomposite hydrogels of poly(N-vinylcaprolactam) prepared by frontal polymerization, *Journal of Polymer Science Part A-Polymer Chemistry,* vol. 50, pp. 4110–4118, 2012.

50. Y. Jiang, D. Chen, J. Song, Z. Jiao, Q. Ma, H. Zhang, et al., A facile hydrothermal synthesis of graphene porous NiO nanocomposite and its application in electrochemical capacitors, *Electrochimica Acta,* vol. 91, pp. 173–178, 2013.

51. M. Monti, M. Rallini, D. Puglia, L. Peponi, L. Torre, and J. M. Kenny, Morphology and electrical properties of graphene-epoxy nanocomposites obtained by different solvent assisted processing methods, *Composites Part A-Applied Science and Manufacturing,* vol. 46, pp. 166–172, 2013.

52. M. Srivastava, M. E. Uddin, J. Singh, N. H. Kim, and J. H. Lee, Preparation and characterization of self-assembled layer by layer $NiCo_2O_4$-reduced graphene oxide nanocomposite with improved electrocatalytic properties, *Journal of Alloys and Compounds,* vol. 590, pp. 266–276, 2014.

53. X. Huang, X. Fang, Z. Lu, and S. Chen, Reinforcement of polysiloxane with superhydrophobic nanosilica, *Journal of Materials Science,* vol. 44, pp. 4522–4530, 2009.

54. Y. Chen, S. Zhou, H. Yang, and L. Wu, Structure and properties of polyurethane/nanosilica composites, *Journal of Applied Polymer Science,* vol. 95, pp. 1032–1039, 2005.

55. A. Al-Mulla, Development and characterization of polyamide-10, 6/organoclay nanocomposites, *International Journal of Polymer Analysis and Characterization,* vol. 14, pp. 540–550, 2009.

56. G. Zuo, C. Liu, H. Luo, F. He, H. Liang, J. Wang, et al., Synthesis of intercalated lamellar hydroxyapatite/gelatin nanocomposite for bone substitute application, *Journal of Applied Polymer Science,* vol. 113, pp. 3089–3094, 2009.

57. S. Tzavalas and V. G. Gregoriou, Infrared spectroscopy as a tool to monitor the extent of intercalation and exfoliation in polymer clay nanocomposites, *Vibrational Spectroscopy,* vol. 51, pp. 39–43, 2009.

58. D. Zhang, A. B. Karki, D. Rutman, D. P. Young, A. Wang, D. Cocke, et al., Electrospun polyacrylonitrile nanocomposite fibers reinforced with Fe_3O_4 nanoparticles: fabrication and property analysis, *Polymer,* vol. 50, pp. 4189–4198, 2009.

59. L. Moghaddam, D. J. Martin, P. J. Halley, and P. M. Fredericks, Vibrational spectroscopic studies of laboratory scale polymer melt processing: application to a thermoplastic polyurethane nanocomposite, *Vibrational Spectroscopy,* vol. 51, pp. 86–92, 2009.

60. J. W. Gilman, S. Bourbigot, J. R. Shields, M. Nyden, T. Kashiwagi, R. D. Davis, et al., High throughput methods for polymer

nanocomposites research: extrusion, NMR characterization and flammability property screening, *Journal of Materials Science,* vol. 38, pp. 4451–4460, 2003.

61. M. Bertmer, M. Wang, M. Krüger, B. Blümich, V. M. Litvinov, and M. van Es, Structural changes from the pure components to nylon 6–montmorillonite nanocomposites observed by solid-state NMR, *Chemistry of Materials,* vol. 19, pp. 1089–1097, 2007.
62. T. M. Arantes, K. V. Leão, M. I. B. Tavares, A. G. Ferreira, E. Longo, and E. R. Camargo, NMR study of styrene-butadiene rubber (SBR) and TiO_2 nanocomposites, *Polymer Testing,* vol. 28, pp. 490–494, 2009.
63. V. Saini, Z. Li, S. Bourdo, E. Dervishi, Y. Xu, X. Ma, et al., Electrical, optical, and morphological properties of P3HT-MWNT nanocomposites prepared by in situ polymerization, *The Journal of Physical Chemistry C,* vol. 113, pp. 8023–8029, 2009.
64. F. Pardal, V. Lapinte, and J.-J. Robin, Modification of silica nanoparticles by grafting of copolymers containing organosilane and fluorine moities, *Journal of Polymer Science Part A: Polymer Chemistry,* vol. 47, pp. 4617–4628, 2009.
65. E. Amerio, G. Malucelli, M. Sangermano, and A. Priola, Nanostructured hybrid materials obtained by UV curing and sol-gel processes involving alkoxysilane groups, *E-Polymers,* 2009.
66. B. Didier, R. Mercier, N. D. Alberola, and C. Bas, Preparation of polyimide/silica hybrid material by sol-gel process under basic catalysis: comparison with acid conditions, *Journal of Polymer Science Part B: Polymer Physics,* vol. 46, pp. 1891–1902, 2008.
67. J. Tronto, F. Leroux, M. Dubois, J. F. Borin, C. F. de Oliveira Graeff, and J. B. Valim, Hyperfine interaction in Zn–Al layered double hydroxides intercalated with conducting polymers, *Journal of Physics and Chemistry of Solids,* vol. 69, pp. 1079–1083, 2008.
68. L.-M. Ai, W. Feng, J. Chen, Y. Liu, and W.-M. Cai, Evaluation of microstructure and photochromic behavior of polyvinyl alcohol nanocomposite films containing polyoxometalates, *Materials Chemistry and Physics,* vol. 109, pp. 131–136, 2008.
69. Y. Miwa, A. R. Drews, and S. Schlick, Detection of the direct effect of clay on polymer dynamics: the case of spin-labeled poly(methyl acrylate)/clay nanocomposites studied by ESR, XRD, and DSC, *Macromolecules,* vol. 39, pp. 3304–3311, 2006.
70. Y. Miwa, A. R. Drews, and S. Schlick, Unique structure and dynamics of poly(ethylene oxide) in layered silicate nanocomposites:

accelerated segmental mobility revealed by simulating ESR spectra of spin-labels, XRD, FTIR, and DSC, *Macromolecules,* vol. 41, pp. 4701–4708, 2008.

71. T. Andritsch, A. Lunding, P. H. F. Morshuis, H. Negle, and J. J. Smit, Permittivity in epoxy based syntactic foam nanocomposites, in *Electrical Insulation and Dielectric Phenomena, 2008. CEIDP 2008. Annual Report Conference on*, pp. 513–516, 2008.

72. P. S. Chua, Dynamic mechanical analysis studies of the interphase, *Polymer Composites,* vol. 8, pp. 308–313, 1987.

73. H. M. L. Thijs, C. R. Becer, C. Guerrero-Sanchez, D. Fournier, R. Hoogenboom, and U. S. Schubert, Water uptake of hydrophilic polymers determined by a thermal gravimetric analyzer with a controlled humidity chamber, *Journal of Materials Chemistry,* vol. 17, p. 4864, 2007.

74. O. Boorman, I. L. Hosier, M. Praeger, R. Torah, A. S. Vaughan, T. Andritsch, et al., *Barium titanate and the dielectric response of polystyrene-based composites*, 2014.

75. A. K. Mishra, S. Allauddin, R. Narayan, T. M. Aminabhavi, and K. V. S. N. Raju, Characterization of surface-modified montmorillonite nanocomposites, *Ceramics International,* vol. 38, pp. 929–934, 2012.

76. H. Jin, T. Andritsch, I. A. Tsekmes, R. Kochetov, P. H. F. Morshuis, and J. J. Smit, Properties of mineral oil based silica nanofluids, *IEEE Transactions on Dielectrics and Electrical Insulation,* vol. 21, pp. 1100–1108, 2014.

77. M. Monthioux and V. L. Kuznetsov, Who should be given the credit for the discovery of carbon nanotubes?, *Carbon,* vol. 44, pp. 1621–1623, 2006.

78. M. Zhang, P. Ding, L. Du, and B. Qu, Structural characterization and related properties of EVA/ZnAl-LDH nanocomposites prepared by melt and solution intercalation, *Materials Chemistry and Physics,* vol. 109, pp. 206–211, 2008.

79. S. Hess, M. M. Demir, V. Yakutkin, S. Baluschev, and G. Wegner, Investigation of oxygen permeation through composites of PMMA and surface-modified ZnO nanoparticles, *Macromol Rapid Commun,* vol. 30, pp. 394–401, 2009.

80. J. Y. Xu, Y. Hu, L. Song, Q. G. Wang, W. C. Fan, and Z. Y. Chen, Preparation and characterization of poly(vinyl alcohol)/graphite oxide nanocomposite, *Carbon,* vol. 40, pp. 450–451, 2002.

81. D. H. Kim, P. D. Fasulo, W. R. Rodgers, and D. R. Paul, Effect of the ratio of maleated polypropylene to organoclay on the structure and properties of TPO-based nanocornposites. Part II: thermal expansion behavior, *Polymer,* vol. 49, pp. 2492–2506, 2008.

82. C.-W. Chiu, C.-C. Chu, W.-T. Cheng, and J.-J. Lin, Exfoliation of smectite clays by branched polyamines consisting of multiple ionic sites, *European Polymer Journal,* vol. 44, pp. 628–636, 2008.

83. Y. Tang, Y. Hu, J. Z. Wang, R. W. Zong, Z. Gui, Z. Y. Chen, et al., Influence of organophilic clay and preparation methods on EVA/montmorillonite nanocomposites, *Journal of Applied Polymer Science,* vol. 91, pp. 2416–2421, 2004.

84. W. Tao, W. Mozhen, Z. Zhicheng, G. Xuewu, and F. Yue'e, Grafting of polymers from clay nanoparticles via high-dose gamma-ray irradiation, *Materials Letters,* vol. 61, pp. 3723–3727, 2007.

85. V. G. Ngo, C. Bressy, C. Leroux, and A. Margaillan, Synthesis of hybrid TiO_2 nanoparticles with well-defined poly(methyl methacrylate) and poly(tert-butyldimethylsilyl methacrylate) via the RAFT process, *Polymer,* vol. 50, pp. 3095–3102, 2009.

86. R. Chai, H. Lian, C. Li, Z. Cheng, Z. Hou, S. Huang, et al., In situ preparation and luminescent properties of CeF_3 and CeF_3:Tb^{3+} nanoparticles and transparent CeF_3:Tb^{3+}/PMMA nanocomposites in the visible spectral range, *Journal of Physical Chemistry C,* vol. 113, pp. 8070–8076, 2009.

87. Y. X. Pan, Z. Z. Yu, Y. C. Ou, and G. H. Hu, A new process of fabricating electrically conducting nylon 6/graphite nanocomposites via intercalation polymerization, *Journal of Polymer Science Part B-Polymer Physics,* vol. 38, pp. 1626–1633, 2000.

88. W. Chen and B. J. Qu, Structural characteristics and thermal properties of PE-g-MA/MgAl-LDH exfoliation nanocomposites synthesized by solution intercalation, *Chemistry of Materials,* vol. 15, pp. 3208–3213, 2003.

89. R. Zhang, Y. Hu, J. Xu, W. Fan, and Z. Chen, Flammability and thermal stability studies of styrene–butyl acrylate copolymer/graphite oxide nanocomposite, *Polymer Degradation and Stability,* vol. 85, pp. 583–588, 2004.

90. G. H. Chen, C. L. Wu, W. G. Weng, D. J. Wu, and W. L. Yan, Preparation of polystyrene/graphite nanosheet composite, *Polymer,* vol. 44, pp. 1781–1784, 2003.

91. A. Houdayer, R. Schneider, D. Billaud, J. Lambert, and J. Ghanbaja, Preparation of new antimony(0)/polyaniline nanocomposites

by a one-pot solution phase method, *Materials Letters,* vol. 61, pp. 171–176, 2007.

92. L. Lei, L. Li, L. Zhang, D. Chen, and W. Tian, Structure and performance of nano-hydroxyapatite filled biodegradable poly((1,2-propanediol-sebacate)-citrate) elastomers, *Polymer Degradation and Stability,* vol. 94, pp. 1494–1502, 2009.
93. J. C. Garrett, I. Sigalas, and M. Helimann, TEM investigation of the interface formation in cubic boron nitride containing alpha-SiAlON composites, *Ceramics International,* vol. 40, pp. 16169–16175, 2014.
94. T. Andritsch, R. Kochetov, Y. Gebrekiros, U. Lafont, P. H. F. Morshuis, and J. J. Smit, Space charge accumulation in epoxy based magnesium oxide and boron nitride nanocomposites, presented at the *International Symposium on High Voltage Engineering,* Cape Town, South Africa, 2009.
95. Q. Jia, S. Shan, Y. Wang, L. Gu, and J. Li, Tribological performance and thermal behavior of epoxy resin nanocomposites containing polyurethane and organoclay, *Polymers for Advanced Technologies,* vol. 19, pp. 859–864, 2008.
96. A. Omrani and A. A. Rostami, Understanding the effect of nano-Al_2O_3 addition upon the properties of epoxy-based hybrid composites, *Materials Science and Engineering: A,* vol. 517, pp. 185–190, 2009.
97. S. Letaief and C. Detellier, Clay-polymer nanocomposite material from the delamination of kaolinite in the presence of sodium polyacrylate, *Langmuir,* vol. 25, pp. 10975–9, 2009.
98. B. B. Marosfoi, S. Garas, B. Bodzay, F. Zubonyai, and G. Marosi, Flame retardancy study on magnesium hydroxide associated with clays of different morphology in polypropylene matrix, *Polymers for Advanced Technologies,* vol. 19, pp. 693–700, 2008.
99. C. D. Green, PhD, ECS, University of Southampton, 2008.
100. T. Tanaka, Y. Ohki, M. Ochi, M. Harada, and T. Imai, Enhanced partial discharge resistance of epoxy/clay nanocomposite prepared by newly developed organic modification and solubilization methods, *Dielectrics and Electrical Insulation, IEEE Transactions on,* vol. 15, pp. 81–89, 2008.
101. L. Xu and M. Yang, In situ compatibilization of polypropylene and polystyrene-grafted nano-sized TiO_2 in the presence of Friedel-Crafts catalyst, *Materials Letters,* vol. 62, pp. 2607–2610, 2008.
102. N. K. Srivastava and R. M. Mehra, Study of structural, electrical, and dielectric properties of polystyrene/foliated graphite nanocomposite

developed via in situ polymerization, *Journal of Applied Polymer Science,* vol. 109, pp. 3991–3999, 2008.

103. A. S. Vaughan, unpublished.

104. H. Zhou, Y. Chen, H. Fan, H. Shi, Z. Luo, and B. Shi, Water vapor permeability of the polyurethane/TiO_2 nanchybrid membrane with temperature sensitivity, *Journal of Applied Polymer Science,* vol. 109, pp. 3002–3007, 2008.

105. M. L. Cerrada, C. Serrano, M. Sanchez-Chaves, M. Fernandez-Garcia, F. Fernandez-Martin, A. de Andres, et al., Self-sterilized EVOH-TiO_2 nanocomposites: interface effects on biocidal properties, *Advanced Functional Materials,* vol. 18, pp. 1949–1960, 2008.

106. S. Filippi, C. Marazzato, P. Magagnini, A. Famulari, P. Arosio, and S. V. Meille, Structure and morphology of HDPE-g-MA/organoclay nanocomposites: effects of the preparation procedures, *European Polymer Journal,* vol. 44, pp. 987–1002, 2008.

107. B. Lin, G. A. Gelves, J. A. Haber, P. Poetschke, and U. Sundararaj, Electrical, morphological and rheological study of melt-mixed polystyrene/copper nanowire nanocomposites, *Macromolecular Materials and Engineering,* vol. 293, pp. 631–640, 2008.

108. J. Loos, A. Alexeev, N. Grossiord, C. E. Koning, and O. Regev, Visualization of single-wall carbon nanotube (SWNT) networks in conductive polystyrene nanocomposites by charge contrast imaging, *Ultramicroscopy,* vol. 104, pp. 160–167, 2005.

109. M. F. Frechette, R. Y. Larocque, M. Trudeau, R. Veillette, R. Rioux, S. Pelissou, et al., Nanostructured polymer microcomposites: a distinct class of insulating materials, *Dielectrics and Electrical Insulation, IEEE Transactions on,* vol. 15, pp. 90–105, 2008.

110. K. Y. Lau, A. S. Vaughan, G. Chen, I. L. Hosier, and A. F. Holt, On the dielectric response of silica-based polyethylene nanocomposites, *Journal of Physics D-Applied Physics,* vol. 46, 2013.

111. S. J. Sutton and A. S. Vaughan, A comparison of fracture and permanganic etching of polypyrrole films, *Journal of Polymer Science Part B-Polymer Physics,* vol. 34, pp. 837–843, 1996.

112. K. Wang, Y. Chen, and Y. Zhang, Effects of organoclay platelets on morphology and mechanical properties in PTT/EPDM-g-MA/organoclay ternary nanocomposites, *Polymer,* vol. 49, pp. 3301–3309, 2008.

113. W. Owpradit and B. Jongsomjit, A comparative study on synthesis of LLDPE/TiO_2 nanocomposites using different TiO_2 by in situ

polymerization with zirconocene/dMMAO catalyst, *Materials Chemistry and Physics,* vol. 112, pp. 954–961, 2008.

114. P. Umek, M. Huskić, A. S. Škapin, U. k. Florjančič, B. Zupančič, I. Emri, et al., Structural and mechanical properties of polystyrene nanocomposites with 1D titanate nanostructures prepared by an extrusion process, *Polymer Composites,* vol. 30, pp. 1318–1325, 2009.

115. J.-M. Yeh, C.-J. Weng, K.-Y. Huang, and C.-C. Lin, Effect of baking treatment and materials composition on the properties of bulky PMMA–silica hybrid sol–gel materials with low volume shrinkage, *Journal of Applied Polymer Science,* vol. 101, pp. 1151–1159, 2006.

116. M. Casciola, A. Donnadio, M. Pica, V. Valentini, and P. Piaggio, Characterization of Zr phosphate/PVDF nanocomposites by vibrational spectroscopy, *Macromolecular Symposia,* vol. 230, pp. 95–104, 2005.

117. M. F. Meléndrez Castro, G. Cárdenas Triviño, J. Morales, J. Díaz Visurraga, and C. Cruzat Contreras, Synthesis and study of photoacoustic properties of (Pd/TiO_2)/polystyrene nanocomposites, *Polymer Bulletin,* vol. 62, pp. 355–366, 2008.

118. A. Choudhury, Polyaniline/silver nanocomposites: dielectric properties and ethanol vapour sensitivity, *Sensors and Actuators B: Chemical,* vol. 138, pp. 318–325, 2009.

119. T. Nakayama, T. A. Yamamoto, Y.-H. Choa, and K. Niihara, Structure and magnetic properties of iron oxide dispersed silver based nanocluster composite, *Journal of Materials Science,* vol. 35, pp. 3857–3861, 2000.

120. A. Vinu, S. Anandan, C. Anand, P. Srinivasu, K. Ariga, and T. Mori, Fabrication of partially graphitic three-dimensional nitrogen-doped mesoporous carbon using polyaniline nanocomposite through nanotemplating method, *Microporous and Mesoporous Materials,* vol. 109, pp. 398–404, 2008.

121. J. C. Sánchez-López, D. Martínez-Martínez, C. López-Cartes, and A. Fernández, Tribological behaviour of titanium carbide/amorphous carbon nanocomposite coatings: from macro to the micro-scale, *Surface and Coatings Technology,* vol. 202, pp. 4011–4018, 2008.

122. G.-X. Chen and H. Shimizu, Multiwalled carbon nanotubes grafted with polyhedral oligomeric silsesquioxane and its dispersion in poly(l-lactide) matrix, *Polymer,* vol. 49, pp. 943–951, 2008.

123. A. Chandra, L.-S. Turng, P. Gopalan, R. M. Rowell, and S. Gong, Study of utilizing thin polymer surface coating on the nanoparticles

for melt compounding of polycarbonate/alumina nanocomposites and their optical properties, *Composites Science and Technology,* vol. 68, pp. 768–776, 2008.

124. G. Z. Li, L. Wang, H. Toghiani, T. L. Daulton, and C. U. Pittman, Viscoelastic and mechanical properties of vinyl ester (VE)/ multifunctional polyhedral oligomeric silsequioxane (POSS) nanocomposites and multifunctional POSS-styrene copolymers, *Polymer,* vol. 43, pp. 4167–4176, 2002.

125. Y. Song, Z. Li, L. Wang, Y. Yao, C. Chen, and K. Cui, One-step preparation of hybrid materials of polyacrylamide networks and gold nanoparticles, *Microscopy Research and Technique,* vol. 71, pp. 409–412, 2008.

126. Y. Rao and S. Chen, Molecular composites comprising TiO_2 and their optical properties, *Macromolecules,* vol. 41, pp. 4838–4844, 2008.

127. P. Maity, S. Basu, V. Parameswaran, and N. Gupta, Degradation of polymer dielectrics with nanometric metal-oxide fillers due to surface discharges, *Dielectrics and Electrical Insulation, IEEE Transactions on,* vol. 15, pp. 52–62, 2008.

128. M. Ganss, B. K. Satapathy, M. Thunga, R. Weidisch, P. Poetschke, and D. Jehnichen, Structural interpretations of deformation and fracture behavior of polypropylene/multi-walled carbon nanotube composites, *Acta Materialia,* vol. 56, pp. 2247–2261, 2008.

129. S. A. Monemian, V. Goodarzi, P. Zahedi, and M. T. Angaji, PET/imidazolium-based OMMT nanocomposites via in situ polymerization: morphological, thermal, and nonisothermal crystallization studies, *Advances in Polymer Technology,* vol. 26, pp. 247–257, Win 2007.

130. H. Zhou, Y. Chen, H. Fan, H. Shi, Z. Luo, and B. Shi, The polyurethane/ SiO_2 nano-hybrid membrane with temperature sensitivity for water vapor permeation, *Journal of Membrane Science,* vol. 318, pp. 71–78, 2008.

131. M. Zubitur, A. Fernandez, A. Mugica, and M. Cortazar, Novel nanocomposites based on poly(p-dioxanone) and organically modified clays, *Physica Status Solidi a-Applications and Materials Science,* vol. 205, pp. 1515–1520, 2008.

132. J. Seiler and J. Kindersberger, Insight into the interphase in polymer nanocomposites, *Dielectrics and Electrical Insulation, IEEE Transactions on,* vol. 21, pp. 537–547, 2014.

133. E. Tamburri, S. Orlanducci, M. L. Terranova, F. Valentini, G. Palleschi, A. Curulli, et al., Modulation of electrical properties in single-

walled carbon nanotube/conducting polymer composites, *Carbon,* vol. 43, pp. 1213–1221, 2005.

134. L. Peponi, A. Tercjak, R. Verdejo, M. Angel Lopez-Manchado, I. Mondragon, and J. M. Kenny, Confinement of functionalized graphene sheets by triblock copolymers, *Journal of Physical Chemistry C,* vol. 113, pp. 17973–17978, 2009.

135. L. Peponi, A. Tercjak, J. Gutierrez, H. Stadler, L. Torre, J. M. Kenny, et al., Self-assembling of SBS block copolymers as templates for conductive silver nanocomposites, *Macromolecular Materials and Engineering,* vol. 293, pp. 568–573, 2008.

136. G. Gramse, I. Casuso, J. Toset, L. Fumagalli, and G. Gomila, Quantitative dielectric constant measurement of thin films by DC electrostatic force microscopy, *Nanotechnology,* vol. 20, p. 395702, 2009.

137. S. Ramasundaram, S. Yoon, K. J. Kim, J. S. Lee, and C. Park, Crystalline structure and ferroelectric response of poly(vinylidene fluoride)/organically modified silicate thin films prepared by heat controlled spin coating, *Macromolecular Chemistry and Physics,* vol. 210, pp. 951–960, 2009.

138. J. A. Smith, M. Josowicz, and J. I. Janata, Polyaniline-gold nanocomposite system, *Journal of The Electrochemical Society,* vol. 150, p. E384, 2003.

139. R. Liu, Imaging of Photoinduced interfacial charge separation in conjugated polymer/semiconductor nanocomposites, *Journal of Physical Chemistry C,* vol. 113, pp. 9368–9374, 2009.

140. Y. Cai, Y. Hu, L. Song, L. Liu, Z. Wang, Z. Chen, et al., Synthesis and characterization of thermoplastic polyurethane/montmorillonite nanocomposites produced by reactive extrusion, *Journal of Materials Science,* vol. 42, pp. 5785–5790, 2007.

141. B. Li, Y. Hu, J. Liu, Z. Chen, and W. Fan, Preparation of poly(methyl methacrylate)/LDH nanocomposite by exfoliation-adsorption process, *Colloid & Polymer Science,* vol. 281, pp. 998–1001, 2003.

142. S. A. Gârea, H. Iovu, A. Nicolescu, and C. Deleanu, A new strategy for polybenzoxazine–montmorillonite nanocomposites synthesis, *Polymer Testing,* vol. 28, pp. 338–347, 2009.

143. Y. Tang, Y. Hu, R. Zhang, Z. Wang, Z. Gui, Z. Chen, and W. Fan, Investigation into poly(propylene)/montmorillonite/calcium carbonate nanocomposites, *Macromol. Mater. Eng.*, vol. 289, 191–197, 2004, DOI: 10.1002/mame.200300157.

144. M. Hernandez, B. Sixou, J. Duchet, and H. Sautereau, The effect of dispersion state on PMMA–epoxy–clay ternary blends: in situ study and final morphologies, *Polymer,* vol. 48, pp. 4075–4086, 2007.

145. R. A. Vaia, R. K. Teukolsky, and E. P. Giannelis, Interlayer structure and molecular environment of alkylammonium layered silicates, *Chemistry of Materials,* vol. 6, pp. 1017–1022, 1994.

146. R. Suresh, S. Vasudevan, and K. V. Ramanathan, Dynamics of methylene chains in an intercalated surfactant bilayer by solid-state NMR spectroscopy, *Chemical Physics Letters,* vol. 371, pp. 118–124, 2003.

147. Q. H. Zeng, A. B. Yu, G. Q. Lu, and R. K. Standish, Molecular dynamics simulation of the structural and dynamic properties of dioctadecyldimethyl ammoniums in organoclays, *Journal of Physical Chemistry B,* vol. 108, pp. 10025–10033, 2004.

148. T. D. Fornes, P. J. Yoon, D. L. Hunter, H. Keskkula, and D. R. Paul, Effect of organoclay structure on nylon 6 nanocomposite morphology and properties, *Polymer,* vol. 43, pp. 5915–5933, 2002.

149. E. M. Moujahid, F. Leroux, M. Dubois, and J.-P. Besse, In situ polymerisation of monomers in layered double hydroxides, *Comptes Rendus Chimie,* vol. 6, pp. 259–264, 2003.

150. D. R. Paul and L. M. Robeson, Polymer nanotechnology:nanocomposites, *Polymer,* vol. 49, pp. 3187–3204, 2008.

151. Z. Sedláková, J. Pleštil, J. Baldrian, M. Šlouf, and P. Holub, Polymer-clay nanocomposites prepared via in situ emulsion polymerization, *Polymer Bulletin,* vol. 63, pp. 365–384, 2009.

152. H. Kim and C. W. Macosko, Morphology and properties of polyester/exfoliated graphite nanocomposites, *Macromolecules,* vol. 41, pp. 3317–3327, 2008.

153. A. Chemtob, C. Croutxé-Barghorn, O. Soppera, and S. Rigolet, Cationic photopolymerization in presence of functionalized silica nanoparticles, *Macromolecular Chemistry and Physics,* vol. 210, pp. 1127–1137, 2009.

154. H. Sawada, N. Naitoh, R. Kasai, and M. Suzuki, Dispersion of single-walled carbon nanotubes in water by the use of novel fluorinated dendrimer-type copolymers, *Journal of Materials Science,* vol. 43, pp. 1080–1086, 2007.

155. A. B. Elmabrouk, T. Wim, A. Dufresne, and S. Boufi, Preparation of poly(styrene-co-hexylacrylate)/cellulose whiskers nanocomposites via miniemulsion polymerization, *Journal of Applied Polymer Science,* vol. 114, pp. 2946–2955, 2009.

156. X.-S. Wang, T. Metanawin, X.-Y. Zhen, P.-Y. Wang, M. Ali, and D. Vernon, Structure-defined C60/polymer colloids supramolecular nanocomposites in water, *Langmuir,* vol. 24, pp. 9230–9232, 2008.

157. N. Osaka, H. Endo, T. Nishida, T. Suzuki, H.-j. Li, K. Haraguchi, et al., Microphase separation in nanocomposite gels, *Physical Review E,* vol. 79, 2009.

158. R. J. Foster, P. J. Hine, and I. M. Ward, Characterisation and modelling of polypropylene/carbon nanofibre nanocomposites, *Polymer,* vol. 50, pp. 4018–4027, 2009.

159. W. Al Rawashdeh, S. Kray, A. Pich, S. Pargen, A. Balaceanu, M. Lenz, et al., Polymeric nanoparticles as OCT contrast agents, *Journal of Nanoparticle Research,* vol. 14, p. 1255, 2012.

160. W. R. Broughton, T. Koukoulas, P. Woolliams, J. Williams, and S. S. Rahatekar, Assessment of nanoparticle loading and dispersion in polymeric materials using optical coherence tomography, *Polymer Testing,* vol. 32, pp. 1290–1298, 2013.

Chapter 6

Theoretical Aspects of Interfaces

Toshikatsu Tanaka[a] **and Alun S. Vaughan**[b]

[a]*IPS Research Center, Waseda University,*
2-7 Hibikino, Wakamatsu-ku, Kitakyushu-shi 808-0135, Japan
[b]*ESC, Faculty of Physical Sciences and Engineering,*
University of Southampton, Southampton SO17 1BJ, UK

t-tanaka@waseda.jp, asv@ecs.soton.ac.uk

This chapter is intended to explain some theoretical aspects of interfaces in polymer nanocomposites. Interfaces are considered to play a crucial role in determining the performances of polymer nanocomposites. Interfaces between inorganic fillers and polymer matrices have long been investigated in colloid science, since the 1930s, and then analyzed with modern quantum mechanics in 1950s. Bonding between two immiscible substances, i.e., organic polymers and inorganic fillers, are needed for homogeneous filler dispersion and the formation of thermodynamically stable composites. Bound polymers were the first concept for interfaces clarified by colloid chemistry. Many factors such as triboelectricity, electrochemical potential, Coulombic attractive force, steric effects, and quantum mechanical forces are involved in the process of determining polymer morphology and the formation of nanocomposites via mutual interactions.

Tailoring of Nanocomposite Dielectrics: From Fundamentals to Devices and Applications
Edited by Toshikatsu Tanaka and Alun S. Vaughan

ISBN 978-981-4669-80-1 (Hardcover), 978-981-4669-81-8 (eBook)
www.panstanford.com

6.1 Introduction

Interfaces are inherently included in composite materials, which will involve interactions between the various components. Associated forces will in general be associated with chemical bonds, hydrogen bonds and van der Waals interactions; in addition, magnetic forces and mechanically structural anchor forces might also be involved. Thus, interaction forces originate from bonding groups, polar groups, surface activity level (specific surfaces), surface roughness, chemical composition and the presence of surface functional groups (=O, –COOH, –OH, etc.) on inorganic fillers. In general, it can be stated that mutual interaction depends on "the degree of wetting" or "effective wetting surface area." A method of maximizing strengthening interactions is therefore to produce as good a dispersion of the filler as possible and to increase the degree of wetting. Surface treatment is an effective technique to improve the level of wetting.

6.2 Classical Models

Nanocomposites can be considered as solidified colloidal dispersions, where colloidal dispersion systems are suspensions of nanoparticles with diameters of the order of one to one hundred nanometers. Interfacial forces are crucial in composites containing two kinds of apparently immiscible substances, such as organic polymers and inorganic fillers. A key concept in such systems is that of an interaction zone. Interface and colloid science had already dealt with nanometer-size substances long before nanotechnology emerged. Gold sols and other nanomaterials with sizes down to 10 nm were studied in the 1920s [1], Langmuir–Blodgett thin films composed of one or more monolayers of organic material have been investigated since the early 1930s [2] and the first measurement of surface forces was made in the early 1950s. The concept of bound polymers was demonstrated in the 1960s [3–7]. The characteristics of interfaces between mutually immiscible substances in colloids were also clarified and colloid science reached a significant level of maturity during the second half of the 20th century. It is important to note that, in colloid science, the forces acting between particles and solvent are well understood for both dilute and condensed phases. Various forces act between colloidal particles, e.g., excluded volume repulsion, electrostatic forces, van der Waals force, entropic forces, and steric forces. Excluded volume repulsion, sometimes called Born repulsion, is caused by the impossibility of overlap between hard particles. Electrostatic forces surrounding a colloidal particle were analyzed in detail and clarified by Derjaguin, Landau, Verwey, and Overbeek, resulting in the DLVO theory [8]. There is a deep minimum in the potential (attractive force) near the particle surface and a shallow minimum (attractive force) far from the particle surface, with a high maximum (repulsive force) in between. Entropic forces act between particles, increasing the entropy until thermal equilibrium is reached. Steric forces arise between polymer-covered surfaces, or in solutions containing non-adsorbing polymer, producing an additional steric repulsive force (which is predominantly entropic in origin) or an attractive force due to depletion of counter ions. There are other more important interfacial phenomena associated with the interaction between particles and their environment. Physical and chemical adsorption is one of them. The former is related to hydrogen, hydrophobic or van der Waals bonding, while the latter is caused by chemical bonding between adsorbate and adsorbent. Surface tension appears on a soft surface, generated by the difference between the free energies of the surface and the bulk. Wetting is the contact of a solid surface by a liquid and includes adhesion, spreading, and immersion wetting. Knowledge of interfacial phenomena in colloids has proved extremely useful in developing polymer nanocomposites.

6.2.1 Bound Polymers: Suggested Interaction Zones

Bound polymer was discovered in the 1960s [3–7]. The concept of a bound layer was suggested in 1969 to explain improvements in mechanical strength and an increase of glass transition temperatures, as shown in Fig. 6.1. The bound layer is immobile, so that matrix molecules can be bound to inorganic fillers to form an interfacial zone, which is considered to become thicker as the interactions between the filler and matrix becomes stronger. Its thickness was estimated as several to several hundred nanometers in polyvinyl chloride (PVC)/xonotlite and PVC/mica composites and several nanometers in rubber/carbon black composites; the thickness was too small to measure in PVC/calcium carbonate composites. The thickness depended on the type of matrix and filler, the method of surface treatment, and the measurement technique, e.g., indirect estimation from the glass transition temperature change, or from the dissipation of vibrational energy through dynamic viscoelasticity, or direct measurement on bound polymer adsorbed on the surface of the filler. The first and second methods give dimensions of the order of 1 to 10 nm as the apparent thickness of the bound polymer layer, while the third method gives the order of 100 nm [9]. A big difference in apparent thickness between measurement methods suggests that the strength of interaction decays not abruptly but gradually over a distance.

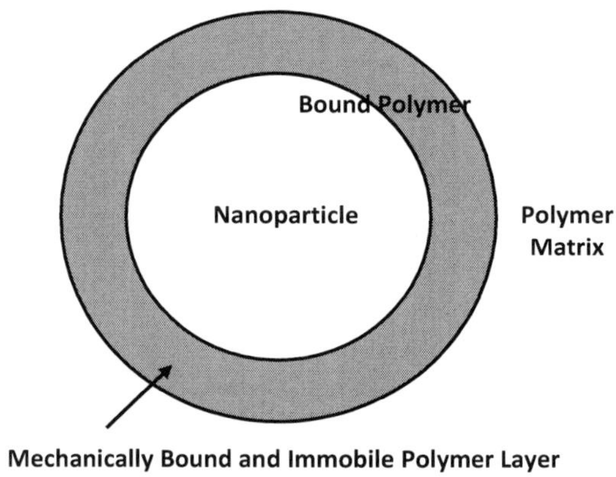

Figure 6.1 Bound polymer model from colloid chemistry.

6.2.2 DLVO Model and Beyond

The stability of colloids is primarily governed by the following two factors.

(1) Inter-particle electrostatic repulsive interactions
(2) Steric repulsion between particles with adsorbed polymer layers

Inter-particle electrostatic repulsion interactions

This effect is associated with surface charge. Electron clouds are distorted in the surface to produce charge, negatively charge in silica and positively charged in metals for instance. When such a surface is in contact with water, water molecules are oriented to form a hydration layer, termed the Helmholtz layer; in the case of polymers, a similar interaction mechanism is possible. For example, acids and bases can be considered as electron acceptors and donors respectively, such that surface charges can appear when particular polymers interact with different solvents. Another possibility is that ions adsorb on the surface of solid polymers in solution.

Surface charges then serves to attract ions of the opposite polarity to form an electric double layer, in which the charge density diminishes to zero far from the interface and where the system as a whole is not charged. The accepted Helmholtz–Stern–Gouy–Chapman double-layer model is shown schematically in Fig. 6.2. A key feature of this model is the presence of a hydrodynamic slip plane, which is associated with the Stern layer at a potential Φ_D. In general, the potential near the slip plane is considered to be equal to ς, the zeta potential obtained from electrophoretic mobility. Surface potential is governed by the surface charge density on particle surfaces, which can be measured by electrophoretic light scattering method.

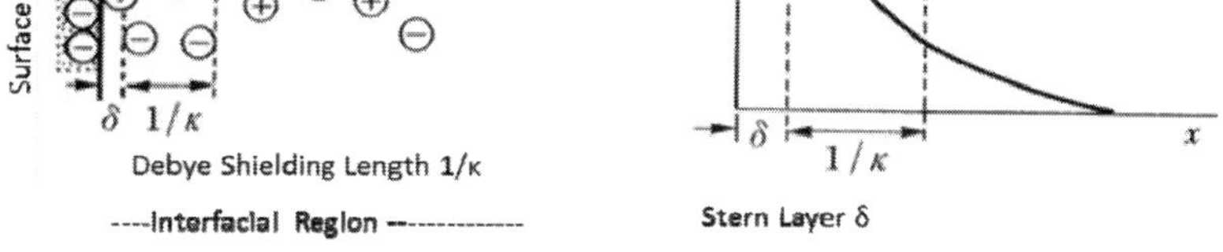

Figure 6.2 Electric double-layer model. (a) Distribution of pair ions. (b) Potential curve. δ: Stern layer length, $1/\kappa$: Debye shielding length.

When steric stability based on entropic effects is neglected, inter-surface interactions consist of van der Waals and Coulombic electrostatic forces. In this case, the DLVO (Derjaguin–Landau–

Verwey–Overbeek) theory can apply. Figure 6.3 shows a DLVO potential energy vs. separation curve. As is evident from this figure, such a plot consists of the primary minimum (attractive potential) comprises a deep valley (hundreds of kT) near the particle surface, the maximum repulsive potential in the middle, and a secondary minimum (several kT) in far distance. The Born repulsion energy near the surface is caused by the overlapping of electron clouds. The secondary minimum appears due to the balance between electrostatic repulsion energy and van der Waals energy; the former decreases more sharply than the latter increases. Since the DLVO potential becomes lower as the separation is reduced, particles are more stable as they aggregate. This is one of the most important characteristics that should be avoided for good particle dispersion.

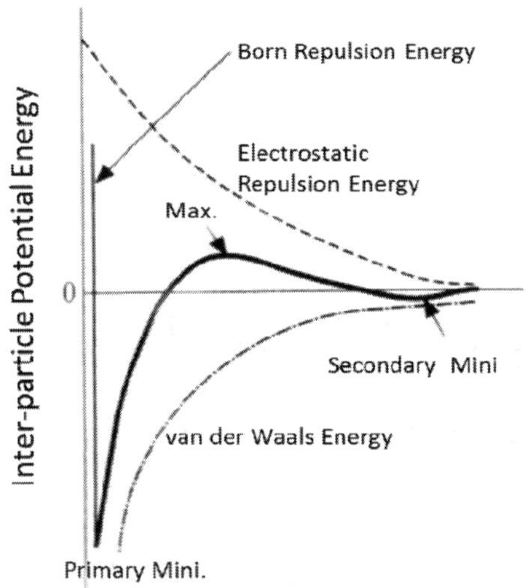

Figure 6.3 DLVO potential energy curve.

The slip plane introduced above does not necessarily exist, for example, when considering coiled chains of polymer electrolytes, microgels, surface graft chains, or particles with surface adsorbed layers. In such cases, the above theory cannot apply; alternative approaches include the Ohshima–Kondo theory and the Hermans–Fujita model, but these fall outside the scope of this book.

Steric repulsion between particles with adsorbed polymer layers

The second mechanism is associated with steric stabilization or repulsion between inorganic particles with adsorbed polymers on their surfaces, as shown in Fig. 6.4. When such two particles approach each other at a distance of less than twice the thickness of the adsorbed layer, interaction of the two layers takes place. The degree of stabilization can then be defined quantitatively in terms of the energy change occurring upon the interaction of the adsorbed layers. The Gibbs free energy change ΔG of the overlap interaction of the adsorbed layers is expressed as $\Delta G = \Delta H - T\Delta S$. If ΔG is negative upon the overlap of the adsorbed layers, flocculation or coagulation will result, whereas if ΔG is positive, stabilization will result. Under isothermal conditions, stability is then a function of the enthalpy change, ΔH and the entropy change, ΔS.

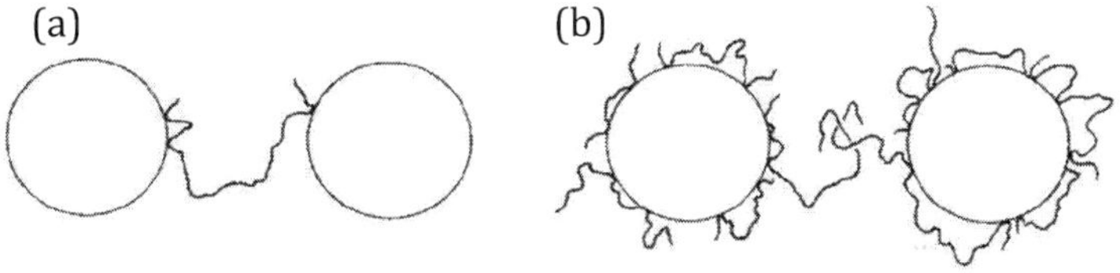

Figure 6.4 Bridging flocculation: (a) Two particles by one polymer molecule. (b) Two particles by two separately adsorbed polymer molecules.

The presence of functional groups such as carboxyl, hydroxyl, amine, and ester groups in the molecular structure of the absorbents leads to positive changes in the Gibbs free energy. Polymers containing carboxyl groups turn out to be the most effective steric stabilizers, because carboxyl groups are thought also to interact strongly with basic sites that are often present on the particle surface. In the case of nonaqueous solvents, long-chain hydrocarbons extending from the surface act as good moieties. Furthermore, copolymers are usually more effective in steric stabilization than homopolymers, due to the presence of more than one type of repeat unit. Then, one repeat unit can act as an anchor to the particle surface, while the other type can act in the moieties extending into the nonaqueous solvent.

When the free energy difference ΔG is negative, flocculation or coagulation tends to take place. Flocculation occurs when the solubility of the disperse phase within the solvent decreases or, in the case of aqueous suspensions, the concentration of the added salt decreases and is represented by CFP (the critical flocculation point), which is closely related to Flory's theta condition for polymeric solutions. This is termed a mixing effect. Coagulation takes place through polymer bridging and occurs when one polymer chain becomes adsorbed onto the surfaces of neighboring two particles. Polymer bridging is characterized by the gradual agglomeration of particles as time goes by. This is termed a cross-linking effect.

Counter repulsion is caused by entropic elasticity since, when interfaces are impenetrable, adsorbed layers are compressed. That is, the polymer segments occupy fewer possible configurations in the compressed state than in the uncompressed state. This reduction in entropy increases ΔG, producing the net effect of repulsion between the particles and thus preventing the particles from flocculating. Thus if a polymer with a radius of gyration that is larger than the average inter-particle distance is adsorbed onto the particles, entropic elasticity is produced by the restraint of conformation and particle agglomeration is suppressed.

6.2.3 Lewis' Concepts: Insight into Dielectrics

Lewis has brought nanocomposites to the world of electrical engineers, emphasizing that the interfaces between two kinds of immiscible substances, such as organic polymers and inorganic particles, play a dominant role in determining the macroscopic properties of polymer nanocomposites [10–13]. His models are physically understandable and are largely based on the electric double-layer model and the DLVO concept that have been considered in the previous section.

Interfaces with Finite Thickness

Specific surface increases as a filler particle decreases in size. Interfaces have a finite thickness, and thereby volume and, in nanocomposites, interfaces are considered to play a major role in determining properties. Figure 6.5 shows representative plots of interfacial volume fraction vs. nanoparticle diameter, for cases where the interface of thickness is considered to be 10, 1 and

0.5 nm, and demonstrates that the total interface contribution can become very significant as the particle diameter is reduced. If the interfaces are determined only by short-range forces and are say 0.5 nm thick, their volume exceeds 50% of the whole when the particle size is reduced below 5 nm. If, however, the interface extends to 10 nm or more, as they might do when the interfaces are charged, they will become dominant when the particle diameter is less than 100 nm. Clearly the properties of the interface can strongly influence the overall dielectric properties of the composite.

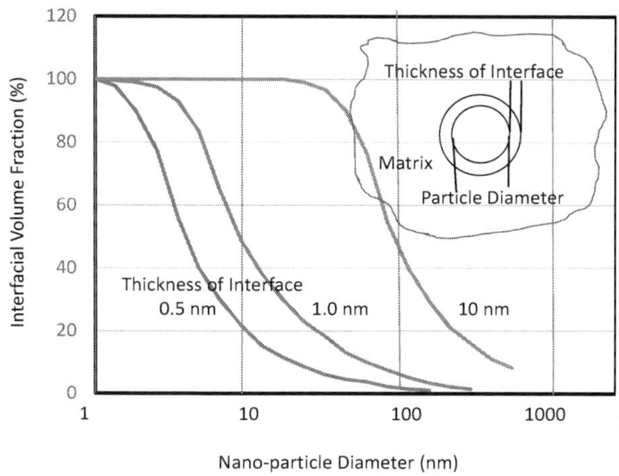

Figure 6.5 Plots of the volume fraction of the interface vs. nanoparticle diameter *d* for the thickness of 10, 1, and 0.5 nm.

Forces Generated in Interfaces [11]

Interfacial forces are of several kinds. There are hard-core forces that are extremely short range, quantum mechanical and repulsive in nature. They arise from overlap interactions between the electron clouds of individual atoms, which correspond to volume exclusion force or Born repulsion force, as discussed in Section 6.2.2. All the other forces are essentially attractive/repulsive electrostatic interactions involving the negative electrons and positive nuclei of the system. The form they take depends on whether the individual atoms or molecules are neutral or possess a net charge. In neutral form any asymmetric distribution of electrons and nuclei in a molecule or atom will give rise to multipole forces of which the dipole component will be dominant. Several interactions are possible. Between ions, the forces are strongly Coulombic and long-range. When permanent dipoles are present, there will be weaker ion dipole and dipole–dipole forces and a range of ion and dipole induced dipole forces. The Debye and Keesom forces [14, 15] fall into this category. All dipole-related polarizing forces are of moderate range and generally attractive in nature and known collectively as van der Waals forces. A particular short-range electrostatic force, but highly directional and therefore of special structural importance, is the hydrogen bond between a donor (a hydrogen atom covalently bonded to an electronegative atom) and an acceptor electronegative atom.

Dielectric phenomena on nanometric scales are governed by the strong long-range Coulombic interactions which occur when charges are present. A charged component will induce both electronic polarization [ion (induced dipole) interaction] and the orientation of any permanent dipoles, [ion–dipole interaction]. The energy associated with these has two components:

(i) an inner one involving quantum mechanical interactions with nearest neighbor atoms and molecules
(ii) an outer one which can be considered classically in terms of permittivity

The change in the latter in passing through an interface can be evaluated in terms of the image law for dielectric interfaces [14, 15]. A frequent situation in nanodielectric systems is one in which the surface or at least a part of the surface of a particle becomes effectively charged and the surrounding polymer matrix responds by establishing a screening counter charge confronting the charge on the particle. The charging of the particle and the development of counter charge in the polymer matrix are distinguishable processes and should be considered separately.

6.3 Physical and Chemical Aspects of Interfaces

6.3.1 Interfaces by Coupling Agents and Wettability

Silane coupling is a convenient means of binding chemical moieties to inorganic substances (nanofillers in this case), as shown in Fig. 6.6. It occurs via the formation hydrogen bonds between two kinds of immiscible substance, i.e., organic and inorganic substances. Covalent, ionic and van der Waals forces are other types of interaction that can occur between different substances and, in the case of silane coupling, hydrogen bonds are converted to covalent bonds by de-hydration. It is important to specify which kind of interfacial force exists in particular cases. The affinity between two kinds of mutually contacted materials is also important in forming wetted interfaces, which can lead to the formation of a bound polymer layer.

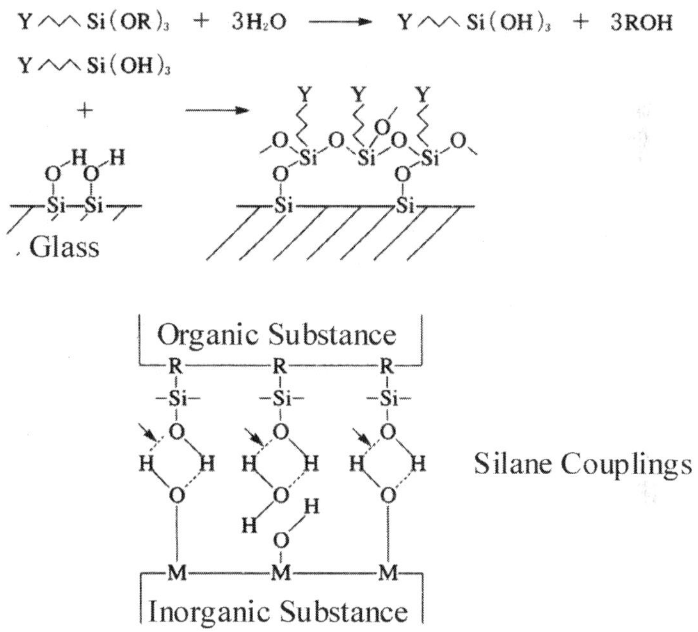

Figure 6.6 Chemical structure of silane couplings.

Wettability or affinity are important concepts to consider when seeking to mix two kinds of materials homogeneously. In order to disperse inorganic nanofillers into organic polymer matrices, the affinity between their surfaces should therefore be taken into consideration. Surface tension is a physical parameter that directly represents the affinity of two substances and can be considered either as a force or as a free energy; it is necessary to reduce the difference in surface tension between the two in order to obtain a good composite. Common practice is to use compatibilizers for polymer alloys, and to use coupling agents for polymer inorganic composites. The solubility parameter (SP) is often utilized as a semi-quantitative measure to represent the affinity (the degree of miscibility) of polymers with organic modifiers and compatibilizers and corresponds to the square root of the cohesive energy density (CED) for the material in question. The CED is the energy needed to remove a molecule from its nearest neighbors, divided by the total volume of the moved molecule. As is well known, the SP value is expressed by SP = $(\Delta E/V)^{1/2}$, where ΔE and V are the evaporation energy per mol to a gas at zero pressure (i.e., infinite separation of the molecules), and the molar volume, respectively [1, 2]. It can be approximately calculated, if the chemical structure is known.

6.3.2 Evidence for Far-Distance Interaction

The glass transition temperature, T_g, is related to molecular mobility and consequently, would be expected to be influenced by the interactions that occur within a system. Variations in this parameter have therefore been examined in two classes of specimen, nanosilica-filled nanocomposites and sandwiched thin films of various thicknesses (~100 nm) [16] that can be considered as constrained model systems. P2VP (poly(2-vinyl pyridine)), PMMA, and polystyrene were evaluated as representative materials. Selected results are shown in Fig. 6.7, which shows the deviation in T_g seen in P2VP nanocomposites (open squares) and PMMA constrained model systems (open circles) together with, right and left, transmission electron micrographs obtained from samples containing 0.4 vol% silica–P2VP (right) and 0.4 vol% silica–PMMA (left) nanocomposites (scale bars = 100 nm). PMMA and P2VP both interact with the silanol groups on the silica surface through hydrogen bonding, whereas polystyrene only interacts weakly with silica. It is therefore considered that the increase in T_g seen in PMMA-silica system can be attributed to the existence of attractive interfacial interactions between the polymer matrices and nanofiller, which serves to suppress cooperative segmental motion in the polymer. The same mechanism can also be postulated to occur in P2VP-silica systems. The interaction distance will depend on the type of interaction force and can be ~100 nm when hydrogen bonding is effective. Interaction distance values of the order of 100 nm have been estimated based on the concept of bound polymer from dynamic viscoelasticity measurement (see Section 6.2.1) and are now considered to be reasonable.

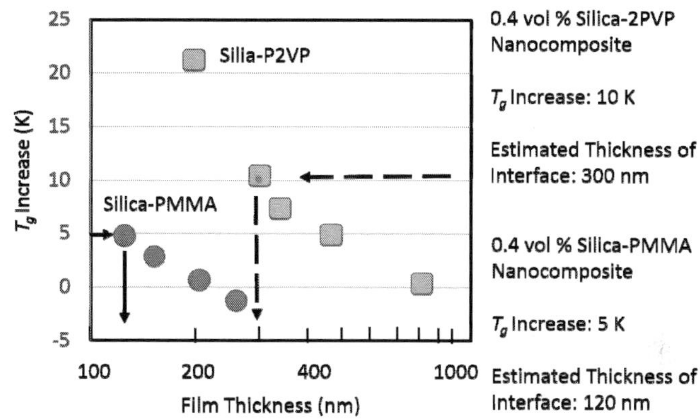

Figure 6.7 Interlayer spacing (film thickness) in "model" nanocomposites that yield the same T_g deviation as 0.4 vol% silica–PMMA and silica–P2VP nanocomposites.

Furthermore, forces other than chemical bonds are also available. For instance, an electrostatic force can be generated due the formation of an electric double layer. This force, being Coulombic is effective over long distances by nature, as shown in Fig. 6.2. This type of force is generated by charge transfer between the two kinds of materials, when they are in contact with each other, so as to equalize the Fermi level or the electro-chemical potential, and is closely related to triboelectric behavior. For instance, it is found experimentally that silicone elastomers and polyamide tend to become positively charged, while polypropylene tends to become negatively charged.

6.3.3 Charge Tails in Interfaces

Charge injection from a filler particle to a polymer matrix is based on electron affinity and the equalization of the Fermi level or electrochemical potential. Charge injection is also related to triboelectricity, as introduced in Section 6.3.2, such that the triboelectric series is useful for providing macroscopic understanding [17, 18]. The triboelectric series indicates that polyolefins tend to attract electrons to become negatively charged, whereas condensation polymers tend to give up electrons to become positively charged. In the absence of any data pertaining to the metal oxides used as nanofillers, it can be assumed that, for the polymers commonly used to prepare nanocomposites, polyethylene, polypropylene, and ethylene vinyl acetate become negatively charged, whereas silicone elastomers, polyamides and epoxy resins become positively charged.

If charge injection takes place, a charge distribution layer termed the Gouy–Chapman diffuse layer is formed, as shown in Fig. 6.2. This charge distribution might affect the local evolution of polymer morphology, as a consequence of the mutual interaction of chemical and electrical forces. However, no such electrochemical study seems to have been undertaken. However, there is the DLVO theory of colloids, which gives the balancing of attractive polarization and repulsive double-layer forces [11].

The Gouy–Chapman diffuse layer, as shown in Fig. 6.2, is associated with the interface (charge distribution) between the nanofiller particle and the polymer matrix in this model. The particle becomes charged possibly due to the equalization of Fermi level or chemical potential or through processes such as ionization of surface groups and adsorption of ions from the polymer matrix. The polymer then responds by establishing an excess counter charge environment, screening the charge on the particle. There are two mechanisms for this. The first is concerned with an induced polarization of the polymer matrix involving both electronic polarizability and the orientation of permanent dipoles. This is an organization process which, in electrochemistry, can be described by the Born expression. The second mechanism that can lead to the formation of a diffuse electrical double layer occurs if the polymer matrix is electrolytic and contains mobile charges. The double layer then consists of positive and negative ions and their distribution can be determined by the Poisson and Boltzmann equations with the help of the Gouy–Chapman or the Debye–Hückel approximation. The Debye shielding length or the electrical interface dimension is of the order of 30 nm for a 10^{-4} M ion concentration, as cited in the reference [6], which suggests that the electrical interface layer could be rather thick due to long range van der Waals forces.

6.4 Multicore Model and Interfacial Structures

6.4.1 Multicore Model

The multicore model [19] shown in Fig. 6.8 consists of

(1) a bonded layer (the first layer);
(2) a bound layer (the second layer);
(3) a loose layer (the third layer);
(4) an electric double layer overlapping the above three layers.

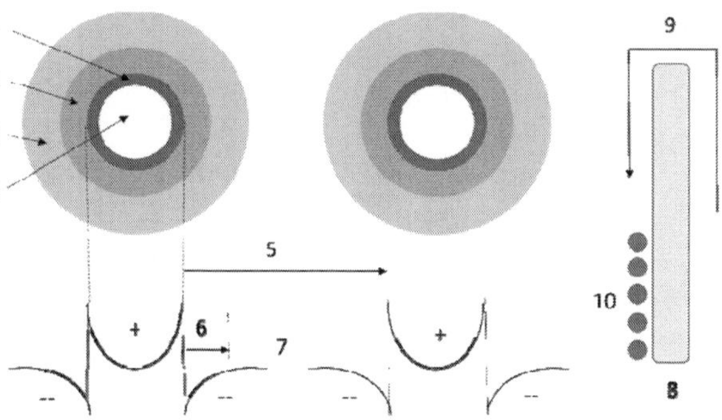

1: The first layer with mutual tight binding: several nm
2: The second layer with deep traps: about 10 nm
3: The third layer with local free volume as ion taps and shallow electronic traps: several tens nm
4: Nano-particle: 20 nm–40 nm in diameter
5: Interparticle distance (surface to surface): 40 to 100 nm
6: Debye shielding length: up to 100 nm
7: Charge is distributed over three layers. Charge tails can be overlapped when particles are closely packed.
8: Electrode facing accumulated charge tails of nano-fillers
9: Charge injection possibly via Schotty emission at high electric field
10: Nano-fillers facing the electrode with collective charge tails will form Coulomb potential possibly to suppress carrier (electron in this case) injection.
+: Positive charge distribution in the case of positively charged nano-particles
–: Counter negative charge distribution with Debye shielding length

Figure 6.8 Multicore model for nanocomposites.

If spherical nanoparticles some tens nm in diameter are homogeneously dispersed within a system, then at reasonable concentrations, their separation becomes comparable to their diameter. That is, both the inter-particle distance and their diameter are in the order of 1 to 10 nm. Inter particle interaction such as a percolation process may take place when characteristic dimensions are of the order of 1 nm. As described previously, interfacial layers exist which are different from both the unperturbed matrix and the bulk nanofiller material; in the multicore model considered here a three layered structure is postulated. The first layer corresponds to a region of chemical bonding between inorganic fillers and organic polymer matrices. The second layer is attributed to a region of special morphology, which forms as a consequence of the influence of the first layer, or a stoichiometrically distinct cross-linked region, depending on the nature of the matrix. The third layer is a region that exists between the second layer and the unperturbed matrix in which local factors such as structure, stoichiometry, cross-link density, etc., are a consequence of the influence of the adjacent regions.

(1) Bonded layer (the first layer)

The first layer corresponds to a transition layer, which is tightly bonded to both the inorganic and organic substances by coupling agents such as silane. The thickness is postulated to be the order of 1 nm. The following results were obtained using small angle X-ray scattering (SAXS) from nanocomposites formulated by the sol-gel method. Sol-gel methodologies include the end-cap method, the blend method, and the copolymerization method.

(a) Silica permeates into the outside polymer matrix, in part, in the covalently bonded region of nanocomposites prepared by the end cap method.

(b) Silica and polymer chains are mutually inter-diffused to form a hydrogen bonded, interpenetrating network region in the case of nanocomposites formulated by the blend method.

A broad peak appears in SAXS in the case of (a), which indicates some important results, as follows. Silica clusters (namely nanoparticles) are homogeneously dispersed with a certain correlation distance (inter nanoparticle distance). Silica domains (nanoparticles) are rich in inorganic substance but are compatibilized with polymer phase in part. A distinct interface exists between the matrix and the silica region. Such peaks have been observed in the PPG (polypropylene glycol)/SiO_2 system, the PEG (polyethylene glycol)/TiO_2 system, and PTMO (polytetramethylene oxide)/SiO_2 system.

In the case of (b), no peak peaks appears in SAXS. Rather, the scattering intensity, $I(q)$ is characterized by the formula that it is proportional to q^{-D}, where q and D represent the wave vector and the fractal dimension, respectively. By investigating various kinds of colloids, it has been ascertained that colloid particles with a smooth surface lead to $D = 4$, for rough surfaces $D = 3$, and where branching leads to the formation of three dimensional network structures $D = 1.9$ to 2.5. In the case of polyvinyl alcohol (PVA) nanocomposites prepared by the blend method, where hydrogen bonding will be significant, D is equal to 1.97. It is presumed that SiO_2 chains and PVA chains are mutually interpenetrated to form a network structure. The same is true for the PTMO/WO_3 system.

In general, the first layer (the bonded layer) is constrained by ionic, covalent, hydrogen, and van der Waals bonding (bond strength decreasing in this order).

(2) Bound layer (the second layer)

The second layer corresponds to a bound polymer layer, which is an interfacial region consisting of a layer of polymers chains strongly bound and/or strongly interacting with the first layer and the surface of the inorganic particle. Cited values for its thickness typically fall in the range 2 to 9 nm. These values, of course, depend critically on the strength of the polymer-particle interactions, the stronger the interactions, the greater the bound polymer fraction.

Chain mobility and crystallinity are deeply involved. The former is directly related to the glass transition temperatures. In addition, the average chain conformation can be affected by the proximity to the nanoparticle surface, such that both the type and degree of crystallinity can change. The chain conformation includes the orientation of constituent groups, local polymer chain stacking at the surface of the nanofiller, the local density of interfaces, the folded structure of polymer chains, and other conformational structures. Crystallinity is also sensitive to interfacial interactions. When curing agents are adsorbed selectively onto nanofillers, the region around the nanoparticles consists of a layer of stoichiometrically cross-linked thermoset region with curing agents. In the case of polyamide/layered silicate nanocomposites, the nucleated region around nanoparticles is considered to correspond to the second layer.

(3) Loose layer (the third layer)

The third layer is a region loosely coupled to and interacting with the second layer. It is generally considered that the loose layer involves different chain conformation, chain mobility, and even free volume or crystallinity from the unperturbed polymer matrix. This layer may extend to several tens nm, being chemically influenced by the second layer. When the preferential adsorption of a curing agent occurs, the region surrounding the nanoparticles consists of a "layer" of stoichiometrically cross-linked thermoset with excess curing agent, surrounded by a "layer" of depleted curing agent and thus less-than-stoichiometric cross-linking. This phenomenon is observed in fiber-filled polymers, epoxy/titania nanocomposites. More subtle changes may also occur, such as preferential adsorption of low molecular weight material to the surface.

(4) Electric double layer

In addition to the above general conclusion concerning interfacial structures based upon chemical and structural factors, superimposed Coulombic interactions will also exist, which will be of great importance where dielectric and electrical insulation properties are concerned. When a nanoparticle is charged and the polymer contains mobile charge carriers, these will become distributed at the interface in such a way that the concentration of counter charge carriers decays away from the contact surface to the Debye shielding length. This corresponds to the Gouy–Chapman diffuse layer, described above, in which the space charge density decays exponentially with distance according to the Born approximation. The Debye shielding length is calculated to be of the order of 30 nm and, consequently, this distribution of displaced positive and negative charges constitutes a long distance dipole, which may have slow time response. This distribution of charge will affect electrical conduction and dielectric properties in the low frequency region. As described previously, the triboelectricity series suggests that polyethylene (PE), polypropylene (PP) and ethylene vinyl acetate (EVA) tend to become positively charged, while silicone elastomer, polyamide and epoxy tend to become negatively charged.

(5) Thickness of each layer in the multicore model

To summarize, the thickness of the first, second, and third layers is considered to be ~1 nm, several nm, and several tens nm, respectively, and it is questionable as to whether or not the extent of the loose layer is the same as that of the Gouy–Chapman diffuse layer; the latter might extend beyond the former. Nevertheless, it seems likely that some far-field effect exists, which results in mesoscopic interactions causing collaborative effects amongst neighboring nanoparticles. Macroscopic phenomena and parameters must be different from material to material, due to the relative difference in thickness and interaction strengths in the multicore model with the far-field effect.

6.4.2 Interfacial Structures

Examples of interfacial structures are shown in Figs. 6.9 and 6.10 for epoxy and polyethylene, respectively. In Fig. 6.9, the interface is represented in an idealized way; in reality, the structure will be much more complicated, as discussed in Section 6.4.1. In the case of polyethylene, it is proposed that LDPE/silica nanocomposite has three layers at its interface.

An additional complexity arise under conditions where the material is exposed to ambient conditions, in that the presence of hydroxyl groups on nanoparticle surfaces will result in hydrogen bonding with water molecules absorbed within the matrix. This has led to the so-called water shell model. Although the solubility of water in many polymers is very low (e.g., polyethylene), the diffusion coefficient can nevertheless be high, which results in water becoming adsorbed within the confined geometries associated with the nanoparticle/matrix interface. The behavior of small aggregates of water in confined spaces is of considerable interest; it has been found that the single file passage of water molecules occurs in 2 nm diameter carbon nanotubes that are hydrophobic. The proposed water shell model is represented schematically in Fig. 6.11 [22]; the water is considered to exist mainly in the second layer of the multicore model. This model has

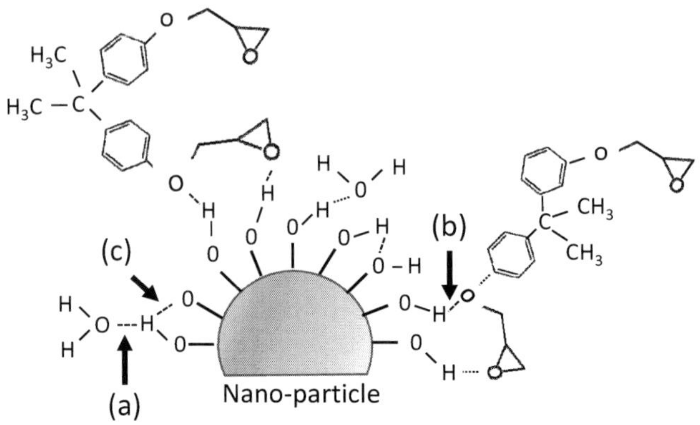

(a) Hydrogen bond between adsorbed water molecule and free surface hydroxyl group
(b) Hydrogen bonds between epoxy and free surface hydroxyl groups
(c) Hydrogen bonds among surface hydroxyl groups

Figure 6.9 Possible structure of interfaces in epoxy/alumina nanocomposite with hydrogen bonding of epoxy rings with surface hydroxyl groups on Al_2O_3 nanoparticles.

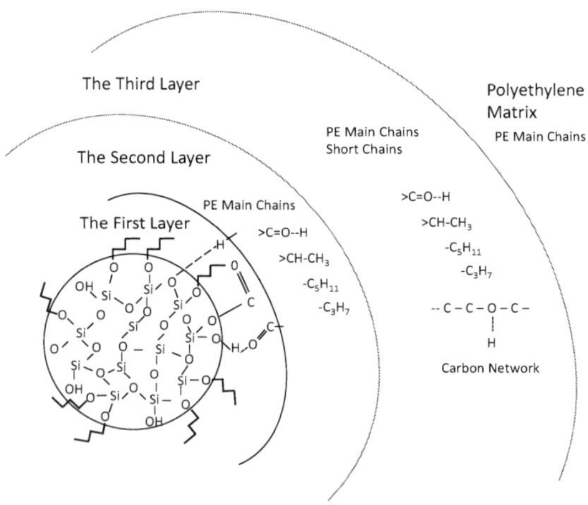

Figure 6.10 Possible structure of interfaces in LDPE/silica nanocomposite with chemical bonds formed in the interface between LDPE and SiO_2 nanoparticles.

been used to explain some dielectric relaxation processes at low frequency in epoxy and polyethylene-based nanocomposites. In the latter case, the dielectric characteristics of the water has been found to be highly sensitive to the nature and degree of surface functionalization. Interfaces are likely to be thermodynamically unstable in the presence of water, especially when they are not surface-treated, which is of considerable significance for the use of nanocomposites is applications where the exclusion of water cannot be strictly guaranteed.

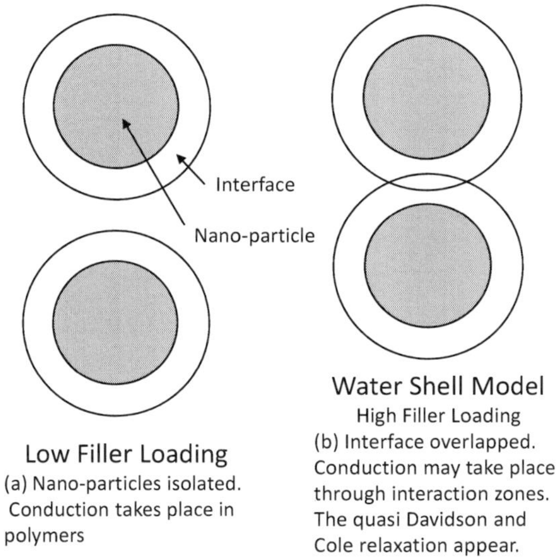

Figure 6.11 Water shell model vs. conventional model.

6.5 Tailored Interfaces

In the most general sense, nanofillers can usually be considered to have a core shell structure when incorporated within a nano-composite. The shell(s) may then include regions of bound polymer, silane-coupled interfaces, or the various structures described by multicore models. While some of these will be generated as a direct consequence of introducing the nanoparticles into the matrix, interfaces can also be tailored by deliberate surface treatment, as described below. Fundamentally, this approach results in the inorganic particle (i.e., oxide, nitride, carbide, montmorillonite) being surrounded by a shell made from an appropriate organic substance (monomer, surfactant, surface active organic molecule, organic dye), such that the surface of the shell interacts with polymer matrices to give the desired material functionality.

6.5.1 Core Shell Structures

The surface modification of silica nanoparticles can be accomplished by traditional methods, through grafting functional groups, or by relatively new methods which give core shell structures. The latter is based upon emulsion polymerization and dispersion polymerization [23–25] and some common approaches are shown schematically in Fig. 6.12 [24].

Recently, controlled radical techniques such as RAFT (reversible addition-fragmentation chain transfer) methods have been developed, as shown in Fig. 6.13. A distinct advantage of using the RAFT approach is the variety of modifications that can be performed on the polymer chain post-polymerization, whereby functionality can be introduced onto the particles that is incompatible with the polymerization conditions.

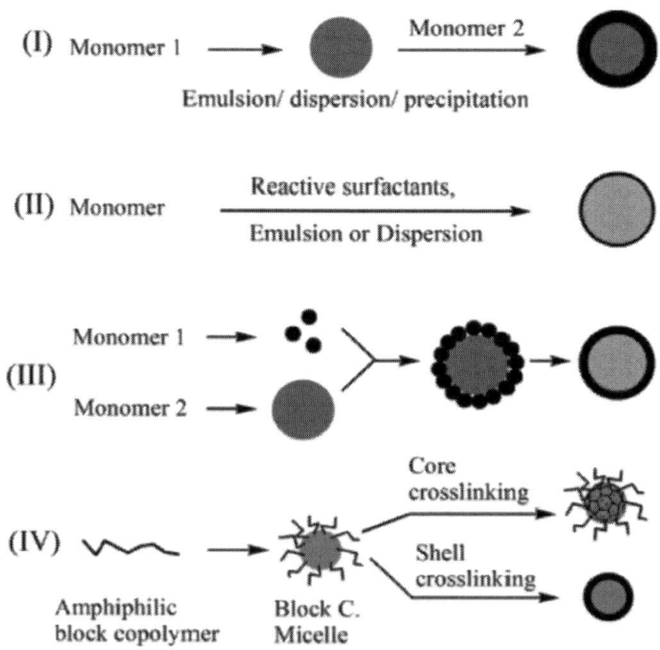

Figure 6.12 Common methods to prepare core shell nanoparticles. Reprinted with permission from ref. [24]. Copyright 2000 American Chemical Society.

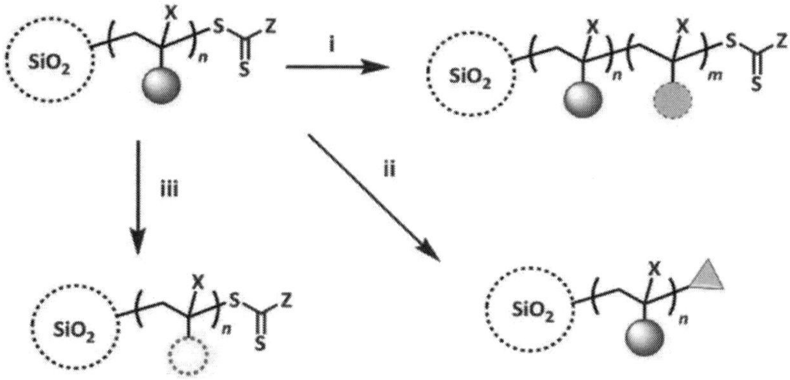

Figure 6.13 Post-polymerization modification of hybrid nanoparticles: (i) chain extension of RAFT-grafted polymer chain; (ii) modification of thiocarbonylthio endgroup; (iii) modification of grafted polymer chain. X represents H or Me, *n* and *m* represent the number of monomer repeat units of the first and second block, respectively. Reprinted with permission from [26]—Published by The Royal Society of Chemistry.

The method termed in situ RAFT polymerization has been used to tailor dielectric properties by preparing core-shell structured polystyrene/$BaTiO_3$ hybrid nanodielectrics. Using this approach, high dielectric constant and low loss materials with weak frequency dependence have been produced [27, 28].

6.5.2 Core–Satellite Structures

Core–satellite structures are particles decorated with ultra-small functional particles (Fig. 6.15). An example of this concept for dielectrics is core–satellite Ag@$BaTiO_3$ nanoassemblies [29], which exhibit multiple performance benefits including high discharged energy density, high breakdown strength and low dielectric loss.

6.5.3 Core Double Shell Structures

If nanoparticles covered by multiple shell structures are available as shown in Figs. 6.14, 6.15, and 6.16, then nanocomposites can, in principle, be produced with excellent overall performance, by overcoming mutually contradictory effects. For example, while an increase in the permittivity of a polymeric substance may be achieved by adding a high permittivity inorganic filler, this generally comes at the price of reduced dielectric breakdown performance. Core double shell structured $BaTiO_3$ (BT)-polymer nanocomposites have been prepared, as shown in Fig. 6.14 [30]. The core is the BT nanoparticles, which exhibits a high dielectric constant, is surrounded first by a shell composed of a hyperbranched aromatic polyamide (HBP) grafted from the surface of the BT nanoparticles. The second layer is PMMA grafted from the terminal groups of the HBP via atom transfer radical polymerization (ATRP).

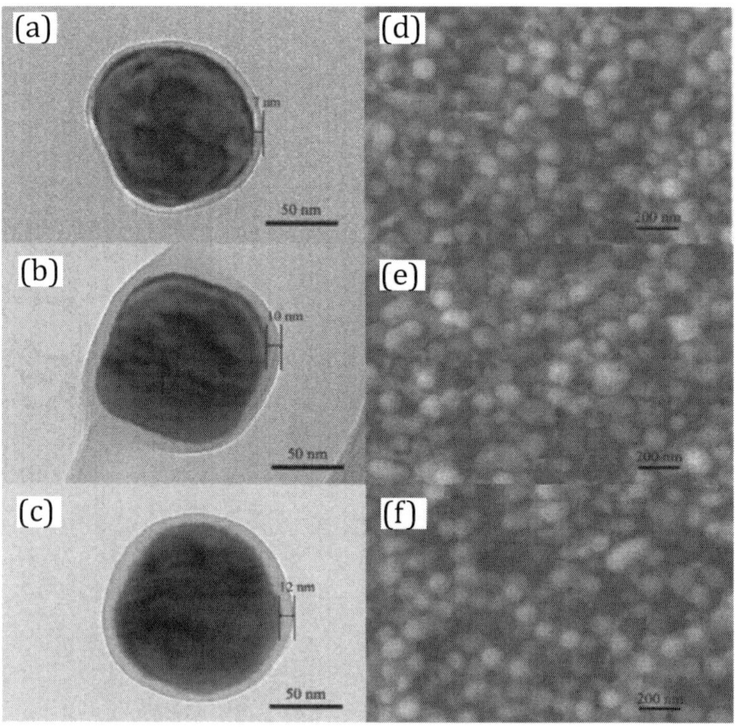

Figure 6.14 TEM images of PS@$BaTiO_3$ core–shell nanoparticles: (a) BT PS1, (b) BT–PS2, and (c) BT–PS3; SEM images of the freeze-fractured cross sections of the nanocomposites films: (d) BT–PS1, (e) BT PS2, and (f) BT–PS3. Reproduced with permission from [27]. Copyright © 2012 WILEY-VCH Verlag GmbH & Co. KGaA, Weinheim.

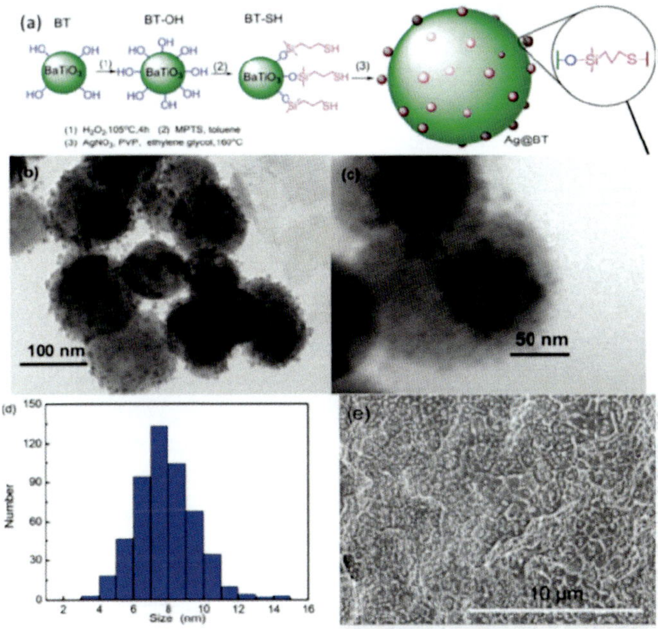

Figure 6.15 (a) Schematic representation of the preparation process for the Ag@BT core–satellite nanoassemblies, (b) TEM image of Ag1%@BT core–satellite nanoassemblies, (c) TEM image of Ag1%@BT–PVDF nanocomposite, (d) the size distribution of Ag nanoparticles and (e) the SEM images of Ag1%@BT–PVDF nanocomposite. Reproduced from [29] with permission from the PCCP Owner Societies.

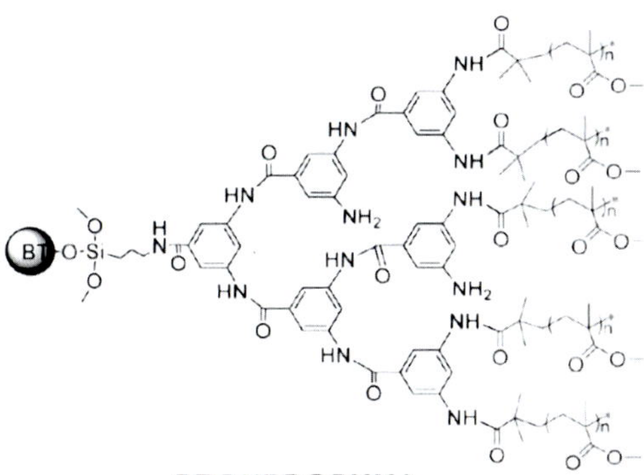

BT@HBP@PMMA

Figure 6.16 An example of a nanoparticle with double shell structures. Reprinted with permission from [30]. Copyright (2013) American Chemical Society.

6.5.4 Bimodal Polymer Brush and Click Chemistry

There are two approaches in the case of polymer brushes, namely, "grafting from" and "grafting to." In the former case, polymer brushes can be grown from the surface of a nanoparticle by an appropriate polymerization reaction. This "grafting from" approach has the advantageous characteristic of achieving high graft densities [31]. In "grafting to," polymer chains are formed prior to reaching the surface of the nanoparticle and, consequently, this reaction is a quick and flexible means of molecular attachment. The "grafting to" approach can be conducted using "Click Chemistry" [32], an example of which is shown in Fig. 6.17. In this case, electroactive oligothiophene molecules were grafted to a silica particle surface in order to confer the desired electroactivity on the nanocomposite. In addition, epoxide groups were also attached, to promote interactions with the epoxy matrix.

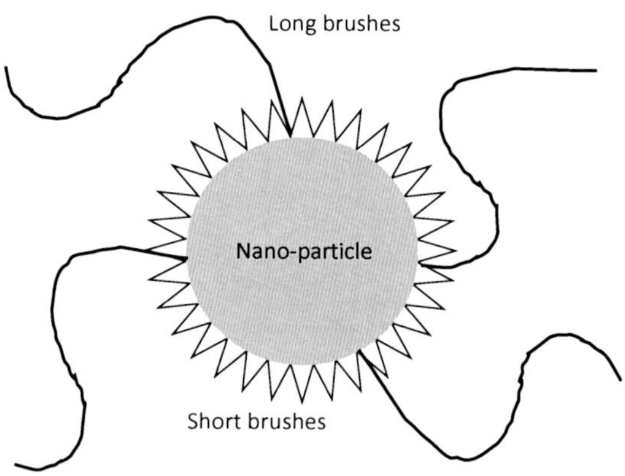

Figure 6.17 Schematic representation of the bimodal brush grafted core functionalized silica nanoparticle. Example: The short brush is oligothiophene and long brush is PGMA (poly(glycidyl methacrylate)). The chemical structure of the entities attached via the azide linker on the silica core by click reaction were: (a) alkyne terminated oligothiophene; (b) ethynylferrocene; (c) alkyne terminated PGMA.

The concept of "Click Chemistry" was introduced in 2001 [32] with the intention of providing a novel and rapid approach selectively to synthesize molecules with desired properties through the C-X-C links. This has opened a new region for combinatorial chemistry that is characterized by the following: (1) modular, of wide scope and starting from materials or reagents readily available; (2) provides high yields; (3) is inert to both oxygen and water; (4) avoids the need for chromatographic purification; (5) has stereospecific properties leading to high-selectivity. A typical click chemistry reaction is the copper-catalyzed 1,3-dipolar cycloadditions between azides and acetylenes [33]. In the case of Fig. 6.15, the first stage is to create a monomodal silica PGMA reference samples based on "grafting from" synthesis known as surface-initiated reversible addition-fragmentation chain transfer (RAFT) polymerization. Then, oligomers or ligands such as (a) alkyne terminated oligothiophene, (b) ethynylferrocene, and (c) alkyne terminated PGMA are attached by a click reaction.

6.6 Computational Simulation

The preceding sections of this chapter have considered a range of theoretical issues associated with systems of nanoparticles. However, with progressive increases in computational power, an area of growing importance concerns the numerical simulation of the behavior of assemblies of atoms or molecules, which can provide a means of both interpreting experimental data and of exploring systems that are not available in the laboratory. For example, computational simulation can be used to explore the behavior of idealized defect-free systems, which can only be approached experimentally. As such, simulations can provide limiting property values and can highlight the impact of defects, impurities, etc. In Chapter 7, computational approaches to charge transport are considered; here, we introduce the application of molecular dynamics techniques for the study of both the structure and properties of nanocomposites.

6.6.1 Molecular Dynamics

Molecular dynamics (MD) concerns the application of classical Newtonian dynamics to the behavior of a many body problem. As such, it is conceptually simple in that it involves the determination of the trajectory of each of the atoms or molecules in the system as a function of time, by solving Newton's equations of motion. Mathematically, the essence of the approach can therefore be expressed in a few simple equations. For a classical system of N particles, then the force F_i on the i-th particle of mass m_i at position r_i results in an acceleration, a_i:

$$F_i = m_i a_i = m_i \frac{d^2 x_i(t)}{dt^2}$$

where the force is obtained from the gradient of the potential energy function, $V(r^N)$ pertinent to the relevant interactions:

$$F_i = -\frac{\partial V(r^N)}{\partial r_i}$$

Evidently, this approach relies critically upon a detailed knowledge of the interactions that are occurring in the system and the adoption of the correct potential energy functions. However, potential energy functions are generally empirical in nature and, in the case of molecular systems, need to incorporate both intramolecular terms (e.g., interatomic separations, bond rotation, etc.) and intermolecular terms (e.g., Coulomb interactions, directional hydrogen bonds, van der Waals interactions, etc.). Nevertheless, assuming that all such interactions can be adequately defined, then the trajectory of the complete system can be determined, time step by time step, provided the initial coordinates and velocities of every atom is known. As in other areas where the progressive evolution of the simulation with time is of interest, it is critically important to employ a time step that is small enough, so that the predictions of the model are independent of this factor. In MD simulations, values of the order of a few femtoseconds are typically adopted. Finally, the behavior of the system is determined by averaging the configuration of the system over many thousands of time-steps. This approach can be computationally demanding and, therefore, the greater the computing power, the larger the system and the more time steps that can be considered. Commonly, simulations consider systems in which the number of particles, the volume and the energy are fixed—the so-called microcanonical ensemble, which equates to the system being isolated. Other scenarios are, however, possible.

6.6.2 MD Simulation of Nanocomposite Structure

A commonly considered problem that has been considered by MD concerns the effect of various material parameters on the tendency of nanoparticles to agglomerate, since enhanced macroscopic properties are generally associated with good levels of dispersion. In view of the importance attached to suitably engineering nanoparticle surface chemistry, Xue et al. [34] recently employed coarse grained MD simulations to explore conformational effects in polydisperse polymer chains grafted onto nanoparticles, the rationale being that by attaching chains that are chemically equivalent to the matrix polymer, they can both shield the nanoparticle and provide steric stabilization to prevent nanoparticle agglomeration. Specifically, this study set out to vary the grafted chain polydispersity, the average chain length, the grafting density and the nanoparticle size and to explore the effect of these parameters on the resulting chain conformations. As highlighted above, a key issue of MD simulation relates to defining the interactions that are occurring in the system. In this study, excluded volume interactions were represented by a truncated and shifted Lennard–Jones potential. Polymer chains were represented as a string of monomers connected together by a nonlinear elastic spring potential. In addition, inter-particle interactions also included pairwise dissipative and random forces. Once each particle was defined, a polydisperse array of the polymer chains, characterized by the unimodal Schulz–Zimm distribution, was grafted onto it, before the remaining space was filled with solvent molecules. Figure 6.18 shows an instantaneous conformation for a nanoparticle carrying a grafted array of polydisperse chains.

The simulation was then conducted at a constant-volume and constant temperature and three dimensional periodic boundary conditions were applied. After relaxing the system, the effect of the chosen input variables on the free chain end density profile, the average radius of gyration of the grafted chains, the average end-to-end distance of the grafted chains, the average thickness of the film of grafted chains and the degree of interpenetration of neighboring chains was determined. While all the results are not easily summarized, the authors conclude that enhanced excluded volume interactions occur for larger nanoparticles, under conditions when the polydispersity is low and the grafting density is high.

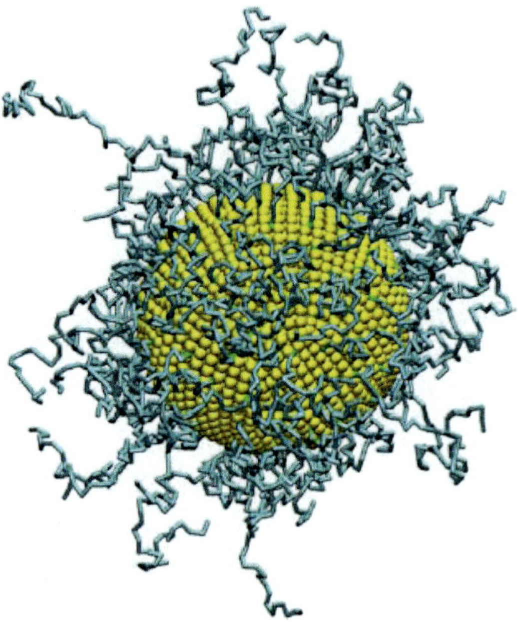

Figure 6.18 Schematic illustration showing a MD snapshot of the conformation of polydisperse array of chains grafted to a nanoparticle. Reproduced from [34] by permission of Taylor & Francis Ltd.

Nanoparticle dispersion and aggregation effects were examined in a non-specific polymer-based nanocomposite by Liu et al. [35]. In this work, each polymer molecule was modeled as a linear sequence of 30 beads of diameter σ. The nanoparticles were defined as spheres of radius 2σ and the various interactions and boundary conditions were defined in a comparable way to the above. Initially, the work examined the effect of varying the nanoparticle/matrix polymer interactions on the nanoparticle dispersion; polymer/filler interactions were defined by an interaction parameter that was varied from $\varepsilon_{\mathrm{np}} = 0.1$ (weak interactions) to $\varepsilon_{\mathrm{np}} = 12.0$ (strong interactions) and the resulting nanoparticle dispersions were defined by the radial distribution function (RDF). When the nanoparticle/polymer interactions were low ($\varepsilon_{\mathrm{np}} < 1$), the simulation led to a peak in the RDF at a spatial separation of 4σ, indicating direct contact of filler particles. At $\varepsilon_{np} > 2$, the degree of interaction between the polymer and the nanoparticles resulted in various structural arrangements being adopted, which include specific peaks in the RDF that indicated a tendency for the system to form agglomerates in which the nanoparticles are separated from one another by intervening layers of polymer. While this tendency for increasing nanoparticle/matrix interactions to favor such conformations is not quantitatively surprising, the absence of strong peaks in the RDF at $\varepsilon_{np} = 2$ is interesting, in that it indicates that the best dispersion is achieved when nanoparticle/matrix interactions are moderate. The study went on to consider the effect of grafting polymer chains onto the nanoparticles; the chains were represented by a sequence of 10 beads and the matrix was assumed to be composed of polymer molecules of the same length. Under conditions where $\varepsilon_{np} = 1.0$ (i.e., direct filler contact in the absence of grafting), only a small peak in the RDF was seen at twice the nanoparticle radius, indicating that, despite the absence of favorable nanoparticle matrix interaction, no direct contact aggregation occurred. Indeed, the presence of a peak in the RDF at 9.5σ, was attributed to repulsion of grafted chains leading to a good dispersion of the nanoparticles in the polymer matrix.

6.6.3 MD Simulation of Nanocomposite Properties

A commonly cited reason for the current interest in nanocomposites concerns the potential of such systems to exhibit unique *combinations* of properties. While the electrical characteristics of nanocomposites for use in insulation applications are, of course, critical, in practice, the addition of nanoparticles has also been discussed as a potential means of enhancing other properties. For example, increasing thermal conductivity would be highly beneficial in many applications, as a means of improving heat dissipation and, therefore, facilitating the design of more compact plant.

A simplistic approach to the topic of heat transfer through a complex multicomponent medium is to adopt an effective medium approach, in which the thermal conductivity of the composite is derived directly from the thermal conductivities of the constituents and the composition. A range of such approaches was described by Progelhof et al. [36]. As with comparable approaches to other parameters, the limiting cases correspond to series and parallel arrangements, such that the thermal conductivity of the composite K_e can be expressed:

$$\left[\frac{(1-\phi)}{K_c} = \frac{\phi}{K_d}\right]^{-1} < K_e < (1-\phi)K_c + \phi K_d$$

where K_d and K_c represent the thermal conductivity of the discrete phase and continuous phase, respectively, and φ is the faction of the discrete phase.

However, this neglects a key factor in heat transfer within composite materials, namely, the influence of the internal interfaces which, phenomenologically, can be considered in terms of an thermal resistance. This is then is defined in terms of the heat flux through the boundary and the temperature drop across it. MD constitutes an effective means of addressing this problem, as described by Cahill et al. [37]. In a common approach to this problem a heat source and a sink are introduced across the interface of interest and the MD simulation is run until a steady state is established. Then, the interfacial thermal conductance can be determined from the temperature change across the interface. Since MD simulations rely on the availability of appropriate interatomic potentials, these must be known for the interface of interest. In addition to the interfacial thermal conductance, MD simulations can also be used to provide detailed understanding of the thermal coupling across the interface, including information relating to coupling between molecular vibrations in the material either side of the interface.

MD thermal conductivity techniques involve the determination of the behavior of the system, including all vibrations. Conversely, phonon dynamics MD techniques have recently been developed in order to provide information on mechanism of heat flow in terms of specific vibrational modes. In this approach, packets of lattice vibrations of defined polarization and with a narrow spread of frequencies are generated and allowed to propagate through the system of interest, such that scattering processes can be explored. For example, it has been shown using this approach that scattering of high frequency phonons at interfaces involves a mixture of acoustic scattering, in which the incident and reflected phonons have the same character, and diffusive scattering, where correlation between the incident and reflected wave packets is lost.

References

1. Sharma, V., Park, K., and Srinivasarao, M. (2009), Colloidal dispersion of gold nanorods: Historical background, optical properties, seed-mediated synthesis, shape separation and self-assembly, *Mater. Sci. Eng.*, vol. R65, pp. 1–38.
2. Petty, M. C. (1983), Langmuir-Blodgett films, *Endeavour*, New Series, vol. 7, no. 2, p. 65–69.
3. Ziegel, K. D. (1969), Role of the interfaces in mechanical energy dissipation of composites, *J. Colloid Interface Sci.*, vol. 20, pp. 72–80.
4. Ziegel, K. D., Frensdorff, H. K., and Fogiel, A. W. (1969), Strain magnification in polymer composites, *J. Appl. Polymer Sci.*, vol. 13, pp. 867–869.
5. Droste, D. H., and Dibenedetto, A. T. (1969), The glass transition temperature of filled polymers and its effect on their physical properties, *J. Appl. Polym. Sci.*, vol. 13, pp. 2149–2168.
6. Pliskin, I., and Tokita, N. (1972), Bound rubber in elastomers: Analysis of elastomer-filler interaction and its effect on viscosity and modulus of composite systems, *J. Appl. Polym. Sci.*, vol. 16, pp. 473–492.
7. Binks, B. P., and Horozov, T. S. (2006), *Colloidal Particles at Liquid Interfaces: An Introduction*, Cambridge University Press, 978-0-521-07131-4.
8. Ninhama, B. W. (1999), On progress in forces since the DLVO theory, *Adv. Colloid Interface Sci.*, vol. 83, pp. 1–17.
9. Souma, I., and Kishibe, M. (1986), Interaction between poly(vinyl chloride) and inorganic fillers, *J. Rubber Inst. Jpn.* (Nihon Gomu Kyokai Shi), vol. 59, pp. 297–301 (in Japanese).
10. Lewis, T. J. (1994), Nanometric dielectrics, *IEEE Trans. Dielectr. Electr. Insul.*, vol. 1, no. 5, pp. 812–825.
11. Lewis, T. J. (2004), Interfaces are the dominant feature of dielectrics at the nanometric level, *IEEE Trans. Dielectr. Electr. Insul.*, vol. 11, no. 5, pp. 739–753.
12. Lewis, T. J. (2004), Interfaces and nanodielectrics are synonymous, *Proc. ICSD*, no. 705, p. 4.

13. Lewis, T. J. (2006), Nano-composite dielectrics: The dielectric nature of the nano-particle environment, *IEEJ Trans. FM*, vol. 126, no. 1, pp. 1020–1030.

14. Evans, D. F., and Wennerström, H. (1999), *The Colloidal Domain*, 2nd ed., Wiley-VCH, New York, NY, USA.

15. Israelachvili, J. N. (1992), *Intermolecular and Surface Forces*, 2nd ed., Academic Press, London, UK.

16. Rüttingstein, P., Preistley, R. D., R. D. Broadbelt, R. D., and Torkelson, J. M. (2007), Model polymer nanocomposites provide an understanding of confinement effects in real nanocomposites, *Nat. Mater.*, vol. 6, pp. 278–282.

17. Tanaka, T., Montanari, G. C., and Mülhaupt, R. (2004), Polymer nanocomposites as dielectrics and electrical insulation-perspectives for processing technologies, material characterization and future applications, *IEEE Trans. Dielectr. Electr. Insul.*, vol. 11, no. 5, pp. 763–784.

18. Jean, M. S., Hudet, S., Guthmann, C., and Berger, J. (1998), Local triboelectricity on oxide surfaces, *Eur. Phys. J. B*, vol. 12, pp. 471–477.

19. Tanaka, T., Kozako, M., Fuse, N., and Ohki, Y. (2005), Proposal of a multi-core model for polymer nanocomposite dielectrics, *IEEE Trans. Dielectr. Electr. Insul.*, vol. 12, no. 4, pp. 669–681.

20. Maity, P., Kasisomayajula, S. V., Parameswaran, V., Basu, S., and Gupta, N. (2008), Improvement in surface degradation properties of polymer composites due to pre-processed nanometric alumina fillers, *IEEE Trans. Dielectr. Electr. Insul.*, vol. 15, no. 1, pp. 63–72.

21. Sangha, S., and Thomas, M. J. (2009), Influence of filler loading on dielectric properties of epoxy-ZnO nanocomposites, *IEEE Trans. Dielectr. Electr. Insul.*, vol. 16, no. 2, pp. 531–542.

22. Zou, C., Fothergill, J. C., and Rowe, S. W. (2008), The effect of water absorption on the dielectric properties of epoxy nanocomposites, *IEEE Trans. Dielectr. Electr. Insul.*, vol. 15, no. 1, pp. 106–117.

23. Wang, Y., Ke, Y., Li, J., Du, S., and Xia, Y. (2007), Dispersion behavior of core-shell silica-polymer nanocomposites, *China Particuology*, vol. 5, pp. 300–330.

24. Li, W-H., and Stöber, H. D. H. (2000), Monodisperse cross-linked core-shell polymer microspheres by precipitation polymerization, *Macromolecules*, vol. 33, pp. 4354–4360.

25. Ramli, R. A., Laftah, W. A., and Hashim, S. (2013), Core-shell polymers: A review, *RSC Adv.*, vol. 3, pp. 15543–15565.

26. Moraes, J., Ohno, K., Maschmeyerc, T., and Perrier, S. (2013), Synthesis of silica–polymer core–shell nanoparticles by reversible addition–fragmentation chain transfer polymerization, *Chem. Commun.*, vol. 49, pp. 9077–9088.

27. Yang, K., Huang, X., Xie, L., Wu, C., Jiang, P., and Tanaka, T. (2012), Core–shell structured polystyrene/$BaTiO_3$ hybrid nanodielectrics prepared by in situ RAFT polymerization: A route to high dielectric constant and low loss materials with weak frequency dependence, *Macromol. Rapid Commun.*, vol. 33, issue 22, pp. 1921–1926. DOI: 10.1002/marc.201200361.

28. Yang, K., Huang, X., Zhu, M., Xie, L., Wu, C., Tanaka, T., and Jiang, P. (2014), Combining RAFT polymerization and thiol-ene click reaction for core–shell structured polymer@$BaTiO_3$ nanodielectrics with high dielectric constant, low dielectric loss, and high energy storage capability, *ACS Appl. Mater. Interfaces*, vol. 6, pp. 1812–1822.

29. Xie, L., Huang, X., Li, B-W., Zhi, C., Tanaka, T., and Jiang, P. (2013), Core-satellite Ag@$BaTiO_3$ nanoassemblies for fabrication of polymer nanocomposites with high discharge energy density, high breakdown strength and low dielectric loss, *Phys. Chem. Chem. Phys.*, vol. 15, pp. 17560–17569.

30. Xie, L., Huang, X., Huang, Y., Yang, K., and Jiang, P. (2013), Core@ double-shell structured $BaTiO_3$–polymer nanocomposites with high dielectric constant and low dielectric loss for energy storage application, *J. Phys. Chem. C*, 117, pp. 22525–22537.

31. Virtanen, S., Krentz, T. M., Nelson, J. K., Schadler, L. S., Bell, M., Benicewicz, B., Hillborg, H., and Zhao, S. (2014), Dielectric breakdown strength of epoxy bimodal-polymer-brush grafted core functionalized silica nanocomposites, *IEEE Trans. Dielectr. Electr. Insul.*, vol. 21, no. 2, pp. 563–570.

32. Kolb, H. C., Finn, M. G., and Sharpless, K. B. (2001), Click chemistry: Diverse chemical function from a few good reactions, *Angew. Chem. Int. Ed.*, vol. 40, pp. 2004–2021.

33. Wang, M., Yuan, Y., and Liang, G. (2012), "Click chemistry" for molecular imaging, *Curr. Mol. Imaging*, vol. 1, no. 1, pp. 9 (1–9).

34. Xue, Y.-H., Quan, W., Qu, F.-H., and Liu, H. (2015), Conformation of polydispersed chains grafted on nanoparticles, *Mol. Simulation*, vol. 41, no. 4, pp. 298–310.

35. Liu, J., Gao, Y., Cao, D., Zhang, L., and Guo, Z. (2011), Nanoparticle dispersion and aggregation in polymer nanocomposites: Insights

from molecular dynamics simulation, *Langmuir*, vol. 27, no. 12, pp. 7926–7933.

36. Progelhof, R. C., Throne, J. L., and Ruetsch R. R. (1976), Methods for predicting the thermal conductivity of composite systems: A review, *Polym. Eng. Sci.*, vol. 76, no. 9, pp. 615–626.
37. Cahill, D. G., Braun, P. V., Chen, G., Clarke, D. R., Fan, S., Goodson, K. E., Keblinski, P., King, W. P., Mahan, G. D., Majumdar, A., Maris, H. J., Phillpot, S. R., Pop, E., and Shi, L. (2014), Nanoscale thermal transport. II. 2003–2012, *Appl. Phys. Rev.*, vol. 1, no. 1, 011305, pp. 1–45.

Chapter 7

Computer Simulation of Nanocomposites at the Molecular Level

David Cubero[a] **and Tatsuo Takada**[b]

[a]*Departamento de Física Aplicada I, Universidad de Sevilla, Escuela Universitaria Politécnica, Calle Virgen de Africa 7, 41011 Sevilla, Spain*

[b]*Tokyo City University, 1-28-1 Tamazutsumi, Setagaya-ku, Tokyo 158-8667, Japan*

dcubero@us.es, takada@a03.itscom.net

7.1 Introduction

Since the 1950s, computer simulation techniques have been successfully applied to gain an understanding and to make verifiable predictions of transport phenomena taking place at the molecular level. Today, they represent an indispensable tool to explore theoretically the world at the nanoscale. They are essential both for the scientist who tries to understand new mechanisms and for the engineer who tries to design new materials.

This chapter is organized as follows. First the current simulation methods available for nanocomposites are briefly discussed. Then, in the following sections we focus on the simulation of electron transport, and more specifically on band theory applied to linear polymers, explicitly considering trapping

Tailoring of Nanocomposite Dielectrics: From Fundamentals to Devices and Applications
Edited by Toshikatsu Tanaka and Alun S. Vaughan

ISBN 978-981-4669-80-1 (Hardcover), 978-981-4669-81-8 (eBook)
www.panstanford.com

mechanisms due to physical and chemical defects. Special emphasis is put on the capabilities and limitations of current computational methods. The last section concludes with some insight into the application of computer simulation methods to nanocomposites.

7.2 Computer Simulation Methods

In most cases, unless one is interested in very low temperatures, the dynamics of atoms and molecules is mainly classical, and therefore, it can be simulated using an algorithm to solve numerically the associated Newton's equations of motion for the positions and velocities of the atoms. Such a technique, known as molecular dynamics, is widely used, with many available codes—both open source and close, commercial, source. Nowadays, these codes are quite fast, allowing the simulation of systems with a large number of atoms with the record in 2013 being a system with more than four trillion (4.125×10^{12}) particles (Eckhardt, 2013). The interactions between the atoms are described using potentials which have been adjusted to experimental data or obtained with ab initio methods.

In contrast, electrons are much lighter particles that, in most cases, cannot be treated classically at the molecular level. The wave character of these quantum particles needs to be explicitly accounted for in a molecular simulation that focuses on electron transport, with crucial quantum properties, such as delocalization, band gaps, or tunneling, that would be left out otherwise. While the interaction of the electrons within the atoms are implicitly accounted for in the atomic potentials of molecular dynamics, the transport properties of charge carriers such as excess electrons or electron holes—an electron vacancy—can only be studied when the quantum character of the electrons is explicitly considered, a fact that inevitably complicates the description.

Density functional theory (DFT) is nowadays the most popular computational method to study electronic properties in materials. It is also widely available, with both open and commercial codes to carry out many tasks, from simple ground-state minimizations to ab initio molecular dynamics. It is usually accurate to describe ground-state properties like bond energies, the total charge density or the geometry of the system, such as bond lengths or crystal cell parameters—though not when van der Waals interactions are important, as we discuss below.

DFT is also frequently used to study electronic transport properties though, then, the accuracy tends to be considerably reduced. One important reason for this can be traced back to the frequent identification of DFT Kohn–Sham energies with the real energy levels of the quasi-particle electronic system. The Kohn-Sham levels are the energies of a fictitious non-interacting system used in all DFT calculations, without real physical meaning, thus being different from the excitation energies of electrons in the material. Even if DFT calculations did not contain approximations, quantities such as band gaps obtained with this procedure would differ drastically from the experimental values. Besides this dubious identification, standard DFT functionals are not capable of producing accurate results, and errors are typically in the range 30–50% (Burke, 2012).

On the other hand, other quantum methods such as Hartree–Fock do provide energy levels that can be interpreted as electron quasi-particle energies, i.e., the levels of an approximate free-particle theory that effectively takes into account the many-body interactions between all electrons and nuclei. However, Hartree–Fock theory contains other limiting approximations and, unlike DFT, bond energies tend to be significantly underestimated. Hybrid functionals, which incorporate a portion of Hartree–Fock exact exchange in DFT calculations, produce a substantial improvement, for example in the predicted band gaps of semiconductors, though is still unreliable for metals and insulators (Paier et al., 2006; Jain et al., 2011).

Therefore, electron transport results obtained this way using non-hybrid DFT functionals, such as LDA/GGA (local density approximation/ generalized gradient approximation), PBE (Perdew–Burke–Ernzerhof), or BLYP (Becke correlation functional; Lee, Yang, Parr electron exchange functional), despite being commonly used in the literature, are usually not expected to be very accurate. A double check using a hybrid functional, such as B3LYP, PBE0 and HSE (Heyd–Scuseria–Ernzerhof), though strictly required, unfortunately is not still frequently carried out.

Another DFT limitation is the size of the system that can be simulated with present computers. Being an expensive technique, typical simulations consider systems with a number of atoms in the order of magnitude of a hundred, which may be too small for the correct modeling of the material in disordered phases. As we discuss below, disordered phases play a key role in electron transport inside polymeric dielectrics.

These accuracy and size limitations can be partially averted by using a one-electron theory in a mixed quantum-classical representation, in which a single electron is treated quantum mechanically, while the atoms interact classically, evolving in time according to the classical equations of motion. This method has been the most successful theoretical approach to explain spectroscopic and dynamics properties of hydrated electrons (Turi et al., 2012), and has already been applied to elucidate excess-electron transport properties in several representative motifs of polyethylene (Cubero et al., 2003a; Cubero et al., 2003b; Cubero et al., 2004; Wang et al., 2014). Polyethylene (PE), possessing a very simple chemical composition, with a single atomic group that is repeated into very long chains, [—$(CH_2)_n$—], and being widely employed in industrial applications, is the simplest example of a semicrystalline polymeric insulator.

A key ingredient of the one-electron approach is the pseudopotential describing the interaction of the excess-electron and the atoms in the material, being obtained by either ab initio calculations (Turi et al., 2012) or fitting experimental data (Cubero et al., 2003a). The use of standard molecular dynamics techniques and the focus on just one excess electron allow the simulation of much larger systems than with DFT. It is also expected to be more accurate in PE due to the use of a semi-empirical pseudopotential. However, the development of a pseudopotential is not always easy to carry out for a generic electron-atom or hole-atom interaction, in contrast to DFT, which is not particular about the specific atoms forming the dielectric.

In the following, we use PE as a case study to highlight the capabilities and limitations of current techniques. In Section 7.6, we study a more complex polymer insulator, showing that even though usual DFT methods may not be accurate, they can still serve as a useful tool to gain qualitative insight of electronic transport in polymeric dielectrics.

7.3 Band Structures and Electrical Conduction in Linear Polymers

The electronic band structure of a crystalline material is a single electron theory (a quasi-particle theory) in which the interactions with the other electrons and nuclei are effectively included in a static, spatially periodic, potential. Other dynamical effects such as phonon (interaction with lattice vibrations) or photon collisions, or even with other single electrons are in principle excluded, but can be accounted for as perturbations.

The continuous bands are a consequence of the spatial periodicity, which implies a homogeneous system of infinite size. For the same reason, every electronic state associated with each energy level is also periodic, and thus delocalized, i.e., with an infinite extension. These states can be thought of as the result of periodically combining the atomic orbitals of all atoms in the system, producing band gaps where no energy levels are allowed.

Figure 7.1 shows a scheme of the relevant band structure of crystalline PE. The highest energy electrons are completely filling the valence band, the highest occupied band. Electrons in this band are in molecular orbitals around the strong covalent bonds that support the PE chains, and thus are of an intra-chain nature. The next band, the conduction band, is completely empty, as is characteristic of insulators.

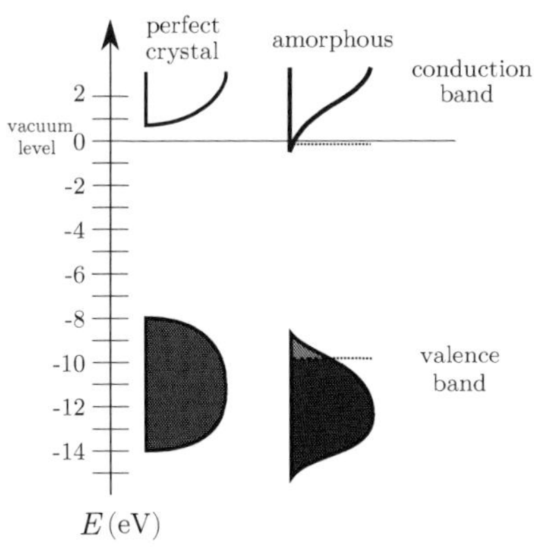

Figure 7.1 Energy bands in crystalline and amorphous phases of polyethylene (schematic). The dotted lines indicate the mobility edges, separating localized and delocalized states for excess electrons and electron holes.

The band gap separating the valence and conduction band has been measured experimentally in the range 8–9 eV (Tanaka, 1973; Delhale et al., 1974). Such a large gap is due to the strength of the covalent bonds in the chains, yielding large ionization energies. DFT calculations typically underestimate this gap by about 2 eV (Righi et al., 2001).

There is, however, agreement between DFT (Righi et al., 2001) and one-electron (Cubero et al., 2003b) methods that the electronic states in the conduction band are of an inter-chain nature. Excess electrons are repulsed by the PE chains and thus stay in the empty regions in between.

There is also agreement that the bottom of the conduction band is above the vacuum level, but the theoretical values differ. While the above DFT calculation (Righi et al., 2001), using the functional LDA/GGA, predicts a value slightly above the vacuum level, 0.1–0.17 eV, the value given by the one-electron approximation, 0.6 eV, is closer to the experimental value 0.4–0.5 eV (Ueno et al., 1986), which was obtained by measuring the onset of conduction in crystalline phases of long n-alkanes. Both theoretical calculations are for a perfect crystal, and thus neglect the disorder effect of a finite temperature, which should further reduce slightly the bottom of the band, as we discuss in the next section.

DFT calculations of the crystal structure of PE using standard functionals fail to provide accurate bound energies, and the predicted lattice parameters typically differ by more than 11% with respect to the experimental values at low temperatures (Kleis et al., 2007). This is due to the famous failure of DFT to describe van der Waals interactions properly. While covalent interactions are responsible for most of the forces within the PE chains, the inter-chain interaction is dominated by van der Waals forces. This problem is usually avoided by using some sort of empirical correction to DFT. Nevertheless, recently the use of a DFT functional specifically designed to deal with van der Waals interactions (Dion et al., 2005) without any empirical input has been shown to produce data for PE (Kleis et al., 2007) that is in better agreement with experiments.

In crystalline PE, both an excess electron at the conduction band and a hole in the valence band should move very fast, with estimated mobilities larger than 1 cm^2/Vs (Cubero et al., 2003b) and 20 cm^2/Vs (Andre et al., 1970), respectively. The fact that the experimental values are much lower, less than 10^{-2} cm^2/Vs (Mizutani, 2006), is an indication that other transport mechanisms are in play in real PE.

Disordered phases, such as amorphous materials, also exhibit band structures. Figure 7.1 illustrates the relevant bands of amorphous PE. Frequently, the mere presence of disorder is able to induce the localization of the lowest (excess electrons) or highest (holes) energy levels in a band, a phenomenon known as Anderson localization. Energy thresholds, called mobility edges, separate the localized and delocalized states. Transport is much faster at delocalized states, the mobility being typically larger by several orders of magnitude when compared to localized states. As we discuss in the next section, the small amount of disorder introduced in the crystalline regions of PE by the thermal fluctuations at room temperature is also able to produce mobility edges in the band ends.

Although the differences between the bands in the crystal and amorphous regions of PE may look small in the scale of Fig. 7.1, they are actually crucial, due to the small magnitude of the thermal energy at room temperature, $k_B T$ = 0.026 eV, which favors the lowest energy levels available.

7.4 Effects of Physical Defects, the Importance of the Underlying Morphology

In crystalline solids, physical defects commonly refer to the deviations of the atom coordinates from their lattice equilibrium positions due to thermal fluctuations. In the crystalline regions of PE at room temperature, they are responsible for the appearance of a tail of localized states, as shown in Fig. 7.3. However, after looking at the whole structure of PE, as well as others linear polymer insulators, physical defects cannot be seen as simple defects, or small perturbations from an idealized crystalline arrangement, but as an essential ingredient of its underlying structure.

Solid PE exhibits a very complicated structure (Dissado et al., 1992), usually presenting a large-scale spherulite structure composed of crystalline lamellae that are formed at the molecular level by folded chains. The details depend highly on the processes that occur during solidification. The interfacial regions between these lamellae typically involve layers of amorphous material located between the folded surfaces of the lamellae. Figure 7.2a shows an image of the surface of a lamella obtained by using atomic force microscopy (Mullin et al., 2011). Generally, solid PE is a semicrystalline polymer where crystalline and amorphous regions are inevitably mixed.

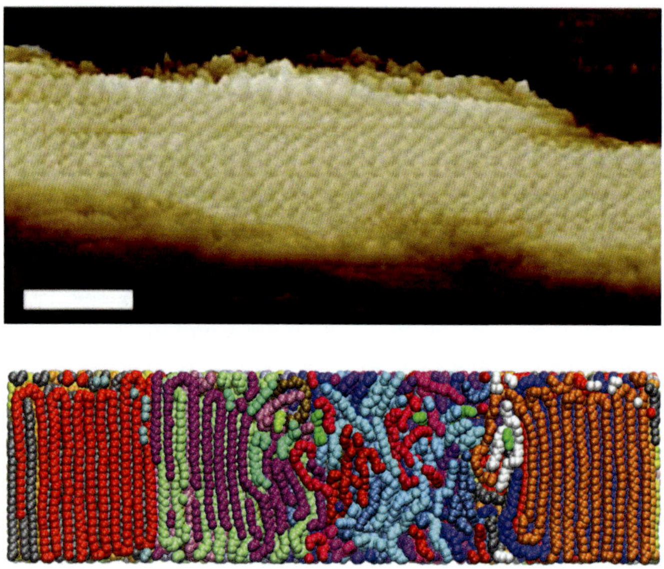

Figure 7.2 Semicrystalline structure of polyethylene. (a) High-resolution image, obtained with torsional tapping atomic force microscopy, showing the (110) surface of a crystalline lamella. Scale bar 5 nm. From Mullin et al. (2011). (b) Molecular dynamics simulation of a lamella-amorphous interface. Individual polymer chains are each colored differently for clarity. Periodic boundary conditions are in place. From Wang et al. (2014).

In order to evaluate the effect on electronic transport of the interfacial structures between lamellar and amorphous regions, molecular configurations modeling these interfaces were generated by Wang et al. (2014) using molecular dynamics. An example, with three lamellae with chains parallel to two interfaces with an amorphous region, is shown in Fig. 7.2b. The amorphous region was obtained by heating a lamella up to 500 K, while keeping the other lamellae at a constant 300 K. The volume and shape of the simulation cell was allowed to change by running the simulations in the NPT ensemble with a Nosé–Hoover algorithm that kept the pressure constant.

The density of states available to excess electrons in several PE motifs are shown in Fig. 7.3. At 1 K, the temperature is so low that the physical defects can be safely neglected in crystalline regions of PE. The atomic positions thus deviate little from the spatially periodic structure associated with the crystalline lattice, yielding electronic states which are also almost periodic. All observed states were indeed delocalized, with the lowest energy at 0.6 eV above the vacuum level. In contrast, thermal fluctuations at room temperature introduce a number of physical defects that are able to induce localization of excess electrons. A tail of localized levels can be seen in Fig. 7.3, down to 0.2 eV. The lowest energy level showing delocalization is slightly below 0.5 eV, shifting down the value for the bottom of the conduction band. This simulation value is in good agreement with the experimental range 0.4–0.5 eV obtained by Ueno et al. (1986) by measuring the onset of conduction in crystalline phases of long n-alkanes—chain polymers with a similar structure to PE, except for the atoms at both chain ends. While it has been shown (Cubero et al., 2003b) that in perfect crystals of long chain n-alkanes there are electronic states that are absent in pure crystalline regions of PE, the energies associated with them are not very different fromthe levels obtained at room temperature.

Further disorder in the atomic coordinates are able to lower even more the energy levels, as shown in Fig. 7.3 for amorphous regions of PE at room temperature. The lowest levels in these regions are found at about −0.4 eV, being localized states, centered at regions with lower atomic densities. Delocalization is observed at about −0.1 eV; below the vacuum level. Again, this value is in agreement with the experimental observations by Ueno et al. (1986), which locate the conduction band level in molten long-chain n-alkanes below the vacuum level. Note that unlike the crystalline regions, where delocalization is due to spatial periodicity, here it occurs entirely through disordered atomic configurations.

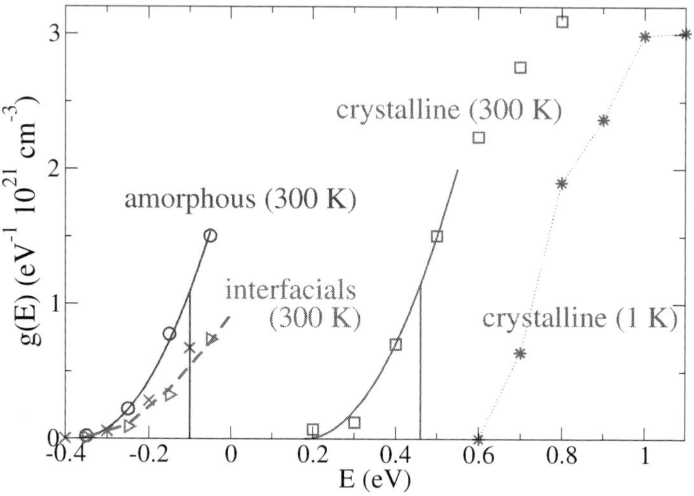

Figure 7.3 Density of states (DoS) for excess electrons in several polyethylene phases, computed from one-electron simulations using a semi-empirical pseudopotential (Cubero et al., 2003a). The vertical solid lines indicate the mobility edges (i.e., the onset of conduction through delocalized states). The crosses and the triangles correspond to lamella-amorphous regions with the chains perpendicular and parallel to the interfacial surface, respectively. The dashed line shows the DoS obtained by weighting the crystalline and amorphous DoS by their volume fraction in the interfacial regions (47%, 53%). From Wang et al. (2014).

Figure 7.3 also depicts the density of states of interfacial configurations, separating lamellae from amorphous regions, as illustrated in Fig. 7.2b. No significant new electronic states were found, though. Indeed, the density of states in the interfacial samples can be estimated by simply weighting the density of states of their crystalline and amorphous regions, as shown in Fig. 7.3.

Although there have been recent attempts to simulate amorphous and interfacial regions of PE using DFT methods (Unge et al., 2012; Unge et al., 2014), the current state of the art does not allow for the accurate simulation of sufficiently large systems using these techniques, and large uncertainties are expected in the results, especially because of the choice of the DFT functional (see our discussion in Section 7.2).

Since real PE contains both crystalline and amorphous regions, the band structure of the whole material is of special interest. Thermal energy is small in the scale of variation of PE bands, thus, the focus is on the lowest, accessible energy levels. For excess electrons it is not difficult to imagine eigenfunctions extended throughout the whole material, formed by joining the wavefunctions within crystal regions with those of the same energy within amorphous regions. This procedure is possible because the mobility edge in the amorphous regions is located below the bottom of the band in the crystal. Therefore, there is an energy band, roughly given by that of the crystal, of delocalized states, extended throughout the material. Conduction through these states is fast, but excess electrons need to gain sufficient energy to be excited to them. Thermal excitation to these states yields an Arrhenius-like factor $\exp(-E_a/k_BT)$ in the electron mobility. see for example, Wang et al. (2014), where E_a is the energy difference between the mobility edge at the crystal and the lowest energy level in the amorphous. The level of crystallinity, determining whether non-crystalline regions percolate the sample, is a relevant parameter here, with implications in the expected mobilities, as discussed in Wang et al. (2014b). In fact, from Fig. 7.3 it is inferred that the typical energy required for an excess electron to penetrate a crystalline region from a state within the amorphous is larger than 0.8 eV. This value is in good agreement with the experimental observations discussed by Mizutani (2006), which associate some physical defects with energy depths in the range of 0.8–1.4 eV.

However, transport through localized states is also possible. In the DFT study of Serra et al. (2002), it was found that, as a result of the dynamical interaction with PE atoms, excess electrons on small crystal regions at room temperature tend to fall spontaneously into a self-trapped state, producing a distortion in the local region surrounding the electron where it buries itself, the so-called polaron. A similar behavior was also observed in the one-electron simulations of amorphous regions reported by Cubero et al. (2004). The self-trap energy is small, about 0.1 eV, so comparable with thermal energy at room temperature where k_BT = 0.026 eV. Therefore, by waiting long enough, it is possible to observe the trapping and detrapping of excess electrons due to phonon interactions, as indeed shown in Fig. 7.4. Here, the center of mass of the excess electron shows that, at the beginning of the simulation, it is trapped near the origin, hopping to another trap after 5 ps, and then again after another 11 ps more. This is a clear indication of transport by hopping induced by phonon interaction.

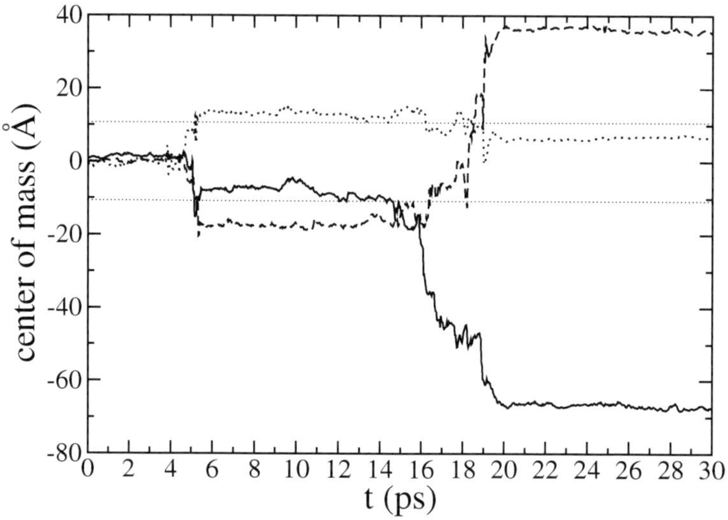

Figure 7.4 Electron hopping between self-trapped states in amorphous polyethylene using the same one-electron method as in Fig. 7.3. The lines depict the components of the center of mass of an electron as a function of time, showing hopping between three self-trapped states. System boundary edges are included as a reference (horizontal dotted lines), though the actual position is arbitrary because of the periodic boundary conditions. From Cubero et al. (2004).

In Serra et al. (2002), the authors did not find a self-trapping state for electron holes in small crystalline regions, which is somehow expected from the intra-chain character, and thus, more extended states of holes. However, the use of larger simulation sizes, larger than the length of the hole, might find a self-trapping behavior also for holes. Furthermore, given their intra-chain nature, in amorphous regions, transport would require hopping between states associated with different chains, most likely facilitated by quantum tunneling, though no computer simulation study has yet analyzed this case.

7.5 Effects of Chemical Defects and Impurities

The tails of localized states discussed in the previous section are not the only traps available for charge carriers. In real materials, the presence of some sort of chemical defect or impurity is inevitable, i.e., a group of atoms with an altered chemical composition. Such structures frequently represent a source of localized states, traps, which critically affects the transport of electrons and electron holes in the material and, thus, its insulating properties. Nowadays, the electrical properties of many semiconductors are routinely controlled by intentionally doping them with impurities.

In PE, having the simple chemical composition [—$(CH_2)_n$—], the most common chemical defects are vinyl groups (i.e., a C=C double bond instead of a single C—C connecting two consecutive carbons) and carbonyl groups (i.e., C=O), the latter frequently introduced by the use of branching agents.

Chemical defects are commonly studied using DFT methods, due to their success in describing covalent interactions. Excess electrons do not covalently attach to PE though, so the accuracy of DFT results here is uncertain. The first DFT calculation of chemical defects in PE was reported in Meunier et al. (2000). By studying the effect of various chemical defects on the Kohm–Sham levels of single molecules of n-alkanes (isolated molecules with up to 13 carbons), a rather crude model of PE, the authors estimated chemical traps for excess electrons with depth values up to 1.6 eV. More elaborate calculations, taking into account the surrounding PE chains, were carried out by Ceresoli et al. (2004) and, more recently, by Huzayyin et al. (2010). The experimentally observed trap depths associated with chemical defects are not much larger than 1 eV (Mizutani, 2006).

However, DFT results in the present context should be interpreted cautiously, because they may lead to inconsistent conclusions, not reproduced with a different functional. An example is given in the reported DFT studies on the stability of excitons (see Ceresoli et al., 2005; Bealing et al., 2013). Excitons are bound states of an electron and a hole, formed by the attractive interaction between both charge carriers. In Ceresoli et al. (2005), the authors, using the functional BLYP with an empirical correction for van der Waals interactions, found no apparent channel for the non-radiative recombination of the exciton in a crystalline region of PE, implying that recombinations only proceed radiatively, i.e., emitting light. In contrast, the DFT calculations of Bealing et al. (2013), using the functional PBE, did find a direct channel for exciton recombination via a non-radiative channel, specifically, via the breaking of a C-H bond and consequently producing appreciable chemical damage in the material. The latter result was confirmed by using the hybrid functional HSE06, thus providing a more convincing case. Experimental evidence also points to non-radiative channels as the dominant recombination mechanism (Le Roy et al., 2005). These results are important for the characterization of the insulating properties of PE, because the mentioned recombination of an electron–hole pair provides a mechanism for degradation of PE under electrical stress.

7.6 Simple Estimations of Trapping Sites in Linear Polymers Using DFT

DFT calculations of energy levels for electrons and electron holes may be off by up to about 1, or maybe 2 eV (judging from the band-gap deviation), sometimes providing the wrong qualitative behavior, as discussed in the previous section for exciton recombination in PE. Despite this, DFT is still a powerful tool that may provide good qualitative insight into the transport mechanisms taking place in the material.

As a case study, let us focus on an ethylene tetrafluoroethylene copolymer (ETFE), a synthetic polymeric insulator with a chemical structure that periodically alternates between that of PE and polytetrafluoroethylene (PTFE), a fluoroalkane with a similar chemical composition to PE, but with all hydrogens replaced by fluorines, i.e., [—$(CF_2)_n$—]. The goal is to obtain some qualitative insight by using simple DFT calculations.

Even though the Kohn–Sham (KS) energies are not exactly the excitation energies of electrons in the material, we are going to regard them as such as a simple approximation. Simplifying further, we might try to account for the electronic traps present in the ETFE sample by looking only at the KS energy levels of PE and PTFE independently, as illustrated in Fig. 7.5. In both cases the system studied was neutral (no excess electrons or holes). The figure shows the higher electronic energy levels of the occupied states, the HOMO (highest occupied molecular orbital), and the lower energy levels that are not occupied, the LUMO (lowest unoccupied molecular orbital). The latter are the relevant, available levels for potential excess electrons, while the former are the energies of electron holes. The wavefunctions associated with these levels are orbitals with appreciable probability near the molecular chains, with the HOMO's (and thus holes) being more intra-chain, as expected from covalent interactions, in which electron pairs are shared between two or more atoms.

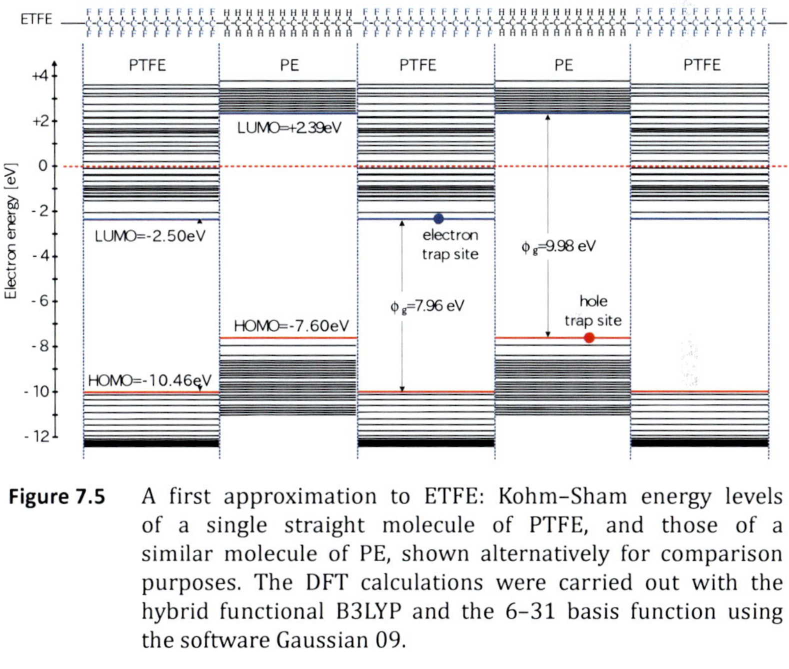

Figure 7.5 A first approximation to ETFE: Kohm–Sham energy levels of a single straight molecule of PTFE, and those of a similar molecule of PE, shown alternatively for comparison purposes. The DFT calculations were carried out with the hybrid functional B3LYP and the 6–31 basis function using the software Gaussian 09.

The energy levels shown in Fig. 7.5 are not exactly those of PE (and PTFE) in a crystalline or amorphous arrangement, but of a single, straight molecule. The energy values are not accurate, but they show some trends that are present in the real materials: The LUMO is above the vacuum level in PE (as it is in crystalline PE, though not in amorphous regions), while is negative in PTFE, in agreement with the well-known positive electron affinity of PTFE. Taking the levels of each material as a model of ETFE, an excess electron in a LUMO of a "PTFE region" will see a neighboring "PE region" as a barrier, because the LUMO in the latter is 4.89 eV above the level it is sitting in. Therefore, a LUMO at a PTFE site must represent an electron trap. Analogously, an electron hole must prefer to sit at a HOMO in a PE site, because the neighboring electrons would prefer to be at the HOMOs in the surrounding PTFE regions, having a lower energy (2.86 eV below). This could be identified with a hole trap. Thus, the conclusion from this simple model is that both electrons and holes have deeps traps in ETFE, which should slow them down drastically, in comparison with transport in PE or PTFE.

Nevertheless, the above model is ignores the fact that the electron orbitals near the connecting atoms between PE and PTFE regions are going to be rather different from those of PE and PTFE separately. It is thus convenient to carry out a DFT calculation of the whole ETFE molecule, the results of which are shown in Fig. 7.6. The LUMO structure is not very different from the simplistic model of Fig. 7.5, with the excess electron showing an appreciable probability density in the PTFE regions only, but the HOMO levels are considerably changed. Here, the electron in the HOMO is occupying a region that expands through several PE regions, centered (as expected) on a PTFE region, implying the absence of deep traps for electron holes. Therefore, the prediction of this more sophisticated model is that holes should move much faster than electrons, which find much deeper traps than in pure PE or PTFE samples.

In order to test these qualitative results, an experiment with a real ETFE sample, subject to a time-periodic external electric field, was carried out. The space charge in the sample was monitored using a Pulse Electro-Acoustic (PEA) system (Takada, 1999). By inserting the ETFE film between the plates of a capacitor and applying a dc voltage between them, a strong electric field of 100 kV/mm was generated in the sample over an interval of 60 min, after which time, the capacitor's plates were shorted for 10 min. This cycle of external dc field application and short-circuit relaxation was repeated several times. Figure 7.7 shows the time evolution of the space charge in the system. After three cycles, the space charge reaches a stationary profile. Apart from positive charges inside the anode and near it, in the stationary state, the charge observed inside the ETFE film is predominantly negative. In the previous cycles, especially during the first cycle, positive charge is observed to move across the ETFE, from the anode to the cathode. The negative charge observed inside ETFE, on the other hand, does not seem to move under the application of the dc field. Rather, it seems to be exposed by the upward movement of the positive charges (also mainly during the first cycle). This behavior is consistent with the predicted picture of slow moving electrons—due to deep traps in the PTFE regions—and much faster moving electron holes.

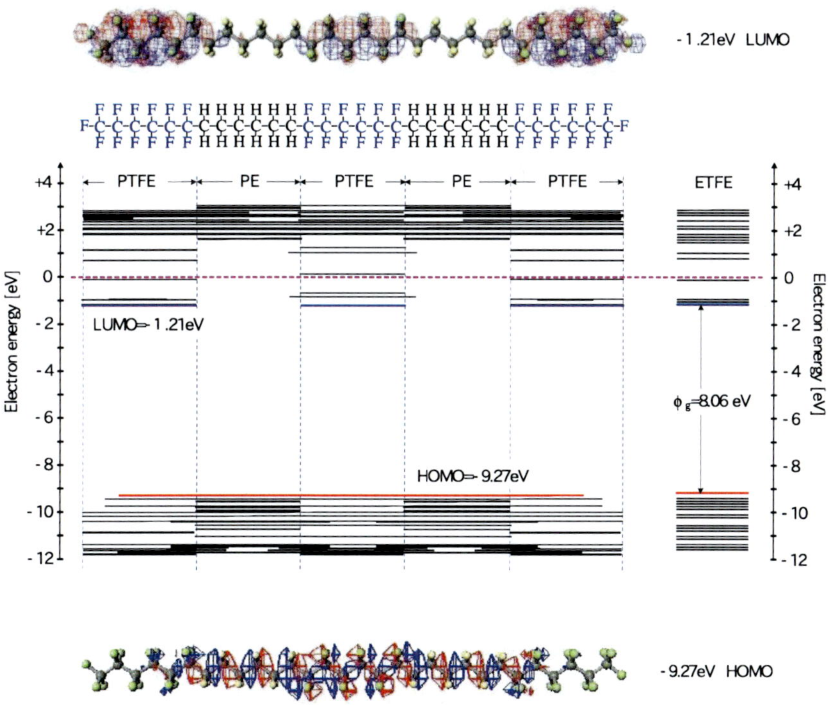

Figure 7.6 Second approximation to ETFE: Kohm–Sham energy levels of a single, straight molecule of ETFE (with 30 carbons, $C_{30}H_{24}F_{38}$). The top panel shows the LUMO, and the bottom panel the HOMO.

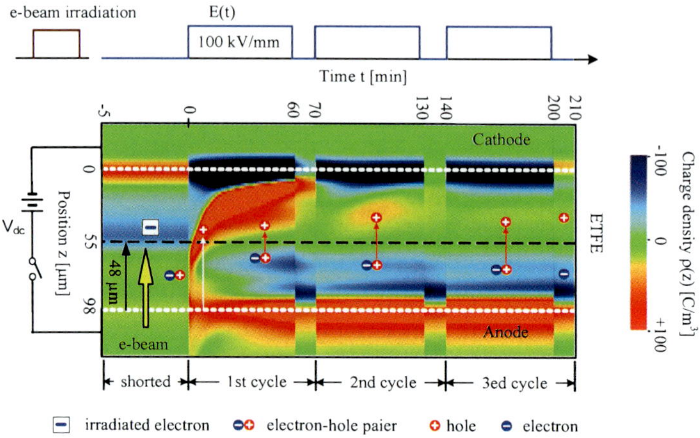

Figure 7.7 Dynamic space charge distribution observed in an ETFE film about 0.1 mm in thickness under the repeated application of a dc field during 60 min and 10 min of short-circuit. Previously, the lower half of the dielectric had been irradiated with an electron beam for 20 min, and then stored for one day, in both cases under a short-circuit condition.

Previously, the EFTE film had been irradiated with a high-energy electron beam, providing a source of excess electrons. The intensity of the beam was 20 nA/cm^2, the electron energy 60 keV, with a total irradiation time of 20 min under short-circuit conditions. The negative charges were observed to be accumulated in the EFTE film immediately after irradiation commenced. A similar procedure using separate samples of PE and PTFE did not show charge accumulation inside the sample during the first 20 min of electron beam irradiation (Nagasawa et al., 2010), which is a further indication that traps for excess electrons are much larger in EFTE than in PE or PTFE, as expected from the simple model described above.

7.7 Insight into Application of Computer Simulation to Nanocomposites

While direct experimental studies of existing and new materials are always required, they have to be complemented with a theoretical understanding of the underlying phenomena at the molecular level. Nowadays, computer simulation methods are important tools for gaining such theoretical insights. By reviewing recent results on simple linear polymers, we have illustrated the potential of these techniques for nanocomposites, showing both their strengths and limitations.

DFT is a very powerful method for the study electronic phenomena in new materials. Being widely available, it is the natural choice for the simulation of nanocomposites with an arbitrary chemical composition. However, it also still contains serious limitations in terms of accuracy and size.

Simulating systems with an appropriate minimum size is important for the adequate description of disordered phases in materials. As has been shown in PE, they might be responsible for important trapping mechanisms. On the other hand, one-electron simulations offer a way to explore larger systems than those that are available with DFT. In new composites, this method requires the development of one-electron potentials using ab initio techniques.

Nevertheless, it should be noted that none of the current simulation methods provides quantitative results with an accuracy that can be systematically controlled. Rather, it is important to contrast the predictions obtained using different techniques and, especially, to compare simulation results with experimental data, in order to check that the relevant qualitative trends are reproduced correctly.

Under many circumstances, a detailed quantitative analysis is not required and only a qualitative description is sufficient. However, the lack of quantitative accuracy in a simulation method sometimes leads to incorrect qualitative conclusions, as we have shown in this chapter. Consequently, checking with a different functional, method, or experiment is essential.

Acknowledgments

DC thanks the funding support from the State Key Laboratory of Electrical Insulation and Power Equipment at Xi'an Jiaotong University under the project EIPE13203.

References

Andre J. M. and Leroy G. (1970), All-electrons band structure of polyethylene in the nearest cell approximation, *Chem. Phys. Lett.*, **5**: 71–74.

Bealing C. R. and Ramprasad R. (2013), An atomistic description of the high-field degradation of dielectric polyethylene, *J. Chem. Phys.*, **139**: 174904.

Burke K. (2012), Perspective on density functional theory, *J. Chem. Phys.*, **136**: 150901.

Ceresoli D., Tosatti E., Scandolo S., Santoro G. and Serra S. (2004), Trapping of excitons at chemical defects in polyethylene, *J. Chem. Phys.*, **121**: 6478–6484.

Ceresoli D., Righi M. C., Tosatti E., Scandolo S., Santoro G. and Serra S. (2005), Exciton self-trapping in bulk polyethylene, *J. Phys. Condens. Matter.*, **17**: 4621–4627.

Cubero D. and Quirke N. (2004), Computer simulations of localized small polarons in amorphous polyethylene, *J. Chem. Phys.*, **120**: 7772–7778.

Cubero D., Quirke N. and Coker D. F. (2003a), Electronic transport in disordered n-alkanes: From fluid methane to amorphous polyethylene, *J. Chem. Phys.*, **119**: 2669–2679.

Cubero D., Quirke N. and Coker D. F. (2003b), Electronic states for excess electrons in polyethylene compared to long-chain alkanes, *Chem. Phys. Lett.*, **370**: 21–25.

Delhale J., Andre J. M., Delhale S., Pireaux J. J., Candano R. and Verbist J. J. (1974), Electronic structure of polyethylene: Theory and ESCA measurements, *J. Chem. Phys.*, **60**: 595–600.

Dion M., Rydberg H., Schroeder E., Langreth D. C., and Lundqvist B. I. (2005), *Phys. Rev. Lett.*, **95**: 109902.

Dissado L. A. and Fothergill J. C. (1992), *Electrical Degradation and Breakdown in Polymers* (Peter Peregrinus, London).

Eckhardt W. and Heinecke A. (2013), SuperMUC boosts the largest molecular dynamics simulation by 4X in Number of Particles, *inSiDE*, 11(1).

Huzayyin A., Boggs S. and Ramprasad R. (2010), Density functional analysis of chemical impurities in dielectric polyethylene, *IEEE Trans. Dielectr. Electr. Insul.*, **17**: 926–930.

Jain M., Chelikowsky J. R., and Louie S. G. (2011), Reliability of hybrid functionals in predicting band gaps, *Phys. Rev. Lett.*, **107**: 216806 (2011).

Kleis J. et al. (2007), Towards a working density-functional theory for polymers: First-principles determination of the polyethylene crystal structure, *Phys. Rev. B*, **76**: 100201.

Le Roy S., Teyssedre G., and Laurent C. (2005), Charge transport and dissipative processes in insulating polymers: Experiments and model, *IEEE Trans. Dielectr. Electr. Insul.*, **12**: 644–654.

Meunier M. and Quirke N. (2000), Molecular modeling of electron trapping in polymer insulators, *J. Chem. Phys.*, 113: 369–375.

Mizutani T. (2006), Behavior of charge carriers in organic insulating materials, *Annual Rept. IEEE-CEIDP*, Whitehead Memorial Lecture, 1–10.

Mullin N. and Hobbs J. K. (2011), Direct imaging of polyethylene films at single-chain resolution with torsional tapping atomic force microscopy, *Phys. Rev. Lett.*, **107**: 197801.

Nagasawa, K. Honjoh, M. Miyake, H. Watanabe, R. Tanaka, Y. and Takada, T. (2010), Charge accumulation in various electron-beam-irradiated polymers, *IEEJ Trans. IEE*, **5**(4): 412–415.

Paier J. et al (2006), Screened hybrid density functionals applied to solids, *J. Chem. Phys.*, **124**: 154709.

Righi M. C., Scandolo S., Serra S., Iarlori S., Tosatti E. and Santoro G. (2001), Surface states and negative electron affinity in polyethylene, *Phys. Rev. Lett.*, **87**: 076802.

Serra S., Iarlori S., Tosatti E., Scandolo S., and Santoro G. E. (2002), Self-trapping vs. non-trapping of electrons and holes in organic insulators: Polyethylene, *Chem. Phys. Lett.*, **360**: 487–493.

Takada, T. (1999), Acoustic and optical methods for measuring electric charge distribution in dielectrics, *IEEE Trans. DEI*, **6**: 519–547.

Tanaka, T. (1973), Optical absorption and electrical conduction in polyethylene, *J. Appl. Phys.*, **44**(5): 2430–2432.

Turi L. and Rossky P. J. (2012), Theoretical studies of spectroscopy and dynamics of hydrated electrons, *Chem. Rev.*, **112**: 5641–5674.

Ueno N., Sugita K., Seki K. and Inokuchi H. (1986), Low-energy electron transmission and secondary-electron emission experiments on crystalline and molten long-chain alkanes, *Phys. Rev. B*, **34**: 6386–6393.

Unge M. and Christen T. (2014), Electron and hole mobility edges in polyethylene from material simulations, *Chem. Phys. Lett.*, **613**: 15–18.

Unge M., Christen T., and Tornkvist C. (2012), Electronic structure of polyethylene: Crystalline and amorphous phases of pure polyethylene and their interfaces, *Annual Rept. IEEE-CEIDP*, 525–530.

Wang Y., MacKernan D., Cubero D., Coker D. F., and Quirke N. (2014), Single electron states in polythylene, *J. Chem. Phys.*, **140**: 154902.

Wang Y., Wu K., and Cubero D. (2014b), Excess electron states and mobility in polyethylene, *Proc. 2014 Inter. Conf. ISEIM*, 393–396.

Part 3

Properties for Dielectric Applications

Chapter 8

Electrical Properties of Polymer Nanocomposites

Alex Tsekmes,[a] Peter Morshuis,[b] and Gary C. Stevens[c]

[a]*Delft University of Technology, Mekelweg 4, 2628 CD Delft, the Netherlands*
[b]*Solid Dielectric Solutions, Herensteeg 17, Leiden, 2311 SG, the Netherlands*
[c]*Gnosys Global Ltd, Frederick Sanger Road,*
The Surrey Research Park, Guildford, Surrey GU2 7YD, UK

peter.morshuis@dielectrics.nl, g.stevens@gnosysgroup.com

8.1 Introduction

Nanodielectrics are promising insulating materials for high-voltage applications. Since the introduction of the term in 1994 [1], significant improvements have been reported of the electrical and dielectric properties of polymers when nanosized fillers are introduced. Nowadays, a tremendous amount of data is available in the literature regarding the electrical/dielectric behavior of polymers reinforced with various types of conducting and non-conducting nanofillers. Nanofillers may have many morphologies and be of different chemical nature; the only requirement is that at least one dimension should be smaller than 100 nm. In this review, we concentrate on non-conducting dielectric

Tailoring of Nanocomposite Dielectrics: From Fundamentals to Devices and Applications
Edited by Toshikatsu Tanaka and Alun S. Vaughan

ISBN 978-981-4669-80-1 (Hardcover), 978-981-4669-81-8 (eBook)
www.panstanford.com

nanoparticles that are also said to form polymer nanocomposites and which may be referred to as nanodielectrics.

One of the main issues in the field of nanodielectrics, in common with nanocomposites in general, is the reproducibility of the experimental results. In many studies, improved electrical/dielectric performance is achieved in nanocomposites compared to the equivalent neat polymers and this has been widely reported. However, adverse or no effects have been noted as well. To what extent are experimental results reproducible and which parameters affect the performance of nanocomposites and in turn the reproducibility of the results?

Many phenomena such as dielectric response, conduction mechanisms, charge trapping and space charge accumulation, occurring in polymers are not yet completely understood, let alone those which occur in polymeric nanocomposites. The large number of involved parameters, including particle properties, sample morphology, and the "ill-defined" interaction between the nanofillers and the host polymer make a deep understanding of their behavior challenging.

Despite the abundance of experimental results and theories trying to explain these results, it appears that only a limited amount of undisputed findings are available. Therefore, in this chapter, we concentrate on the relative permittivity, DC-conductivity and space charge behavior of polymeric nanocomposites. The main goal is to evaluate the existence of trends and the role of interfaces, as interfacial effects are one of the most important factors that influence the performance of nanodielectrics. In this case, the very large specific internal surface area that nanofillers produce in nanocomposites results in a material whose bulk properties are, primarily, controlled by its internal interfaces and the interfacial region close to the nanoparticles. John Lewis stated that nanodielectrics are synonymous with interfaces [2].

8.2 Relative Permittivity

8.2.1 Anomalous Behavior

The relative permittivity of nanocomposites is a thoroughly investigated property, which provides convincing evidence that structural changes of polymers take place in the vicinity of nanofillers. The behavior of the relative permittivity of nanocomposites is often described in the literature as "anomalous", "unusual" or "unexpected". This refers to the fact that nanocomposites often exhibit a lower relative permittivity than neat materials, which cannot be explained using the conventional rules of composite mixing, given that nanofillers usually have a higher permittivity than the neat polymer host materials. Andritsch [3] used a number of different approaches to fit the experimental results of the relative permittivity of epoxy-nanocomposites without succeeding. It was proposed that the "anomalous" behavior of nanocomposites arises from the interfaces and interfacial contributions that are currently poorly understood.

Many experimental data verify that nanocomposites exhibit lower permittivity values than the base materials. Roy et al. investigated the dielectric properties of cross-linked polyethylene (XLPE) reinforced with 5 wt% of nanosilica (SiO_2) [4]. Nanocomposites either with surface-treated or untreated nanoparticles exhibit lower permittivity values than the base material over a broad frequency range. The authors attributed this behavior to reduced polymer chain mobility as a result of nanoparticle incorporation. A decrease in relative permittivity of nanocomposites based on low-density polyethylene (LDPE) has also been reported by Ciuprina et al. [5]. Nanocomposites containing 2 wt% of titanium dioxide (average particle size of 15 nm) showed a lower permittivity than the base polymer. In this case, the nanoparticles were modified with maleinized polyethylene.

Krivda et al. [6] showed that the addition of 2.5 wt% of nanosilica (average particle size of 25 nm) to epoxy decreases its relative permittivity over the whole frequency range studied, i.e., 10^{-3} to 10^5 Hz (Fig. 8.1). Iyer et al. [7] also demonstrated that the addition of nanosized silica (average particle size of 20 nm) decreases the relative permittivity of epoxy. In this example, nanocomposites with 2.5 wt% filler concentration exhibited lower permittivity values than neat epoxy in the frequency range of 10^{-3} to 10^5 Hz.

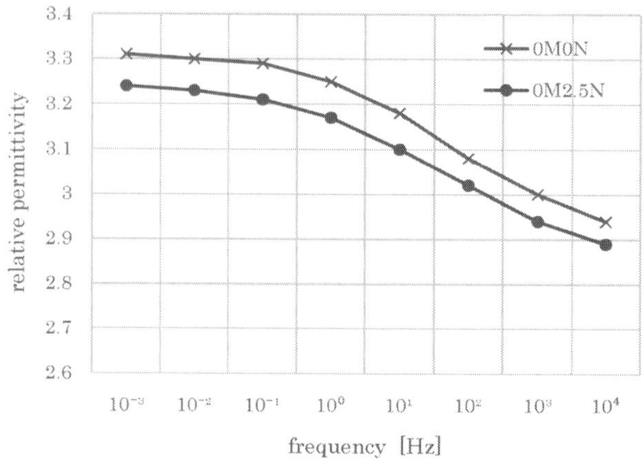

Figure 8.1 Relative permittivity of epoxy nanocomposite with 2.5 wt% of SiO_2 (0M2.5N) and neat epoxy (0M0N) as a function of frequency. After [6].

8.2.2 The Role of Interfaces

The high surface-to-volume ratio of nanofillers results in an enormous interfacial area, which inevitably plays a significant role in the behavior and performance of nanodielectrics. The significant contribution of interfaces has also been experimentally verified. Schuman et al. [8] showed that the relative permittivity of epoxy-titanium dioxide nanocomposites significantly decreases with the use of suitable surfactants. They treated titanium dioxide nanofillers (average particle size of 32 nm) in different ways and demonstrated the importance of adhesion between the filler and the polymer regarding the relative permittivity of the resulting nanocomposites. Chemical bonding between fillers and matrix improves their compatibility and adhesion, resulting in a more effective re-organization of the polymer matrix in the interfacial region [9].

Since interfaces determine to a great extent the behavior of nanodielectrics, the surface properties of the fillers are of major importance and critical to the performance of polymeric nanocomposites. At the nanoscale, surface properties of fillers become more important than bulk properties. Hydrophilicity is one of the main issues that should be taken into account [10].

When nanofillers have a hydrophilic nature, due to their high surface-to-volume ratio, they result in significant water adsorption. The use of suitable surfactants has multiple advantages as they may influence the degree of hydrophilicity and hydrophobicity [11, 12] and increase the adhesion between the host polymer matrix and fillers via chemical bonding. This will tend to reduce particle-to-particle interactions [13] and positively contribute to the degree of dispersion and spatial distribution of particles by reducing the difference in surface tension between organic polymer matrices and inorganic fillers [14].

The importance of surface properties makes the comparison between other parameters such as size and shape, extremely difficult, especially when the relative permittivity is investigated. The main reason is that different types of fillers do not only have different bulk properties and geometrical characteristics but also different surface properties. However, it is commonly observed that there is a consistent trend mainly related to the filler concentration rather than other properties such as size and shape. Thus, regardless of the chemical nature and morphology of the particles, nanocomposites often exhibit a lower relative permittivity than the base material, particularly at low filler concentrations. As the concentration of nanofiller increases, the relative permittivity of the samples increases as well.

8.2.3 Influence of the Filler Type and Fill Grade

The dependence of the relative permittivity on the fill grade has been demonstrated by a number of researchers. Andritsch [3] investigated the dielectric response of epoxy filled with aluminum oxide (average particle size of 25 nm), aluminum nitride (average particle size of 60 nm), and magnesium oxide (average particle size of 22 nm). The nanofillers were treated with a silane coupling agent. He found that in all three cases, epoxy nanocomposites show a consistent trend, which mainly depends on the filler concentration. A characteristic example is presented in Fig. 8.2, where the relative permittivity of epoxy-magnesium oxide nanocomposites is shown at concentrations from 1.5% to 10%. Note, however, that the lowest permittivity occurs at 2% and then increases progressively from 5% to 10%.

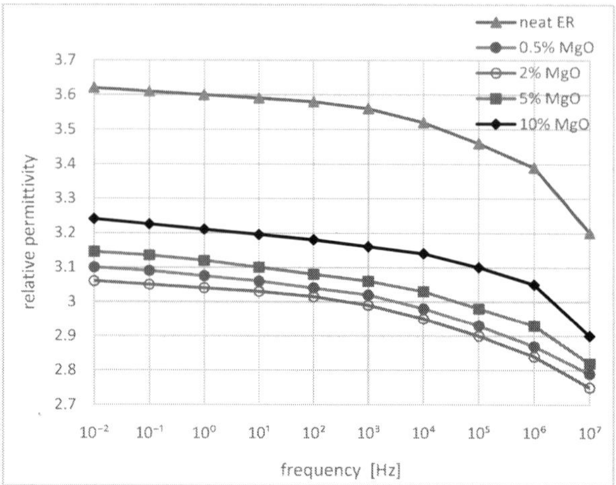

Figure 8.2 Relative permittivity of epoxy-magnesium oxide nanocomposites at 20 °C compared to neat epoxy as a function of frequency [3].

Kochetov [15] also studied the relative permittivity of epoxy nanocomposites with modified fillers and he found a similar trend, which depends on the fill grade. Therefore, he showed that epoxy-aluminum oxide (average particle size of 30 nm), epoxy-aluminum nitride (average particle size of 60 nm) and epoxy-magnesium oxide (average particle size of 22 nm) exhibit a critical filler concentration at which the relative permittivity is lowest. Above this concentration, the relative permittivity of the nanocomposites starts increasing with the concentration. These findings are summarized in Fig. 8.3.

Singha and Thomas [16] investigated the relative permittivity of epoxy-zinc oxide (average particle size of 45–70 nm) and epoxy-titanium dioxide (average particle size of 50 nm) nanocomposites. The same trend was also found in this research. Thus, at low filler concentrations, the nanocomposites have a lower relative permittivity than the neat material. However, as the filler content increases, the relative permittivity increases as well, surpassing that of the unfilled epoxy matrix. These results are presented in Fig. 8.4. Tsekmes et al. [17] found the same trend for epoxy-hexagonal boron nitride (average particle size of 70 nm) and epoxy-cubic boron nitride composites (average particle size of 150 nm). These results are included in Fig. 8.3.

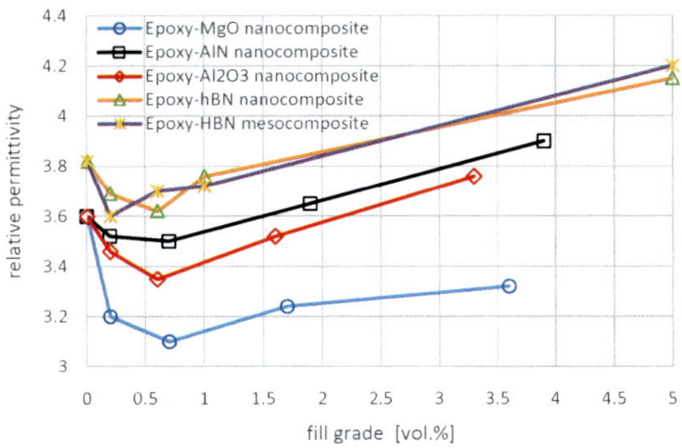

Figure 8.3 Relative permittivity of epoxy nanocomposites as a function of nanofiller concentration (fill grade) @1 Hz and 20 °C [17].

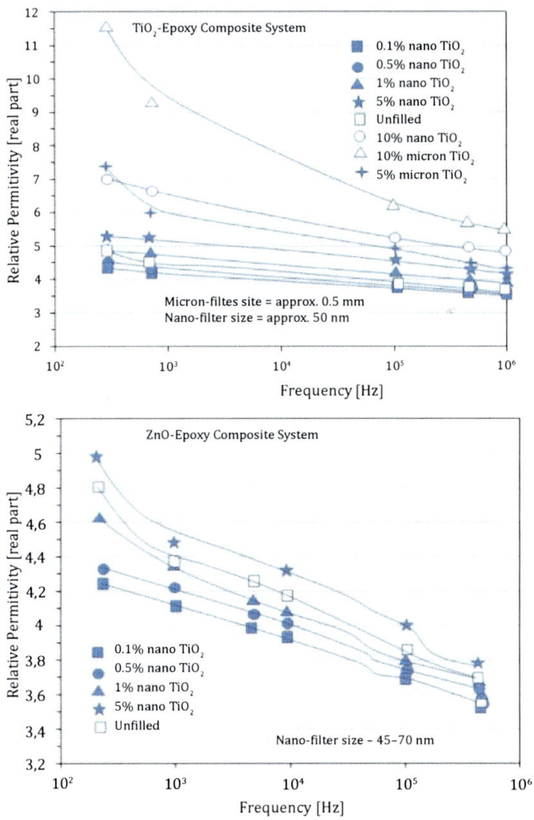

Figure 8.4 Relative permittivity of epoxy-zinc oxide and epoxy-titanium dioxide nanocomposites. After [16].

Ishimoto et al. [18] investigated the dielectric response of LDPE reinforced with nanosized magnesium oxide particles (several tens of nm) and they found that an increase in nanofiller concentration results in a higher relative permittivity. Finally, the dielectric response of poly(vinylidene fluoride) (PVDF)-barium titanate nanocomposites was investigated by Yu et al. [19]. The particles had an average particle size of 100 nm and they were surface treated. The relative permittivity of the produced composites exhibits higher values when the filler concentration increases.

8.2.4 Parameters Affecting the Relative Permittivity of Nanocomposites

It has been shown so far that a polymeric nanocomposite can exhibit a lower relative permittivity than the base material. This occurs at relatively low nanofiller concentrations and it is often attributed to interactions between the polymer matrix and filler, which change the polymeric structure and chain dynamics locally by restraining the mobility of the polymer chains in the vicinity of filler surfaces [3, 4, 15, 20]. On the other hand, it was found by different research groups that the relative permittivity may then increase with a further increase of the filler concentration.

What can cause the increase of the relative permittivity? A potential mechanism can be related to the nanofiller permittivity. As the filler concentration increases, the permittivity of filler starts playing a more important role, overshadowing the decrease that polymer re-organization can cause [3]. However, in many cases the high permittivity values cannot be explained based only on the higher relative permittivity of the filler. Another potential mechanism is operative and this is likely to be related to the presence of water.

Zhang and Stevens [21] suggested that in both polar and non-polar nanodielectrics, absorbed water is located at the polymer-nanoparticle surface. Tsekmes et al. showed that water uptake positively depends on the filler concentration by means of thermogravimetric analysis [22]. A critical issue to be considered is the extent to which water uptake affects the dielectric response of nanocomposites.

Lau et al. [11] investigated the effect of water uptake on the dielectric response of LDPE-silica nanocomposites. The average particle size ranged from 10 to 20 nm. The samples were immersed in distilled water and it was found that water uptake significantly influences the relative permittivity over the whole frequency spectrum studied, i.e., 10^{-1} to 10^{5} Hz (Fig. 8.5).

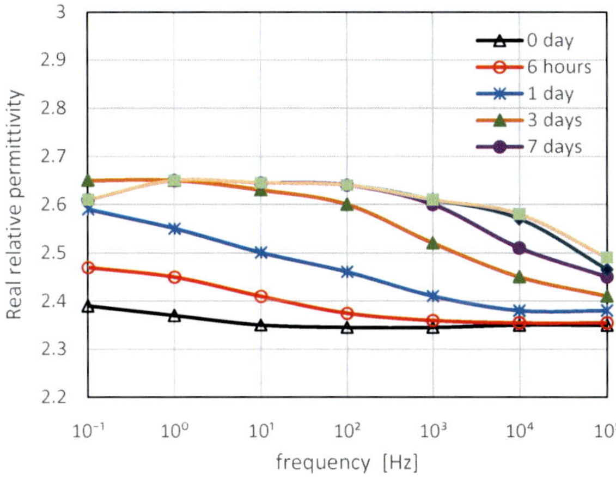

Figure 8.5 Relative permittivity of nanocomposites with 2 wt% of untreated nanosilica which were immersed in water for different time periods. After [11].

Kochetov [15] investigated the effect of thermal treatment on the dielectric response of epoxy nanocomposites. It was found that thermal treatment of epoxy-magnesium oxide nanocomposites at 160 °C for two days leads to a significant reduction in the relative permittivity and dielectric losses within the frequency range of 10^{-2} to 10^{7} Hz. The decrease was attributed to the evaporation of water and further polymerization of the sample.

Tsekmes et al. [22] found that epoxy nanocomposites, after a long period of vacuum storage, exhibit lower permittivity values and dielectric losses over a wide frequency spectrum, ranging from 10^{-2} to 10^{6} Hz. The effect was attributed to the removal of water.

Summarizing the results obtained by different researchers, it can be stated that the inclusion of nanofillers within polymers affects their structure and sometimes leads to a reduction in the relative permittivity. Decreased permittivity values are usually observed at relatively low filler concentrations, i.e., ≤5 wt%. At higher filler concentrations, polymeric composites often exhibit higher permittivity values than the base material.

It has also been shown that water uptake has a dramatic influence on the relative permittivity of nanocomposites. Attention is therefore required to account for absorbed water when comparing the relative permittivity of nanocomposites with that of other composites or neat polymers. Increased permittivity values and high dielectric losses at low frequencies are often attributed to low frequency dispersion [23] or quasi-DC [24] phenomena. However, the high permittivity values and dielectric losses are most often caused by a significant amount of water absorption, which significantly affects the dielectric response and obscures the relaxation peaks. Both the real and imaginary part of the complex permittivity decrease when the samples are relatively "dry". A slight increase of dielectric loss at low frequencies is expected due to interfacial polarization. This increase greatly depends on the nanofiller concentration, i.e., a higher filler concentration leads to a larger interfacial area and usually to higher dielectric losses [3, 19, 22].

8.3 DC-Conductivity

The electrical conductivity of polymeric nanocomposites is reviewed in an attempt to evaluate the existence of consistent trends and behavior in a similar way to the approach taken in addressing the relative permittivity of polymeric nanocomposites. A thorough analysis of the potential conduction mechanisms that can take place in polymeric nanocomposites can be found in the review by Nelson [25]. The intention here is to assess the consistency of the results found in the literature and the extent to which interfaces contribute to the electrical conductivity of nanocomposites. Most studies report an increase of the electrical conductivity when nanofillers are introduced to a polymer. However, nanocomposites with lower conductivity values than the base polymer have also been reported. It is worth noting that the intrinsic electrical conductivity of the nanofillers may be higher than that of the base polymer but it is also possible that ionic impurities on the surface or in the bulk of the nanofiller may also contribute to the behavior of the nanocomposite and it is clear from the literature that assessment of impurities is not commonly undertaken.

Choudhury et al. [26] found that the addition of aluminum nitride particles, with a size smaller than 100 nm, to epoxy leads to higher conductivity values. In the same study, the authors evaluated the influence of surface treatment of the filler on the DC-conductivity. They found that nanocomposites with treated particles exhibited lower conductivity values than nanocomposites containing untreated particles, but still, the values were higher than the neat material. The authors attributed the improved performance of treated nanofillers to good interfacial adhesion between the polymer and the nanoparticles.

Singha and Thomas [16] investigated three different types of epoxy nanocomposites. All of them showed higher DC-conductivity values than neat epoxy. More specifically, epoxy-titanium dioxide with a particle size of 50 nm, epoxy-aluminum oxide with a particle size of 45 nm and epoxy-zirconium oxide with a particle size of 45–70 nm were tested. Nanocomposites with different filler concentrations were produced, namely 0.1, 0.5, 1, and 5 wt% and all of them exhibited a higher conductivity than the neat material. Andritsch et al. [27] also found a higher DC-conductivity than the base material for three different types of epoxy nanocomposites, namely epoxy-aluminum oxide, epoxy-aluminum nitride, and epoxy-magnesium oxide. The nanofillers were surface treated, with an average particle size of 25, 60, and 22 nm, respectively.

Wang et al. [28], produced nanocomposites based on LDPE. The polymer was reinforced with treated silica particles with an average particle size of nearly 30 nm. The authors reported that the DC-conductivity depends on the filler concentration. Thus, at concentrations lower than 1 wt%, the nanocomposites exhibited a lower DC-conductivity than the base material while, at higher filler concentrations, the conductivity of the nanocomposites surpassed that of the neat LDPE.

Ishimoto et al. [18] reported that the electrical conductivity of LDPE-magnesium oxide nanocomposites is lower than that of the neat material and almost independent of the filler concentration. They produced nanocomposites with untreated magnesium oxide nanoparticles with an average particle size of several tens of nanometer and all them had lower conductivity values compared to the base material. The conductivity of the samples was measured at three temperatures, i.e., 50, 70, and 90 °C at 25 kV/mm. A quasi steady state value was considered after 50 min of voltage application. Fleming et al. [29] produced nanocomposites with nanosized zinc oxide particles and dispersants. They found that the conductivity of the nanocomposites decreases by 1–2 orders of magnitude compared to neat LDPE.

Kozako et al. [30] investigated the electrical conductivity of epoxy-based systems containing microsized particles coated with nano-aluminum hydroxide ($Al(OH)_3$—also referred to as aluminum trihydroxide). They found that the application of the microparticles significantly decreased (approximately 7 orders of magnitude) the DC-conductivity of the epoxy composites. These results clearly demonstrate that the nature of interfaces plays a major role in the electrical conduction of polymeric composites.

Finally, Krivda et al. [6] found that the addition of 5 wt% of nanosized silica to a heavily loaded microcomposite based on epoxy (65 wt% of microsized silica) resulted in a higher DC-conductivity. The microsized silica particles had undergone an epoxy-silane treatment while there was an unknown treatment for the nanosized silica particles. The results were consistent at a number of electric field levels.

It seems to be difficult to determine a specific trend that governs the electrical conductivity of nanocomposites. However, it is important to consider that comparison of different published studies requires significant attention to the detail of both the materials, the measurement methods used and the conditions of measurement. Parameters such as temperature, electric field, polarization time, and glass transition temperature (relative to the measurement temperature) play a critical role in the determination of the electrical conductivity. Very few researchers use the same methods and conditions for determining the electrical conductivity of nanocomposites and few account for impurities and the presence of water; so caution must be exercised when comparing results. Being aware of all the difficulties and uncertainties, it can be stated that, usually, the addition of nanofillers increases the DC-conductivity of polymers, and this appears to be the case for both thermoplastic materials with a low glass transition temperatures where charge transport is primarily controlled by a mobile amorphous phase, as well as thermosetting resins with high glass transition temperatures.

The increase in DC-conductivity might be anticipated due to the higher intrinsic electrical conductivity of some nanofillers, but since the low concentrations at which nanofillers are commonly used fall well below the percolation threshold, the polymer matrix should still dominate nanocomposite conductivity. This is in contrast to nanofillers that may contain labile impurities at the surface or in the bulk of the nanoparticles, which may gain access to the polymer matrix and increase the bulk conductivity of the polymer—in this case, we would expect to see a conductivity that would increase with increasing nanofiller concentration. In contrast, surface treatment of nanofillers results in lower conductivity values, reflecting the potential importance of adhesion between the nanofiller and matrix for the behavior of nanocomposites.

Recent work by Stevens and Vaughan [31] on surface functionalized nanosilicas in epoxy resin matrices with low intrinsic conductivity of both the polymer and the nanosilicas has shown that nanocomposite electrical conductivities can be lower than that of the unfilled polymer. This result, obtained on highly dispersed nanophase nanocomposites, would be expected from interfacial trapping of mobile charge, which effectively lowers the effective charge carrier concentration and mobility in the nanocomposite.

As far as the role of water uptake is concerned, it is expected to contribute slightly to the electrical conductivity of nanocomposites, particularly in relatively "dry" samples. Water ions do not necessarily lead to ionic conduction [32]. Even in the case that water is assumed to lead to ionic conduction, this may not be the dominant conduction mechanism [32]. For sufficiently high applied electric fields and temperatures, water uptake can indirectly influence the DC-conductivity of nanocomposites by leading to space charge formation, field distortion and changes to the injection/extraction processes at the electrodes. Within the Ohmic region, however, the role of water uptake in the DC-conductivity is expected to be negligible.

Finally, an important parameter that significantly affects the electrical conductivity of nanocomposites and often is not taken into account is sample homogeneity. Particle dispersion and distribution are often used to describe the morphology of nanocomposites. The former refers to the separation of particles from each other while the latter is used to describe the way that the nanofillers are incorporated in the polymer matrix. So far, there is no objective way of determining the degree of homogeneity in nanocomposites. Researchers usually use a visual inspection or some further analysis with the help of algorithms. However, regardless of the means of analysis, the evaluation of particle dispersion and distribution is often based on a limited number of TEM or SEM images. It is of great importance to develop an objective way to characterize the morphology of nanocomposites. In this case, correlation between different studies will be more straightforward, giving us the chance to understand more deeply the mechanisms that dictate conduction processes within polymeric nanocomposites.

8.4 Space Charge

LDPE and its cross-linked analogue (XLPE) are widely used as insulation compounds for cables operated at ac voltage. Under high dc voltages, however, LDPE is known for its tendency to accumulate large amounts of space charge, which may result in severe ageing and degradation [33, 34]. In general, it is reported that the introduction of nanoparticles into polyethylene can significantly reduce space charge accumulation [34–42]. Although a considerable amount of literature is now available on the results of space charge measurements, the precise mechanisms of how space charge dynamics are affected by the nanophase in nanocomposites are still debated. The space charge behavior is reviewed here in an attempt to summarize those aspects on which some common agreement has been achieved.

8.4.1 General Findings

While a variety of different nanoparticles have been used, many researchers focus on inorganic materials and the metal oxides in particular, such as magnesium oxide, MgO, and silica, SiO_2.

Terashima et al. [43] reported on the successful development of a 250 kV DC cable incorporating inorganic particles in the XLPE compound. They showed that, in the composite insulation system, space charge accumulation was strongly suppressed, resulting in improved breakdown strength. No details were mentioned apart from the fact that small weight fractions of about 1% were used.

In 2006, Hayase et al. [35] published the results of space charge measurements on LDPE-MgO thin films under ultra-high poling fields, i.e., 50–250 kV/mm. Plain, unfilled LDPE specimens were used as a reference. It was found for the reference specimens that, at poling fields above 150 kV/mm, a positive charge packet was injected at the anode, subsequently moving deep into the bulk of the specimen. For the 0.5 and 1 wt% (weight percentage) nanocomposites, hardly any space charge accumulation was observed (Fig. 8.6).

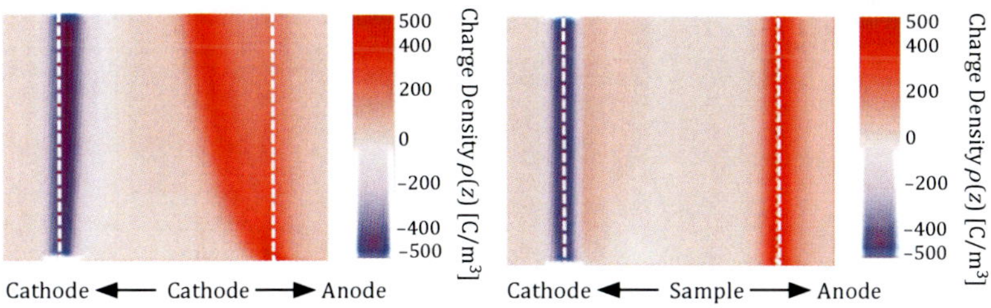

Figure 8.6 Space charge density profile; horizontal axis: position in the test specimen, vertical axis: time. Left: Space charge profile for unfilled LDPE, 250 kV/mm poling field. Right: Space charge profile for 1 wt% MgO LDPE nanocomposite, 250 kV/mm. After [35].

In [44], an attempt was made to correlate the non-existence of positive space charge packets to a reduction in the high field conductivity brought about by the inclusion of MgO nanoparticles. It was found that the conductivity of the LDPE-MgO composite was up to 2 orders of magnitude lower than for the reference LDPE specimens, the difference being larger at 50 °C than at room temperature.

In [45] and [46], the effect of MgO particle size on the space charge behavior in LDPE-MgO was studied. Nanoparticles with an average diameter of 50, 500, and 1000 nm were used for creating 0.5, 1, and 4 wt% XLPE-MgO composites in [45]. The maximum reduction of space charge accumulation was obtained for the composite with 1–4 wt% of 50 nm particles.

In Fig. 8.7, the calculated maximum field enhancement is plotted for the reference XLPE specimen and for the 1 wt% composites containing nanoparticle of different size. For an applied electric field of 80 kV/mm, the field enhancement is smallest for the XLPE with 50 nm particles, i.e., a maximum electric field was calculated of 93.4 kV/mm compared to 137.2 kV/mm for the reference XLPE.

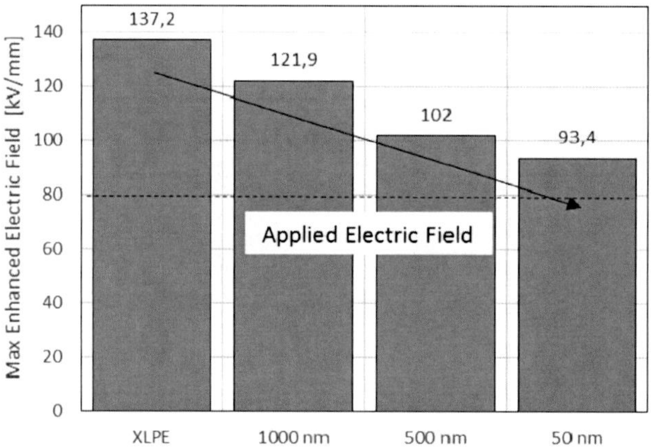

Figure 8.7 Calculated maximum enhanced electric field for different MgO/XLPE composites (1 wt%). After [45].

These results are confirmed in [46], where space charge results obtained from LDPE-based nanocomposites containing 30, 100, and 500 nm MgO particles are presented. The composite with 1 wt% of 30 nm particles exhibited the largest space charge reduction.

Huang et al. [47] studied the effect of surface modification of SiO_2 on space charge suppression in LDPE-SiO_2 composites. In their study, the nanocomposites containing untreated SiO_2 showed a significant amount of heterocharge accumulation, whereas the specimens based upon surface-modified SiO_2 showed a stark reduction in the accumulation of space charge. The TEM images presented in this paper show agglomerates in the specimens containing untreated SiO_2, whereas the surface-treated SiO_2 appears to be much better dispersed.

Another approach to study the effect of surface modification was taken by Wei et al. [48], who improved the SiO_2/epoxy resin host interaction using plasma polymerization to apply a polyethylene oxide coating onto the nanoparticles. This surface modification led to a mitigated space charge build-up.

8.4.2 Mechanisms of Space Charge Suppression

A common feature of most approaches to explaining space charge suppression in nanocomposites is the assumption that a high density of deep traps is introduced by the nanoparticles. Some papers report the results of measurements made to determine the trap distribution in polyethylene nanocomposites. In [39], isothermal discharge current measurements and space charge measurements were made on LDPE and on LDPE-ZnO nanocomposites. It was shown that the addition of 7 wt% ZnO significantly increased the deep trap level density (Fig. 8.8), leading to a strong suppression of space charge accumulation.

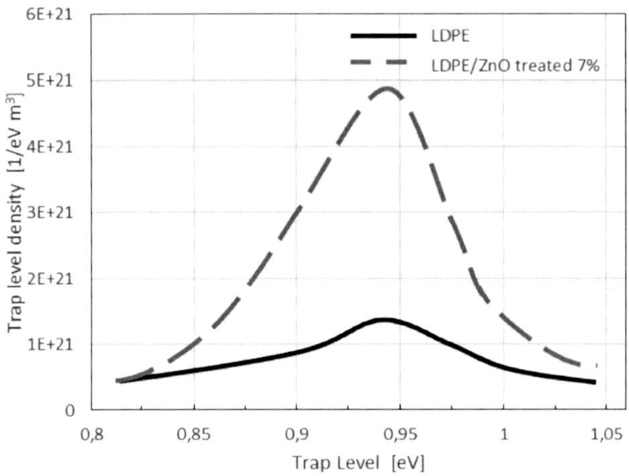

Figure 8.8 Trap level distribution LDPE and LDPE/ZnO (7 wt%, treated). After [39].

In [40], composites of LDPE with a Na exchanged Y-zeolite (so-called NaY zeolite) nanoparticles were studied. The space charge accumulation in the composite was strongly suppressed, which the authors attributed to the existence of deep traps in the porous surface of the zeolite. Thermally stimulated current measurements were performed to show a sharp increase of the deep trap level density, shown in Fig. 8.9, similar to that reported in [39].

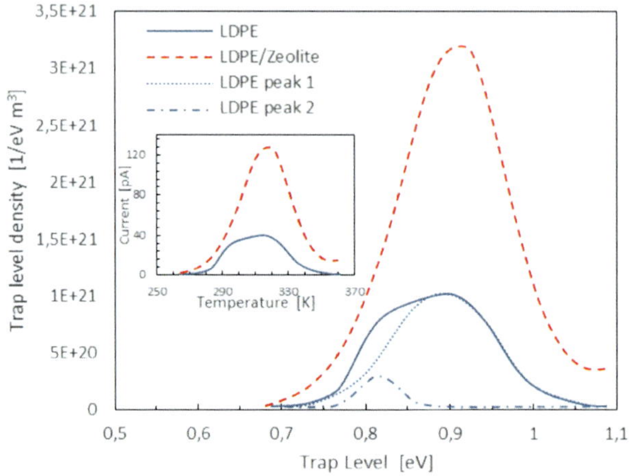

Figure 8.9 Trap level distribution in LDPE and LDPE/zeolite 3% doped; the inset provides the thermally stimulated current spectrum. After [40].

For the explanation of the suppression of space charge in polyethylene nanocomposites, different approaches have been proposed. One approach focuses on the reduction of charge injection by deep trapping of a thin layer of homocharge close to the electrodes [49–51], while another approach assumes a modification of the bulk nanocomposite that affects charge migration and recombination [36, 52]. Finally, some authors suggest that both approaches are of importance [39, 40].

In [49], the authors made an attempt at modeling the suppression of space charge accumulation in LDPE-MgO nanocomposites by assuming a bipolar transport model. They showed that, for a concentration of 5 wt% MgO, the concentration of deep traps increased significantly. This, in return, would lead to a large amount of homocharge just in front of the electrodes, not detectable by PEA space charge measurements due to the limited spatial resolution. Assuming a Schottky-type injection law, the reduction of the electric field at the electrode-dielectric interface would suppress charge injection.

The reduction of charge injection is consistent with the results of experimental work showing that surface modification of LDPE samples can significantly reduce space charge accumulation. In [53], it is shown that surface oxyfluorination of LDPE leads to a significant suppression of space charge injection without changing the bulk properties of the LDPE specimens. Also, in this case, the behavior was explained by the reduction of the electric field at the electrode-dielectric interface. An alternative approach was made in [51], where a barrier layer of silver nanoparticles embedded in an organosilicon matrix was used between the electrode and LDPE. A strong reduction of space charge accumulation in the LDPE specimen was realized through the creation of deep trapping sites at the silver nanoparticles.

To explain the findings in [35], Takada et al. [52] proposed a model of deep trapping introduced by the MgO nanoparticles. The trapping depth of chemical defects in LDPE, such as carbonyl groups (0.45 eV), was considered insufficient to explain the trapping of positive charge carriers close to the anode-dielectric interface. They attributed deep trapping centers to the induced dipole moments of the high permittivity nanoparticles.Trap depths between 1 and 5 eV were calculated for MgO nanoparticles subjected to high electric fields. Recombination of charge carriers of opposite polarity was also considered, with the nanoparticles acting as recombination centers.

An attempt to investigate the relative importance of both approaches (interface and bulk) was made in [50, 54]. To separate interface and bulk effects, test specimens were created consisting of three layers, with either a layer of SiO_2 filled LDPE at both electrode interfaces and a layer of unfilled LDPE in between or a layer of SiO_2 filled LDPE between two layers of unfilled LDPE. The authors concluded that the nanoparticles do reduce charge injection at the electrode/dielectric interface, but also that changes in the insulation bulk are required to explain the reduction of space charge. It was shown that under isothermal temperature conditions, a band theory model can explain the results obtained for these multi-layer specimens if it is assumed that the introduction of the nanoparticles results in deep traps that move the Fermi level closer to the transport states.

In conclusion, current literature points to charge accumulation being strongly suppressed if:

- an interface layer containing a substantial amount of deep traps is introduced between the electrode and dielectric; such a layer can be created by the introduction of nanoparticles (the smaller the nanoparticles at a fixed concentration—the larger the internal surface area, the larger the effect);

- the nanoparticles are appropriately functionalized to make them compatible with the polymer host to ensure an electrically intimate interface.

Further studies on surface-modified nanofillers in nanocomposites with known internal surface area for varying degrees of dispersion will enable a more complete assessment of the mechanisms of space charge suppression.

References

1. Lewis, T. J. (1994). Nanometric dielectrics, *IEEE Trans. Dielectr. Electr. Insul.*, **1**, pp. 812–825.
2. Lewis, T. J. (2004). Interfaces and nanodielectrics are synonymous, *2004 International Conference on Solid Dielectrics*, IEEE, pp. 792–795.
3. Andritsch T. (2010). Epoxy based nanocomposites for high voltage DC applications, synthesis, dielectric properties and space charge dynamics, PhD thesis, Delft University of Technology.
4. Roy, M., Nelson, J. K., MacCrone, R. K., Schadler, L. S., Reed, C. W., Keefe, R. (2005). Polymer nanocomposite dielectrics: the role of the interface, *IEEE Trans. Dielectr. Electr. Insul.*, **12**, pp. 629–643.
5. Ciuprina, F., Plesa, I., Nothingher, P. V., Tudorache, T., Panaitescu, D. (2008). Dielectric properties of nanodielectrics with inorganic fillers, *2008 Ann. Rep. CEIDP*, IEEE, pp. 682–685.
6. Krivda, A., Tanaka, T., Fréchette, M., Castellon, J., Fabiani, D., Montanari, G. C., Gorur, R., Morshuis, P. H. F., Gubanski, S., Kindersberger, J., Vaughan, A., Pelissou, S., Tanaka, Y., Schmidt, L. E., Iyer, G., Andritsch, T., Seiler, J., Anglhuber, M. (2012). Characterization of epoxy microcomposite and nanocomposite materials for power engineering applications, *IEEE Electr. Insul. Mag.*, **28**, pp. 38–51.
7. Iyer, G., Gorur, R., Richert, R., Krivda, A., Schmidt, L. E. (2011). Dielectric properties of epoxy based nanocomposites for high voltage insulation, *IEEE Trans. Dielectr. Electr. Insul.*, **18**, pp. 659–666.
8. Schuman, T. P., Siddabattuni, S., Cox, O., Dogan, F. (2012). Improved dielectric breakdown strength of covalently-bonded interface polymer-particle nanocomposites, *Composite Interfaces*, **17**, pp. 719–731.
9. Andritsch, T., Kochetov, R., Morshuis, P. H. F., Smit, J. J. (2011). Proposal of the polymer chain alignment model, *2011 Ann. Rep. CEIDP*, IEEE, pp. 624–627.

10. Zou, C., Fothergill, J. C., Rowe, S. W. (2008). The effect of water absorption on the dielectric properties of epoxy nanocomposites, *IEEE Trans. Dielectr. Electr. Insul.*, **15**, pp. 106–117.

11. Lau, K. Y., Vaughan, A. S., Chen, G., Hosier, I. L. (2012). Dielectric response of polyethylene nanocomposites: the effect of surface treatment and water absorption. *2012 Ann. Rep. CEIDP*, IEEE, pp. 275–278.

12. Shokoohi, S., Arefazar, A., Khosrokhavar, R. (2008). Silane coupling agents in polymer-based reinforced composites: a review, *J. Reinforced Plastics Composites*, **27**, pp. 473–485.

13. Rong, M. Z., Zhang, M. Q., Ruan, W. H. (2006). Surface modification of nanoscale fillers for improving properties of polymer nanocomposites: a review, *Mater. Sci. Technol.*, **22**, pp. 787–796.

14. Tanaka, T., Imai, T. (2013). Advances in polymer nanodielectrics over the past 50 years, *IEEE Electr. Insul. Mag.*, **29**, pp. 10–23.

15. Kochetov, R. (2012). Thermal and electrical properties of nanocomposites, including material processing, PhD thesis, Delft University of Technology.

16. Singha S., Thomas, M. J. (2008). Dielectric properties of epoxy nanocomposites, *IEEE Trans. Dielectr. Electr. Insul.*, **15**, pp. 12–23.

17. Tsekmes, I. A., Kochetov, R., Morshuis, P. H. F. (2014). A unified model for the permittivity and thermal conductivity of epoxy based composites, *J. Phys. D: Appl. Phys.*, **47**, p. 415502.

18. Ishimoto, K., Tanaka, T., Ohki, Y., Sekiguchi, Y., Murata, Y., Gosyowaki, M. (2008). Comparison of dielectric properties of low-density polyethylene/MgO composites with different size fillers, 2008 *Ann. Rep. CEIDP*, IEEE, pp. 208–211.

19. Yu, K., Niu, Y., Xiang, F., Zhou, Y., Bai, Y., Wang, H. (2013). Enhanced electric breakdown strength and high energy density of barium titanate filled polymer nanocomposites, *J. Appl. Phys.*, **114**, p. 174107.

20. Todd, M. G., Shi, F. G. (2003). Characterizing the interphase dielectric constant of polymer composite materials: effect of chemical coupling agents, *J. Appl. Phys.*, **94**, pp. 4551–4557.

21. Zhang, C., Stevens, G. C. (2008). The dielectric response of polar and non-polar nanodielectrics, *IEEE Trans. Dielectr. Electr. Insul.*, **15**, pp. 606–617.

22. Tsekmes, I. A., Kochetov, R., Morshuis, P. H. F., Smit, J. J. (2015). The role of particle distribution in the dielectric response of epoxy-boron nitride nanocomposites, *J. Mater. Sci.*, **50**, pp. 1175–1186.

23. Jonscher, A. K. (1999). Dielectric relaxation in solids, *J. Phys. D: Appl. Phys.*, **32**, pp. 57–70.
24. Dissado, L. A., Hill, R. M. (1984). Anomalous low-frequency dispersion. Near direct current conductivity in disordered low-dimensional materials, *J. Chem. Soc. Faraday Trans. 2: Mol. Chem. Phys.*, **80**, pp. 291–319.
25. Nelson, J. K. (2010). *Dielectric Polymer Nanocomposites*, Springer.
26. Choudhury, M., Mohanty, S., Nayak, S. K., Aphale, R. (2012). Preparation and characterization of electrically and thermally conductive polymeric nanocomposites, *J. Miner. Mater. Char. Eng.*, **11**, pp. 744–756.
27. Andritsch, T., Kochetov, R., Morshuis, P. H. F., Smit, J. J. (2010). DC conduction in epoxy based nano- and mesocomposites, *2010 Ann. Rep. CEIDP*, IEEE, pp. 1–4.
28. Wang, W., Li, S., Tang, F., Li, J. (2012). Characteristics on breakdown performance of polyethylene/silica dioxide nanocomposites, 2012 *Ann. Rep. CEIDP*, IEEE, pp. 521–524.
29. Fleming, R. J., Ammala, M., Casey, P. S., Lang, S. B. (2008). Conductivity and space charge in LDPE containing nano- and micro-sized ZnO particles, *IEEE Trans. Dielectr. Electr. Insul.*, **15**, pp. 118–126.
30. Kozako, M., Hikita, M., Sato, M., Valdez-Nava, Z., Diaham, S. H., Lebey, T. (2014). Effects of nano-alumina hydrate coating for conductive fillers on dielectric properties of epoxy composite materials, 2014 *Int. Symp. Electr. Insul. Mat.*, IEEE, pp. 77–80.
31. Stevens, G. C., Freebody, N., Hyde, A., Perrot, F., Szkoda-Giannaki, I., Vaughan, A. S., Virtanen, S., Baker, P., Bon, S. A. F., Coles, S. R., Medlam, J. A. (2015). Balanced nanocomposite thermosetting materials for HVDC and AC applications, *2015 Electr. Insul. Conf.*, IEEE, pp. 193–196.
32. Dissado, L. A., Fothergill, J. C. (1992). *Electrical Degradation and Breakdown in Polymers*, Peter Peregrinus Ltd.
33. Yewen, Z., Lewiner, J., Alquie, C., Hampton, N. (1996). Evidence of strong correlation between space-charge buildup and breakdown in cable insulation, *IEEE Trans. Dielectr. Electr. Insul.*, **3**, pp. 778–783.
34. Murakami, Y., Okuzumi, S., Nagao, M., Fukuma, M., Sekiguchi, Y., Goshowaki, M., Murata, Y. (2010). Space charge measurement in MgO/LDPE nanocomposite up to breakdown under DC ramp voltage, *IEEJ Trans. Electr. Electron Eng.*, **5**, pp. 395–399.

35. Hayase, Y., Aoyama, H., Tanaka, Y., Takada, T., Murata, Y. (2006). Space charge formation in LDPE/MgO nano-composite thin film under ultra-high DC electric stress, *8th Int. Conf. Prop. Appl. Dielectric Mater.*, IEEE, pp. 159–162.

36. Hayase, Y., Tanaka, Y., Takada, T., Murata, Y., Sekiguchi, Y., Reddy, C. C. (2009). Space charge suppression effect of nano-size fillers added to polymeric materials, *J. Phys.: Conf. Ser.*, **183**(1).

37. Andritsch, T., Kochetov, R., Lennon, B., Morshuis, P. H. F., Smit, J. J. (2011). Space charge behavior of magnesium oxide filled epoxy nanocomposites at different temperatures and electric field strengths, *2011 Electr. Insul. Conf.*, IEEE, pp. 136–140.

38. Tanaka, T., Bulinski, A., Castellon, J., Fréchette, M., Gubanski, S., Kindersberger, J., Montanari, G. C., Nagao, M., Morshuis, P., Tanaka, Y., Pélissou, S., Vaughan, A., Ohki, Y., Reed, C. W., Sutton, S., Han, S. J. (2011). Dielectric properties of XLPE/SiO_2 nanocomposites based on CIGRE WG D1.24 cooperative test results, *IEEE Trans. Dielectr. Electr. Insul.*, **18**, pp. 1482–1517.

39. Tian, F., Lei, Q., Wang, X., Wang, Y. (2011). Effect of deep trapping states on space charge suppression in polyethylene/ZnO nanocomposite, *Appl. Phys. Lett.*, **99**, p. 142903.

40. Han, B., Wang, X., Sun, Z., Yang, J., Lei, Q. (2013). Space charge suppression induced by deep traps in polyethylene/zeolite nanocomposite. *Appl. Phys. Lett.*, **102**, p. 012902.

41. Huang, X., Li, Y., Liu, F., Jiang, P., Iizuka, T., Tatsumi, K., Tanaka, T. (2014). Electrical properties of epoxy/POSS composites with homogeneous nanostructure, *IEEE Trans. Dielectr. Electr. Insul.*, **21**, pp. 1516–1528.

42. Murakami, Y., Suzuki, H., Kawashima, T., Nagao, M. (2014). Electrical characteristics of PVDF/BTO nanocomoposites under DC voltage application, *2014 Ann. Rep. CEIDP*, IEEE, pp. 683–686.

43. Terashima, K., Sukuki, H., Hara, M., Watanabe, K. (1998). Research and development of ±250 kV DC XLPE cables, *IEEE Trans. Pow. Del.*, **13**, pp. 7–16.

44. Kanegae, E., Ohki, Y., Tanaka, T., Sekiguchi, Y., Murata, Y., Reddy, C. C. (2010). Space charge behavior in multi-layered dielectrics with LDPE and LDPE/MgO nanocomposites, *2010 ICSD*, IEEE, pp. 1–4.

45. Kim, D. S., Kim, D. H., Nam, J. H., Ha, S. T., Lee, S. H. (2013). Investigation of space charge distribution of MgO/XLPE composites depending on particle size of MgO as inorganic filler, *Appl. Mech. Mater.*, **481**, pp. 108–116.

46. Qiongxia, Z., Li, L., Jiandong, W., Qiaohua, W., Yi, Y. (2014). Effect of nano-filler grain size on space charge behavior in LDPE /MgO nanocomposite, *2014 ISEIM*, IEEE, pp. 429–432.

47. Huang, X., Jiang, P., Yin, Y. (2009). Nanoparticle surface modification induced space charge suppression in linear low density polyethylene, *Appl. Phys. Lett.*, **95**, p. 242905.

48. Wei, Y., Phung, B. T., Zhao Jun, H., Ostrikov, K. (2014). Plasma polymer-coated on nanoparticles to improve dielectric and electrical insulation properties of nanocomposites, *IEEE Trans. Dielectr. Electr. Insul.*, **21**, pp. 548–555.

49. Zhang, L., Zhou, Y., Tian, J., Sha, Y., Zhang, Y., Wu, H., Wang, Y. (2014). Experiment and simulation of space charge suppression in LDPE/MgO nanocomposite under external DC electric field, *J. Electrostatics*, **72**, pp. 252–260.

50. Li, S., Ni, Z., Yongjie, N., Xia, W., Chen, G., Teyssèdre, G. (2015). Space charge characteristics of LDPE nanocomposite/LDPE insulation system, *IEEE Trans. Dielectr. Electr. Insul.*, **22**, pp. 92–100.

51. Milliere, L., Makasheva, K., Laurent, C., Despax, B., Boudou, L., Teyssèdre, G. (2014). Effects of a modified interface by silver nanoparticles/SiOC:H barrier layer against space charge injection under HVDC, *2014 Ann. Rep. CEIDP*, IEEE, pp. 883–886.

52. Takada, T., Hayase, Y., Tanaka, Y. (2008). Space charge trapping in electrical potential well caused by permanent and induced dipoles for LDPE/MgO nanocomposite, *IEEE Trans. Dielectr. Electr. Insul.*, **15**, pp. 152–160.

53. An, Z., Xie, C., Jiang, Y., Zheng, F., Zhang, Y. (2009). Significant suppression of space charge injection into linear low density polyethylene by surface oxyfluorination, *J. Appl. Phys.*, **106**, p. 104112.

54. Wang, X., Lv, Z., Wu, K., Chen, X., Tu, D., Dissado, L. A. (2014). Study of the factors that suppress space charge accumulation in LDPE nanocomposites, *IEEE Trans. Dielect. Electr. Insul.*, **21**, pp. 1670–1679.

Chapter 9

Dielectric Breakdown of Polymer Nanocomposites

Shengtao Li, Liuqing Yang, Wenfeng Liu, and Weiwang Wang

State Key Laboratory of Electrical Insulation and Power Equipment, Xi'an Jiaotong University, 28 Xianning West Road, Xi'an, Shaanxi 710049, China

sli@mail.xjtu.edu.cn

9.1 Introduction

Bulk electrical breakdown is a phenomenon that occurs within a very short time, of the order of 1 ms and—since it results in the destruction of insulating materials—is a significant parameter for electrical engineering design. In addition, discharge along a material surface, surface flashover, is also an important discharge phenomenon in electrical engineering. Often times, the interface between the solid insulating material and the gaseous phase acts as the weak point in an electrical insulation system. Therefore, both bulk breakdown and surface flashover in polymer nanocomposites are discussed in this chapter.

Tailoring of Nanocomposite Dielectrics: From Fundamentals to Devices and Applications
Edited by Toshikatsu Tanaka and Alun S. Vaughan

ISBN 978-981-4669-80-1 (Hardcover), 978-981-4669-81-8 (eBook)
www.panstanford.com

Since the 1930s, synthetic polymers have been developed and widely used as electrical insulating materials in the electrical engineering field due to their high breakdown strength of up to $\sim 10^9$ V/m and high resistivity of 10^{16} $\Omega \cdot$m. Tremendous efforts have been made to understand electrical breakdown in polymers. In general, the breakdown strength of polymers depends on a number of factors amongst which are the thickness, temperature and pressure. These factors interact with the complicated structures and morphology of polymeric materials, which makes breakdown processes very difficult to understand. For applications in the electrical engineering, polymer composites are widely used to improve mechanical, thermal or electrical properties. However, the breakdown strength of the polymer composites are usually decrease due to the fillers, which can be regarded as the defect in the polymer matrix.

Electrical breakdown of polymer nanocomposites not only depends on the polymer matrix, but also is significantly influenced by the introduced nanofiller. Consequently, polymer nanocomposites can show encouraging dielectric breakdown properties [1–3]. A majority of experimental results indicate that many factors influence the breakdown properties of nanocomposites, such as the modification of nanofillers [4, 5], nanofiller loading and type [1, 6], cohesive energy density (CED) and the glass transition temperature (T_g) of the polymer [7, 8]. For the nanofillers, the physicochemical characteristics is very crucial for the breakdown performance of the nanocompostis. The interaction of nanoparticles (NPs), which have different polar or non-polar functional groups on their surface, with the polymer matrix influences the breakdown properties. In addition, changes in morphology due to the incorporation of NPs, particularly for semicrystalline polymers, are of great importance to the breakdown performance. It is well acknowledged that the interface between nanofiller and polymer dominates the dielectric breakdown performance [9–11]. Therefore, many researchers pay considerable attentions to the physical and chemical properties of the interface and try to tailor the interface by physical and chemical methods in order to obtain enhanced dielectric breakdown properties.

The introduced NPs occupy the amorphous regions or are located at the amorphous-crystalline interface, leading to changes in morphology in nanocomposites. Because the charge transport characteristics are closely related to morphology, the dielectric breakdown properties, which depend strongly on the charge transport characteristics, can be tailored by incorporating NPs. Based on the electric breakdown model [12], the role of the interfacial region on charge carrier transport and carrier multiplication properties is discussed to deepen the understanding on the dielectric breakdown mechanisms in nanocomposites [7, 13, 14].

9.2 Dielectric Breakdown Performance in Polymer Nanocomposites

From an engineering point of view, based on the traditional dielectric breakdown requirement of a solid insulating material, the breakdown strength is the most crucial parameter for the dielectric materials. Usually, Weibull statistics and Weibull plots with two parameters are employed to exhibit the breakdown properties (IEEE Standard 930-2004: 2004). Additionally, surface flashover voltage is used to evaluate the flashover properties and is usually tested using plain or finger shape electrodes.

The electrical properties of nanocomposites are mainly determined by the dispersion, nanoparticle and polymer matrix characteristics. The physical and chemical characteristics of the interface can be influenced by these factors, leading to changes in micro scale features such as traps, carrier mobility and free volume. These features are essential to the breakdown process in nanocomposites. Additionally, the surface flashover processes are also influenced by these factors. Many experimental results have indicated that both the breakdown and surface flashover properties of polymers are improved by incorporating a small amount of nanofiller.

9.2.1 Influence of Preparation on Dielectric Breakdown

Nanoparticles are known to be difficult to disperse homogeneously in a polymer matrix due to the large specific surface area and the strong surface free energy. The dispersion highly dependents on the preparation method. In addition, surface modification of

NPs not only improves the dispersion, but also significantly changes the physicochemical properties of the interface, resulting in the improvement of breakdown properties of nanocomposites.

9.2.1.1 Dispersion and surface modification of nanoparticle

The dispersion of NPs in a polymer matrix significantly influences the breakdown properties of nanocomposites. Nanocomposties EP/TiO_2, composing of a commercial epoxy (EP) resin and 2.5 wt% rutile TiO_2 NPs, were prepared by in situ and ex situ methods, respectively [15]. As shown in Fig. 9.1, a noticeable improvement in breakdown strength was observed in the in situ nanocomposites compared to the ex situ one. This improvement is attributed to an enhanced dispersion of the NPs in the epoxy resin.

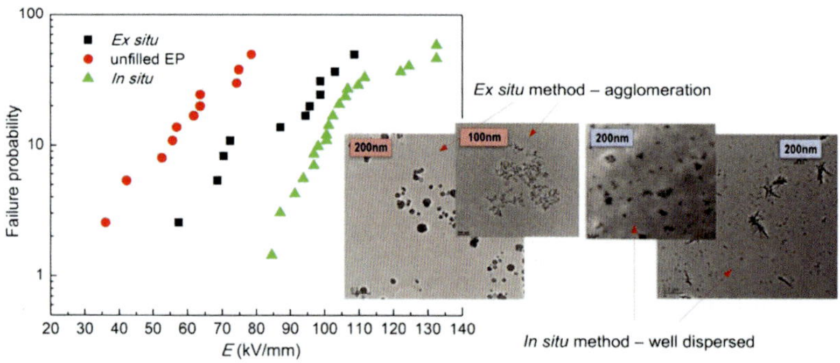

Figure 9.1 Two-parameter Weibull plots of EP and its nanocomposites [15].

Besides the preparation method, the surface functionalization of NPs can not only alter the dispersibility of NPs, but also change the interactions that occur between the NPs and the matrix, eventually influencing the breakdown behavior [4]. A significant improvement in the breakdown strength was observed in LDPE/TiO_2 composites after modification of TiO_2 by *N*-(2-aminoethyl)-3-aminopropyl-trimethoxysilane (AEAPS). As shown in Fig. 9.2, the breakdown strength was enhanced by nearly 40% compared to equivalent systems produced using the as-received NPs. It was considered that the surface modification strengthens the electron scattering, which is mainly caused by the presence of polar groups in the interfacial regions by AEAPS. A similar result was obtained in LLDPE/silica nanocomposites [16]; Figure 9.3 shows that the breakdown strength of the nanocomposites is slightly lower than that of the LLDPE. However, it is relatively higher in the nanocomposites with silane surface-treated silica NPs compared to that of the untreated NPs. It should be noted that surface functionalization of NPs does not always enhance breakdown strength. For example, the improvement of dispersion of $BaTiO_3$ by phosphonic modification decreased the breakdown strength of nanocomposites compared to that of the base polymer [17]. In general, surface modification changes the interfacial region structure, leading to the improvement of breakdown strength of nanocomposites. However, there is still a controversy on the mechanism by which interfacial modification influences dielectric breakdown.

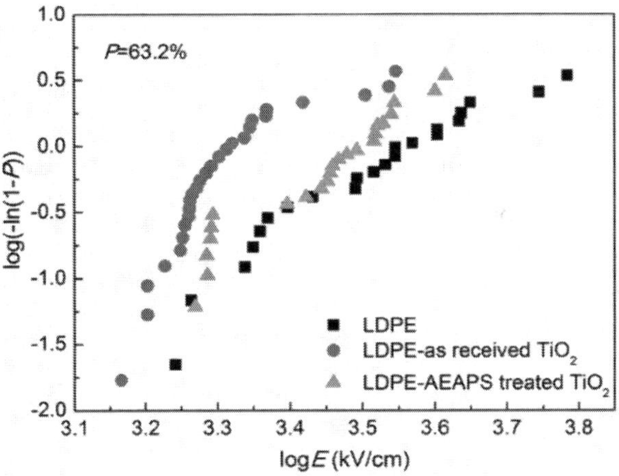

Figure 9.2 Weibull plots of dielectric breakdown strength for neat LDPE and its composites [4].

9.2.1.2 Loading and type of nanoparticle

(1) Breakdown property

Incorporating NPs generally enhances the breakdown strength of polymers. A lot of work has focused on the breakdown property of nanocomposites with various NPs and loading levels. It is notable that a small amount of NPs is beneficial for the breakdown property of composites. As shown in Fig. 9.4, the ac breakdown strength was enhanced at low nano-Al_2O_3 loading [18], but significantly decreased beyond a certain value. Figure 9.5 shows Weibull plots of the ac breakdown strength of LDPE/TiO_2 nanocomposite [19] and associated parameters are presented in Table 9.1. From this, it is evident that incorporating the nano-TiO_2 improved breakdown strength; an increase of 13.6% was observed in the sample containing 1 wt% nanofiller.

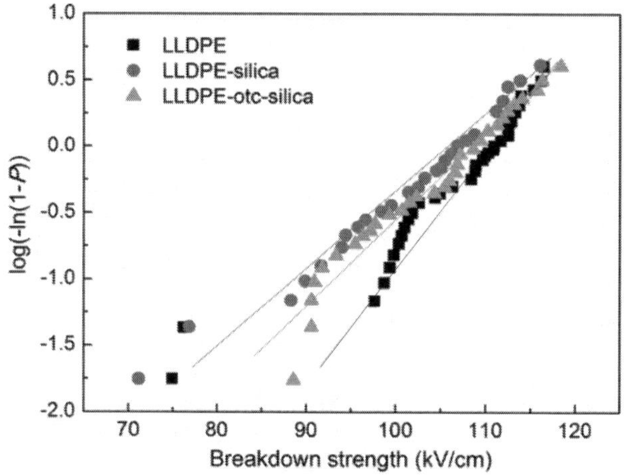

Figure 9.3 Weibull plots of dielectric breakdown strength for neat LLDPE and its composites [16].

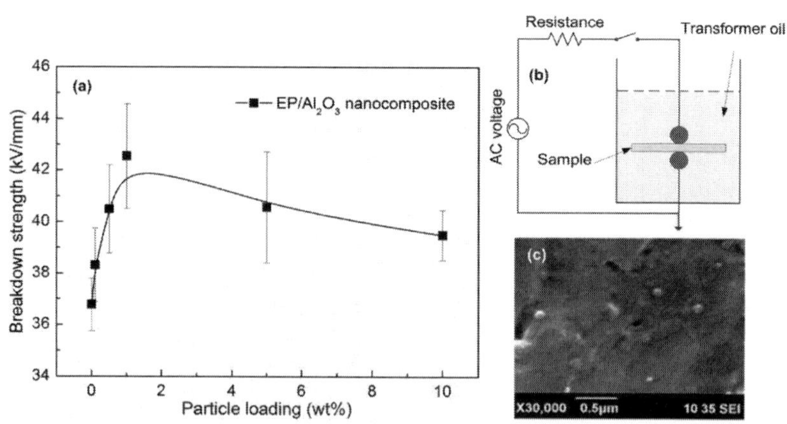

Figure 9.4 (a) ac dielectric breakdown strength of EP/Al_2O_3; (b) schematic diagram of the breakdown test; (c) SEM image with loading of 1 wt% [18].

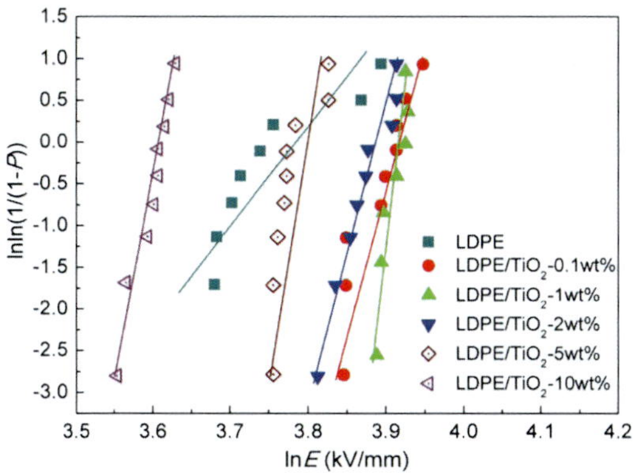

Figure 9.5 Weibull plots of ac breakdown for LDPE/TiO_2 nanocomposite [19].

Table 9.1 Calculated Weibull parameters of LDPE/TiO_2 nanocomposites [19]

NP loading (wt%)	Shape parameter β	Scale parameter α(kV/mm)
0	9.81	44.12
0.1	30.00	49.90
1	63.60	50.34
2	32.80	48.92
5	34.64	44.61
10	45.16	37.04

Note: Weibull cumulative probability of the electrical breakdown.

$$F(E;\alpha,\beta) = 1 - \exp\left(-\left(\frac{E}{\alpha}\right)^{\beta}\right)$$

Furthermore, the type of NPs is also important for the breakdown behavior of nanocomposite. As shown in Fig. 9.6 [19], at low NP loadings (<2 wt%), both nano-Al_2O_3 and nano-TiO_2 composites exhibit comparable breakdown performance. At such low loading levels, the type of NPs has little effect on the breakdown performance; rather, the interfacial region between the NPs and the polymeric matrix dominates the breakdown behavior. However, at higher NP loadings (>2 wt%), the breakdown strength of the Al_2O_3/LDPE systems remains high, in contrast to the significant decrease seen in the TiO_2/LDPE nanocomposites. It is likely that the characteristics of the NPs then play a dominant role, due to the difference in their physical properties:

nano-TiO_2 exhibits strong polarity and high permittivity, while nano-Al_2O_3 has weak polarity and low permittivity. The ac breakdown property of polyimide (PI) nanocomposites (Fig. 9.7) [1] shows that the addition of nano-Al_2O_3 and nano-SiO_2 increased the breakdown strength of PI for NPs loading of 1 wt% up to 10 wt%, while other NPs, i.e., $BaTiO_3$, TiO_2, ZnO, and SiC, decrease the breakdown strength. Here it should be noted that the remarkable decrease of breakdown strength of the SiC nanocomposites is probably caused by the aggregation of SiC NPs.

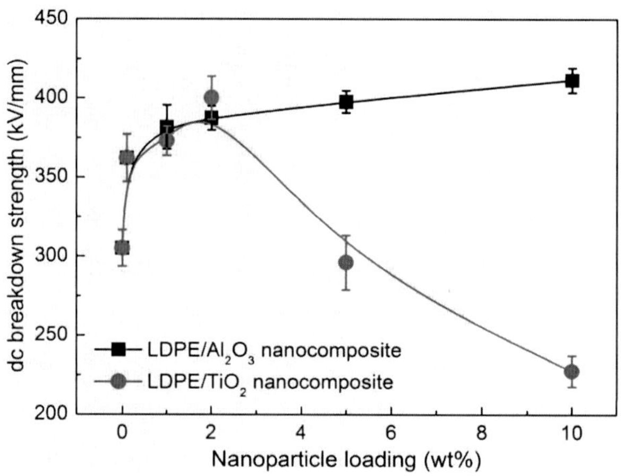

Figure 9.6 dc breakdown strength of LDPE nanocomposites [19].

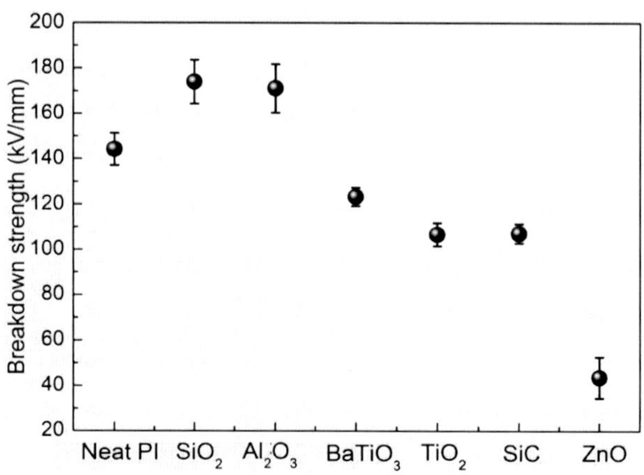

Figure 9.7 ac breakdown strengths for different PI composites (5 wt%) [1].

Plenty of experimental results support the conclusion that small amounts of NPs can enhance the breakdown strength of nanocomposites. To make this point clear, we define a normalization constant *k*, which is defined as the ratio of the breakdown strength of systems containing NPs and that of the associated unfilled polymer [2]. The variation of *k* as a function of NPs loading is shown in Fig. 9.8. Small amounts of NPs (i.e., loadings from 0.05 to 2 wt%) enhanced the ac breakdown strength, while excessive addition of NPs resulted in a reduction in breakdown strength. Consequently, there is an optimal loading of NPs, which will result in the best breakdown performance.

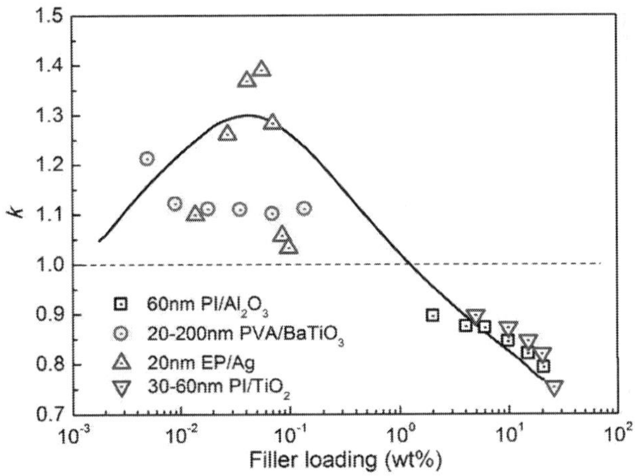

Figure 9.8 Ratio *k* as a function of nanofiller loading [2].

It is interesting that even metal NPs can increase the breakdown strength, as shown in Fig. 9.8. The addition of 6 wt% of silver NPs increased the breakdown strength by about 40% compared to the base resin, whereas, excessive silver NPs led to a significant reduction in breakdown strength [20].

In addition to spherical NPs, NPs with other nanostructures, such as layered and network configurations, have also been employed to improve the breakdown strength of nanocomposites.

(a) Layer-structured nanocomposites

Since the research group at Toyota reported the enhancements in various properties of nylon by the inclusion of layered silicate [21], layer-structured additives have attracted a lot of interest.

Imai et al. [22] have studied the dielectric properties of systems based upon an epoxy resin and a layer-structured silicate organically pre-treated using alkylammonium ions. The time to breakdown of samples tested under ac voltage at different temperatures is shown in Fig. 9.9a. Evidently, the nanocomposite exhibited superior breakdown times, particularly at higher temperatures. In particular, the breakdown time of the epoxy-silicate nanocomposite is over 20,000 min at 145 °C, which is outstanding among the polymers. Such improvement can be ascribed to the electric charge carriers introduced by the layer-structured silicates. As shown in Figs. 9.9b,c, tree propagation is relatively straight in the base resin, while frequent branching occurs in the nanocomposite. It is suggested that a decrease in the electric field around the tips leads to slower propagation of electrical trees and, consequently, longer breakdown time in nanocomposites.

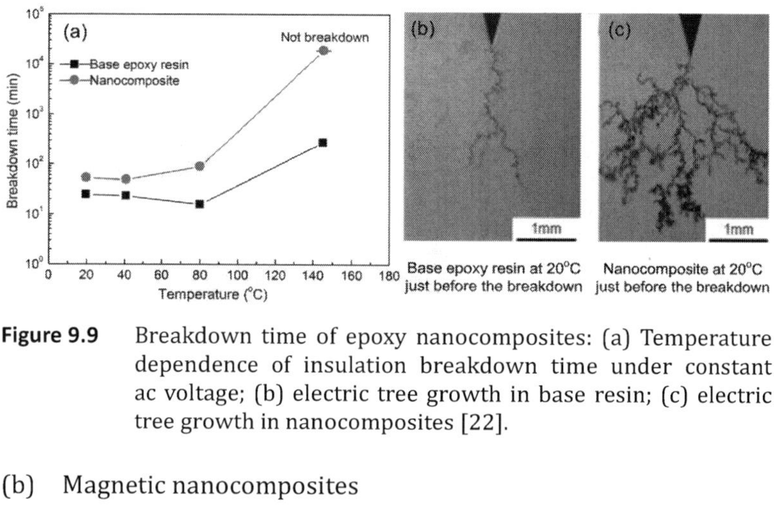

Figure 9.9 Breakdown time of epoxy nanocomposites: (a) Temperature dependence of insulation breakdown time under constant ac voltage; (b) electric tree growth in base resin; (c) electric tree growth in nanocomposites [22].

(b) Magnetic nanocomposites

Magnetic NPs can also affect the dielectric property of nanocomposites. The breakdown behavior of poly(methyl methacrylate) (PMMA)/$CoFe_2O_4$ has been studied in a liquid nitrogen bath [23] and resulting two parameters Weibull distribution plots corresponding to systems containing different loading levels of $CoFe_2O_4$ are shown in Fig. 9.10. This indicates that the addition of $CoFe_2O_4$ NPs increases the dielectric breakdown strength of the PMMA from 153 to 180 kV/mm, as shown in Table 9.2. In this case, the polymer modified with magnetic NPs ($CoFe_2O_4$) can be utilized at low temperature (e.g., liquid nitrogen temperature) for high voltage application. It is anticipated that the electric field would not be high enough to cause a breakdown between NPs. This is likely the reason for the higher breakdown strength of the nanodielectrics.

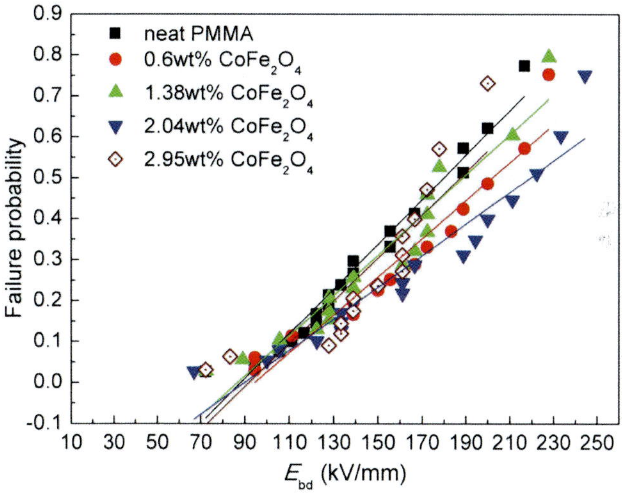

Figure 9.10 Dielectric breakdown of unfilled and $CoFe_2O_4$ filled PMMA samples [23].

Table 9.2 Calculated two-parameters of Weibull distribution plots [23]

Samples	Shape parameter α (kV/mm)	Scale parameter β
Neat PMMA	152.9	4.28
0.6 wt%	169.0	4.00
1.38 wt%	160.2	3.88
2.04 wt%	180.3	3.74
2.95 wt%	162.5	4.09

(c) Nanostructured composites

Polyhedral oligomeric silsesquioxane (POSS) is another option to increase the breakdown strength of thermoplastic polymers. Figure 9.11 shows the breakdown strength of polypropylene (PP)/POSS composites containing octamethyl POSS and isooctyl POSS [24]. It can be concluded that both PP/POSS nanocomposites show enhanced breakdown strength compared to that of the neat PP; the isooctyl POSS/PP nanostructured composites present higher ac breakdown strength and the octamethyl PP/POSS composites exhibit higher lighting impulse (LI) breakdown strength. In this case, POSS acted as an electron scavenger by absorbing charge at the boundaries of the spherulites in the polymer and consequently prevented the charge from going through the bulk insulating material. In addition, the melting point of octamethyl POSS is higher than that of neat PP. This is another possible reason for the enhancement of breakdown strength. Further research should focus on the mechanical, thermal and long-term aging property of nanostructured POSS composites.

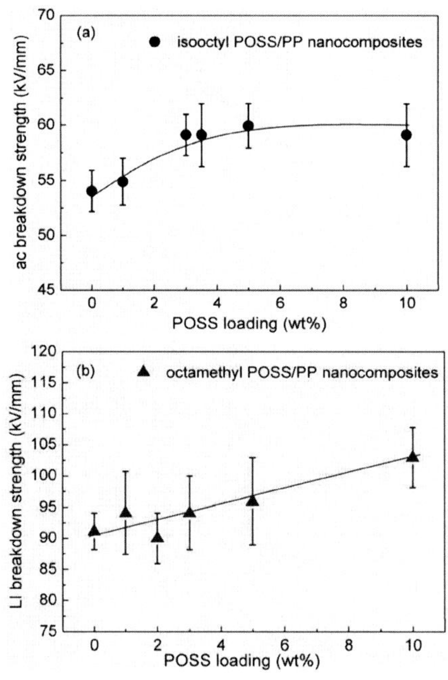

Figure 9.11 (a) ac breakdown strength of isooctyl POSS nanocomposites; (b) lightning impulse breakdown strength of octamethyl POSS nanocomposites [24].

(2) Surface flashover behavior

Surface flashover of insulation system is crucial for many power equipments, especially in vacuum, since it happens at lower electrical field than breakdown. In order to improve the surface flashover performance of insulators in vacuum, NPs and microparticles (MPs) have been employed [25–30].

Previous studies indicated that NPs improved flashover performance, while MPs decreased it at the same loading level. Figure 9.12 presents surface flashover data obtained from EP composites containing MPs and NPs obtained under impulse voltage conditions (40/300 ns) in vacuum [25]. The EP/$Al(OH)_3$ nanocomposite exhibits raised flashover voltage with increasing NPs loading; the flashover voltage is increased by about 9% with a 5 wt% nano-$Al(OH)_3$ loading. However, the EP/$Al(OH)_3$ microcomposites show contrary behavior in that the inclusion of MPs lowers the flashover voltages: A low loading of MPs causes a serious reduction, while higher loadings have less of a detrimental effect. Such differences may be ascribed to the traps introduced by the doping particles [31]; MPs can introduce shallow traps while NPs introduce deep traps. The deep traps are crucial in improving flashover performance by trapping the charge carriers and thereby weakening secondary electron multiplication processes.

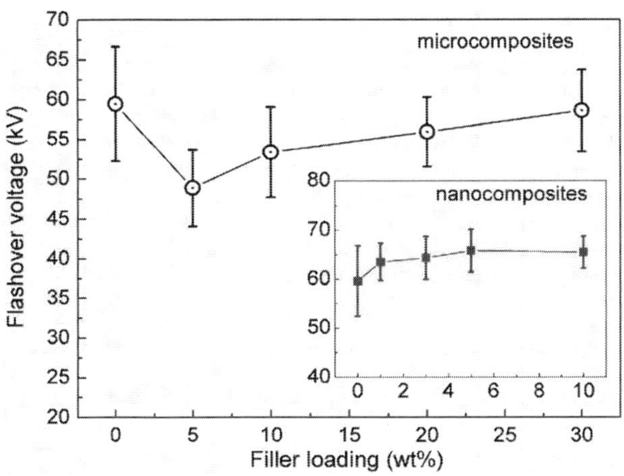

Figure 9.12 Flashover voltage as a function of filler loading in EP/$Al(OH)_3$ [25].

Experimental results obtained from other nanocomposites also support the viewpoint that small amounts of NPs improve flashover performance. Figure 9.13a indicates that a small amount of nano-Al_2O_3 NPs (up to 3 wt%) increased the flashover voltage by over 58% compared with neat polystyrene (PS). Further increasing the loading of NPs decreases the flashover voltage [32]. Equally, nano-Al_2O_3 NPs results in the same effect in LDPE composites [33]. A small amount of nano-Al_2O_3 NPs (up to 1 wt%) improved the flashover performance of LDPE nanocomposites while excessive loadings caused a reduction.

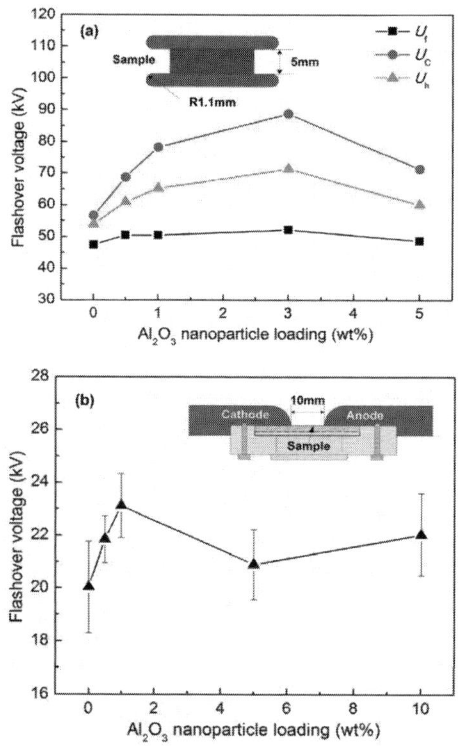

Figure 9.13 Flashover voltage as a function of nanofiller loading: (a) EP/Al_2O_3 nanocomposites [18]; (b) LDPE/Al_2O_3 nanocomposites [33].

It can be observed that NPs change the dielectric characteristics of polymers, such as permittivity and surface conductivity, which in turn affect the flashover performance. Previous studies reported that small amounts of NPs reduce the permittivity and increase the surface conductivity of nanocomposites. On one hand, the lower permittivity may weaken the electric field distortion near the triple junction and thereby improve the flashover performance. On the other, a higher surface conductivity would tend to relieve the accumulation of surface charges, which may also benefit flashover performance [18].

Hybrid doping of NPs and MPs can also improve the surface flashover performance of polymers. As shown in Fig. 9.14, the EP composites containing both micro- and nano-Al_2O_3 exhibit improved flashover performance under both dc and 20 kHz ac voltages. Such enhancement is due to modified surface conditions caused by the embedded particles. On one hand, the embedded particles can generate multiple triple points, which prevent the flashover by scattering the charge carrier transport between the gap. On the other, they may also mitigate surface charge accumulation on the sample surface, leading to the enhancement of the flashover voltage.

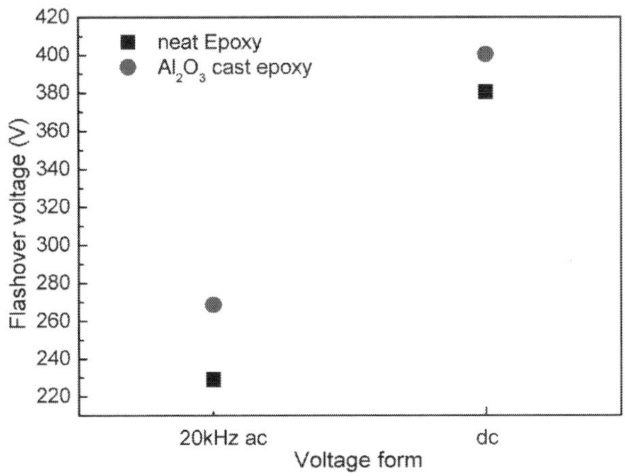

Figure 9.14 ac and dc flashover voltages for neat and nano/micro-Al_2O_3 composite in nitrogen [28].

Conventionally, the two key properties of short-term breakdown performance (i.e., bulk breakdown and surface flashover) are regarded as independent of each other, since bulk breakdown is controlled by bulk characteristics while surface flashover is dominated by the surface conditions. However, recent work has revealed that NPs may modify both the bulk breakdown and surface flashover in comparable ways. As shown in Fig. 9.15, different NPs (i.e., nano-Al_2O_3 and nano-TiO_2) have the same effect on both short-term bulk breakdown and surface flashover performance under dc conditions; a small amount of NPs is beneficial to both while excessive loading has the opposite effect [32].

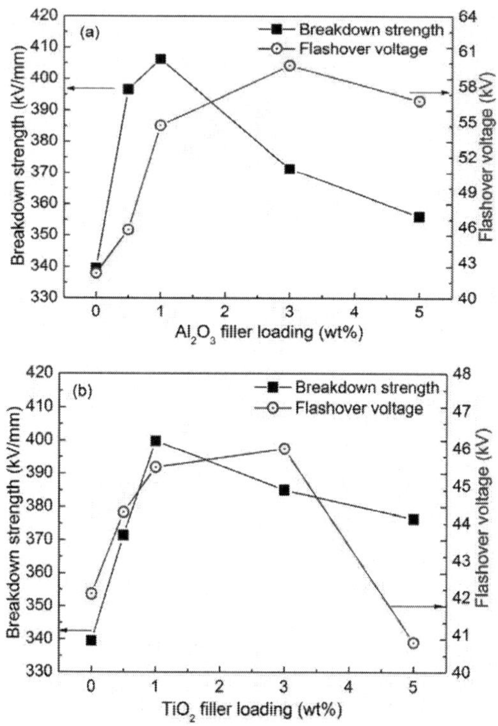

Figure 9.15 Relationship between dc surface flashover voltage and dc bulk breakdown strength in PS nanocomposites: (a) PS/Al_2O_3 nanocomposites; (b)PS/TiO_2 nanocomposites [32].

9.2.1.3 Polymer characteristics

The cohesive energy density (CED) is a crucial characteristic of polymers, since it can evaluate the inter-molecular forces, reflect the interaction between chemical groups and represent the flexibility of molecular chains. For neat polymer, it is believed that high CED yields high electric strengths [34]. However, the situation for nanocomposites is more complex; the breakdown strength of nanocomposites with increasing CED first increases and then decreases [2, 35]. Hence, it is difficult to link the physical and chemical properties of nanocomposites only by the CED parameter.

Interestingly, recent work has revealed that the breakdown strength of nanocomposites is definitely related to T_g. As another crucial parameter of polymers, T_g reflects chain dynamics. As shown in Fig. 9.16, on increasing T_g the parameter k decreases, indicating a decrease of the breakdown strength. Generally, polymers with low T_g are characterized by highly flexible chains, which are likely to twine around NPs and form a strong interacted interface, the so-called "soft interface." Strong interacted soft interfaces are beneficial to breakdown performance. Conversely, the less flexible chains in polymers with high T_g tend to form weak interfaces with NPs, which can be defined as a "hard interface"; this decreases the breakdown performance. Hence, in order to enhance breakdown performance, it is better to employ polymers with low T_g. When, however, a polymer of high T_g is necessary, some techniques, i.e., appropriate surface fictionalization of NPs, should be employed to strengthen the interaction in the interface region.

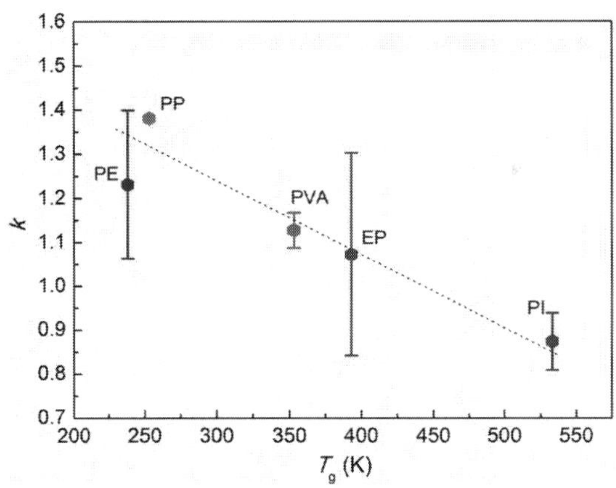

Figure 9.16 Ratio k as a function of T_g for polymers (data from [2]).

9.2.2 Effect of Ambient Condition

The breakdown strength of polymers is sensitive to the ambient conditions, particularly temperature and pressure. Conventionally, breakdown strength refers to the maximum value obtained at low temperature. On increasing the temperature, the breakdown strength first remains constant or may slightly increase. Above a certain critical temperature, it then monotonically decreases

[36]. The critical temperature is closely related to T_g. In nanocomposites, the effect of temperature on breakdown strength is similar to that of the neat polymer.

In addition, mechanical stress also affects breakdown behavior. Since stress can influence molecular chain movement, crystallinity, and free volume, it then leads to changes in breakdown strength [37]. As evidence, LDPE/Al_2O_3 nanocomposites subjected to different mechanical treatment exhibited distinct differences in breakdown strength [38]. Li et al. performed hydrostatic treatment with different pressures of 50, 100, and 200 MPa for 1 h and immediately measured the breakdown strength of each samples without pressure loading. As shown in Fig. 9.17, the breakdown strength of the nanocomposites with different Al_2O_3 loadings increased after the pressure treatment; specifically, the breakdown strength increases by 46% for the LDPE-Al_2O_3-1 wt% system after being subjected to a pressure of 200 MPa. This can be ascribed to a decrease in the free volume as a consequence of applying the pressure.

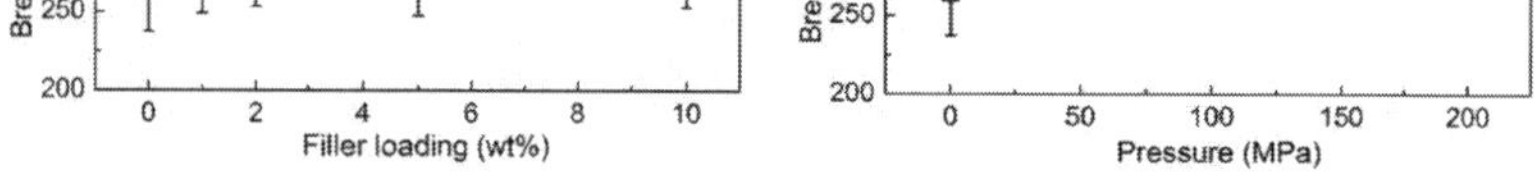

Figure 9.17 dc breakdown strength with respect to particle loadings in LDPE/Al_2O_3 nanocomposites: (a) Breakdown strength as a function of nanoparticle loading; (b) breakdown strength as a function of pressure [38].

9.2.3 Linking Macroscopic Breakdown Performance to Microscopic Characteristics

From the above discussion, it can be concluded that the introduction of NPs will result in the formation of nanoscaled interfaces; these interfaces and interactions between the individual NPs lead to significant changes in macroscopic breakdown strength. Apparently, such modifications fall in the nano- to micro-scale. Hence, discussing the breakdown performance in terms of the microscopic characteristics would lead to a better understanding.

9.2.3.1 Free volume

Free volume, defined as the inter-molecular distance in the amorphous regions of the polymer, is a key microscopic characteristic that affects the mobility of side groups, the segmental motion of the molecule chains, and the charge carriers involved in charge transport [39, 40]. Experimentally, free volume is usually obtained from the *P–V–T* equation [41] or by the positron annihilation method. Although the positron annihilation method is of high accuracy, the small number of available positron sources limits its application. Instead, free volume is usually calculated from the *P–V–T* equation, despite the time consuming nature of this for polymers [42].

Up to now, there is still no consistent conclusion on how NPs influence free volume. Nelson et al. [43] reported that the free volume of TiO_2/EP nanocomposites obtained from the *P–V–T* equation increased; the relative free volume increased by 2–3% over the temperature range from 370 to 550 K. This is contrary to result obtained from microcomposites, where the free volume decreases on the addition of MPs, as shown in Fig. 9.18. Uchaki et al. also reported a decrease in free volume in clay nanocomposites with different loadings, which they ascribed it to a decrease in segmental chains mobility [41].

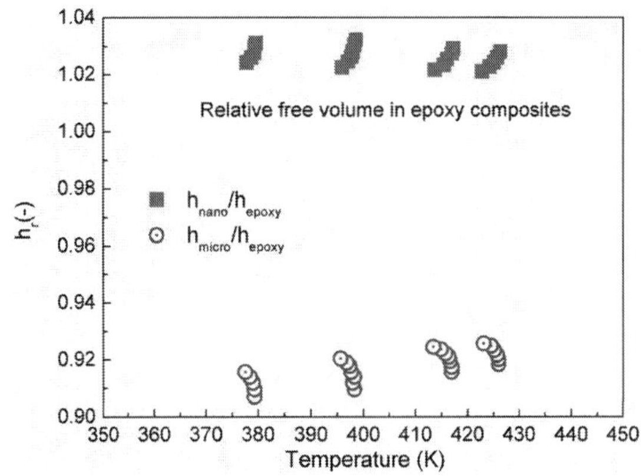

Figure 9.18 The relative free volume for TiO_2–epoxy nano- and microcomposites extracted from *P–V–T* measurements [43].

Conventionally it is believed that breakdown strength would be enhanced by decreasing the free volume [44] and, experimentally, Part et al. [37] supported such a viewpoint. These workers reported an enhanced breakdown strength on applying a compressive stress, which may result in a reduction in free volume.

9.2.3.2 Trap

It is well accepted that for polymers, reduced crystallinity would result in enhanced breakdown strength, since the sufficient crystal boundaries introduce more deep traps and benefit the breakdown strength [36, 45–47]. However, for nanocomposites, NPs may act as nucleating agents and thereby improve the crystallinity. Contrary to the conventional viewpoint, this can enhance breakdown strength. The key issue for such an apparent contradiction is that the intrinsic reason for enhancing breakdown strength is the traps rather than the crystallinity itself.

Considering nanocomposites, NPs act as trap centers and introduce new traps. Experimentally, a new high temperature TSC peak identified in XLPE/nano-silica composites has verified the deep traps introduced by NPs [13]. These deep traps can capture charge carrier, resulting in a decrease of charge mobility and the average free path that the electrons can accelerate over. Macroscopically, nanocomposites, despite higher crystallinity, exhibit enhanced breakdown strength.

Figure 9.19 illustrates the relationship between deep traps and the breakdown strength of LDPE/Al_2O_3 nanocomposites. It can be seen that both the depth of the trap level and the trap density affect breakdown strength. At the low loading level of 0.5 wt%, the nanocomposites exhibit a deep trap level energy of 1.65 eV, a high trap density of 6×10^{16} m^{-3} and the consequently high breakdown strength of 354 kV/mm, as discussed above. Further increasing the loading results in a decrease in both the trap leve and the trap density; at the same time, the breakdown strength decreases. The change of deep trap can be ascribed to changes in the spherulitic structure and the amorphous-crystalline interface that result from the introduction of a small amount of NPs.

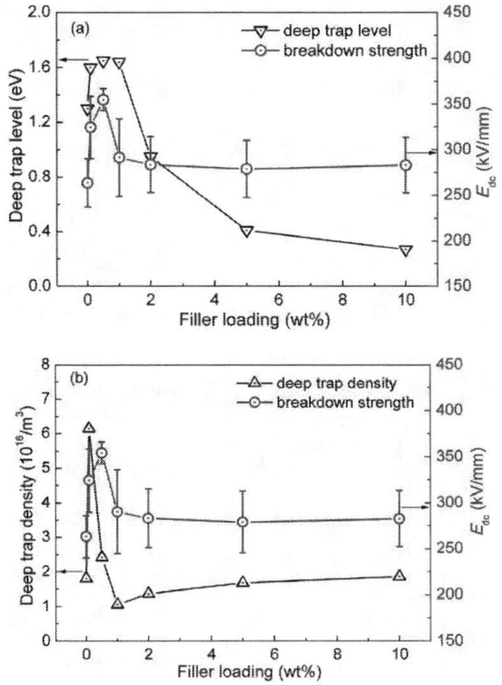

Figure 9.19 Relationship between deep traps and breakdown strength: (a) Deep trap level and breakdown strength; (b) deep trap density and breakdown strength.

9.3 Breakdown Mechanism of Polymer Nanocomposites

9.3.1 Conventional Breakdown Mechanism in Unfilled Polymers

An avalanche breakdown mechanism, one of the electrical breakdown types, is often employed to understand the breakdown process in solid dielectrics [48]. In this, the existence of energetic electrons with sufficient kinetic energy is necessary to trigger impact ionization, which is crucial to the avalanche breakdown process. For polymer dielectrics, the large band gap (~8 eV), the low carrier mobility (10^{-10}~10^{-14} m^2/(V·s)), the small carrier mean free path and the large concentration of localized gap states make it very hard for carriers to gain enough energy from the applied field to generate sufficiently energetic electrons.

Previous research has suggested that, although very hard, it is still possible for polymers to contain energetic electrons. Zeller et al. [49] have estimated that the chemical structure of saturated PE may provide band mobility of the order of 10^{-4} $m^2/(V \cdot s)$. Such high mobility state can be required for avalanche formation. If the applied field is high enough, the injected electrons from the cathode can become energetic and may subsequently collide with the polymer chains, resulting in impact ionization, degradation or breakdown. Hence, an avalanche breakdown mechanism is still available to understand the electrical breakdown process.

However, the breakdown process is more complicated in many polymers, because of their complex hierarchical structures. Based on the free volume breakdown theory [44], it is impossible for the carriers to acquire enough energy to trigger avalanche breakdown, since the free volume is considered to be very small (less than 1 nm) and cannot therefore provide a sufficiently large mean free path (l_E) for the carriers. However, free volume measurements are often obtained at low temperature and cannot apply to the room temperature case. At room temperature, the l_E can reach a few tens of nanometers due to the dramatically increased free volume. This enables the electrons to acquire enough energy to cause structural deformations, chain scission and eventually breakdown [12].

Considering the statistical nature of breakdown, it is insufficient to take only the avalanche process into account. Besides the charge carriers discussed in the avalanche process, defects and local breakdown channels also contribute to the cumulative breakdown [50]. Here, pre-existing point defects may extend under the electric field to form "clusters" of defect and subsequently grow into a discrete breakdown channel along the direction of the electric field. Figure 9.20 shows the relationship between the electric field, E, the free path of charge carriers l and the energy acquired by the carriers in a free path. This indicates that impact ionization and secondary electron emission occur at a relatively long free path. The free volume plays an important role in the carrier multiplication and energy accumulation during the above processes. Breakdown can therefore be understood not only by reducing the "threshold" energy for continuing damage formation but also by accelerating carriers over significantly longer free path.

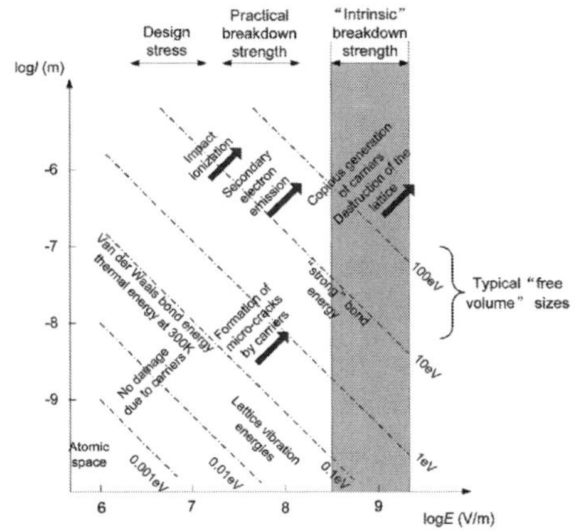

Figure 9.20 Relationship between electric field *E*, free path of charge carriers in insulators *l* and energy acquired by the carriers in a free path [50].

9.3.2 Mechanisms to Understand Breakdown in Nanocomposites

As discussed above, the interface plays the dominant role in the dielectric properties of nanocomposites [10]. Figure 9.21 contains a schematic representation showing a multi-region structured interface around a spherical NP. Here, the interfacial region has been divided into three zones from interior to exterior [2].

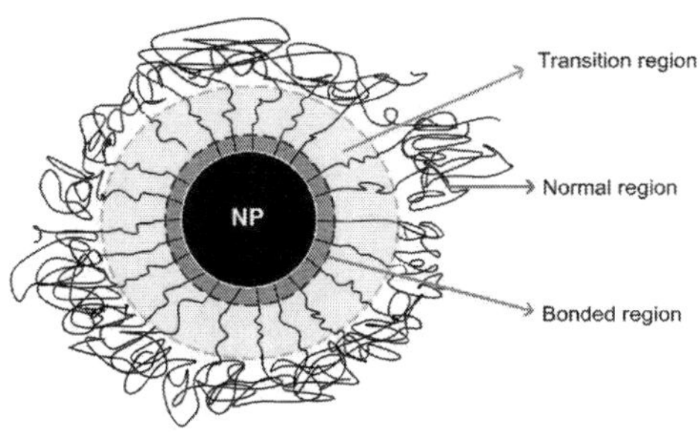

Figure 9.21 Schematic of multi-region structure around spherical NP [2].

(1) Bonded region

According to colloid chemistry, a Stern layer is formed due to a portion of counter ions adsorbed by electrostatic Coulomb force around NPs and other forces, since the Fermi level differs between inorganic particles and a polymer matrix. It is speculated that a vast distortion of potential occurs in the bonded region and that deep traps with the highest density are introduced. Moving outwards from the surface of the NP, the density of traps decreases and the shallow traps gradually take a dominated position. Space charge is accumulated under the applied field such that the electric field increases sharply, which leads to breakdown occurring first in the bonded region. In general, the bonded region plays a key role in the short-term breakdown behavior of nanocomposites.

(2) Transitional region

In the transitional region, the molecular chains of the polymer matrix are strongly associated with the bonded region and NP surface. The molecular chains may adopt an orderly arrangement in this region, such that each chain experiences an average force under electromechanical stress. The characteristics of this region are strongly influenced by the CED of the polymer matrix; flexible molecular chains equate to a small value of CED. Thus, more energy is needed to break the chain up. In this way, the service life of the material will be effectively extended.

Changes in both the mobility of molecular chains and the overall crystallinity may affect the behavior of nanocomposites. The former directly influences T_g, while the presence of the NPs may elevate the latter above the level seen in the unfilled matrix. The transitional region is considered to be crystalline. The thickness of the transitional region is thicker than that of the bonded region and affected by the surface condition of NPs.

(3) Normal region

The property of the normal region is similar to that of the matrix. The polymer chains enwind around the NPs randomly.

In the following sections, possible mechanism behind the breakdown phenomena described above will be discussed in terms of the multi-region structured interface.

9.3.2.1 Potential barrier model

A new potential barrier model was proposed to explain both the short-term breakdown and the long-term failure behavior of nanocomposites [18]. As discussed above, the short-term breakdown of nanocomposites is closely related to NP loading, such that there is an optimal value of loading for short-term breakdown improvement. The effect of the interfacial region on charge transport and breakdown properties differs at low and high NP loadings.

As shown in Fig. 9.22, charge carriers in polymer can easily hop the barrier Δf_3 into the transition region due to the applied field. Two probable effects make the carriers stay in the transition region, one being the reduction of carrier mobility due to the process of collision with defects in the transition region and the other is that local traps can capture charge carriers to decrease the energy obtained from the electric field to hop the barrier Δf_1 since the thickness d_2 is shorter than the mean free path. These two factors can decrease the density of mobile carriers, resulting in high breakdown strength.

When the NP loading increases, the distance between neighboring particles is reduced, leading to the overlap of transition regions, as shown in Fig. 9.22. The thickness d_2 largely extends to d_3, which is longer than the mean free path. In this way, carriers can gain enough energy from the electric field to hop the barrier Δf_1. When particle loading exceeds the percolation threshold, many conductive paths are formed through the overlap of the transition region in the bulk of the nanocomposite. Carriers move easily through these conductive paths. The mobility and the density of mobile carriers increase greatly, leading to an increase in conductivity and a decrease in the breakdown strength.

The free volume can be changed by incorporating NPs [51]. The fact that the free volume increased in epoxy nanocomposites (10 wt% nano-TiO_2) [52] indicated that the breakdown strength decreased. This is contrary to what is expected. It can be inferred that the free volume depends on the NP loading and polymer matrix. Hence, it is necessary to investigate the influence of free volume on dielectric breakdown performance to better understand the dielectric breakdown mechanism in nano-composites.

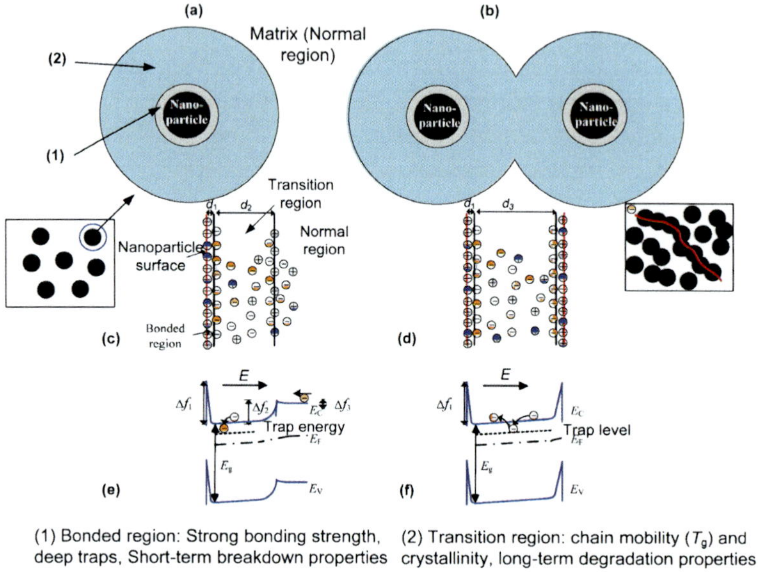

Figure 9.22 Potential barrier model around NPs and short-term breakdown and long-term failure properties in nanodielectrics [18]: (a) Schematic of the charge distribution around an isolated particle. (b) Schematic of the charge distribution when the interfacial region overlaps. (c) Potential barrier around an isolated particle without applied field. E_C, E_V, E_F, and E_g are the bottom edge of conduction band, the top edge of valence band, the Fermi level and the forbidden band gap of the interaction zone, respectively. Δf_1 is the barrier caused by the bonded region. Δf_2 and Δf_3 are the barriers between NP and polymer matrix. (d) Potential barrier without an applied field when the transition region overlaps between two neighboring particles; the thickness of d_3 has a great increase. (e) Carriers restraint around an isolated particle with an applied field. Charge carriers are restrained in the transition region since d_2 is shorter than the mean free path of carriers, even when the field E is applied. (f) Carrier transport through the transition region with an applied field when they overlap between two neighboring particles. In this case, carriers can obtain enough energy from the applied field E to hop the barrier Δf_1, resulting in carrier transport.

The increase in breakdown strength caused by a reduction of free volume is probably due to a consequent decrease in the mean free path of charge carriers under the electric field.

The effects of hydrostatic pressure on dielectric properties of LDPE/Al_2O_3 nanocomposites have been investigated [38]. Figure 9.23 shows a simple model of the effect of pressure on molecular chain movement and free volume in LDPE/Al_2O_3 nanocomposites. Free volume exists in the polymeric matrix and in the interaction zone. It should be pointed out that the interaction/miscibility between the NPs and the matrix is weak. The overall free volume can be changed because of the existence of an interaction zone. During hydrostatic pressure treatment, the molecular chains are pressed to reduce the inter-molecular distance, leading to a decrease in free volume. It is supposed that the size distribution of the free volume becomes uniform and, in this way, it can be easily seen how breakdown strength would be enhanced by hydrostatic pressure treatment.

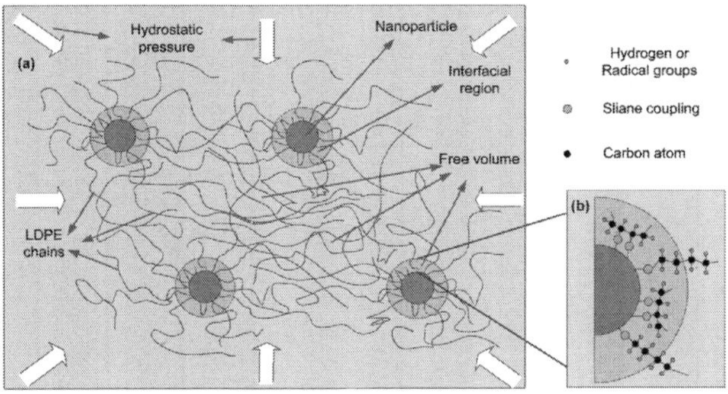

Figure 9.23 The simple model of pressure effect on molecular chains movements and free volume in LDPE/alumina nanocomposites: (a) Local distribution of NPs in LDPE and the interactions between NPs and LDPE chains; (b) polymer chain segments boned to NP surface in an interfacial region [38].

Free volume breakdown can be regarded as the "intrinsic" breakdown of polymers. Hence, the breakdown strength is usually high based on the free volume breakdown model. Actually, the breakdown strength of polymers is lower than the "intrinsic" breakdown strength, due to various secondary effects. One of the most important secondary effects concerns the influence of space charge on the breakdown process [36]. Therefore, charge injection and accumulation behaviors need to be considered in the dielectric breakdown processes.

9.3.2.2 Charge transport model

A negative space charge accumulation near the cathode is known as "homocharge." This can decrease the electric field at the cathode to reduce electron injection, resulting in the increased breakdown strength, as shown in Fig. 9.24. The large negative space charge accumulation adjacent to the anode, known as "heterocharge," conversely causes field enhancement at the anode. This can increase charge injection from this electrode, leading to a decrease in breakdown strength. The temperature and the rate of voltage rise may strongly affect the nature of the accumulated charges, whether homocharge or heterocharge. Particularly in polymers, charge transport and subsequent accumulation effects are considered in terms of electron energy band and localized state models [12]. Nelson et al. pointed out that the presence of traps in amorphous regions determines electrical conduction [53].

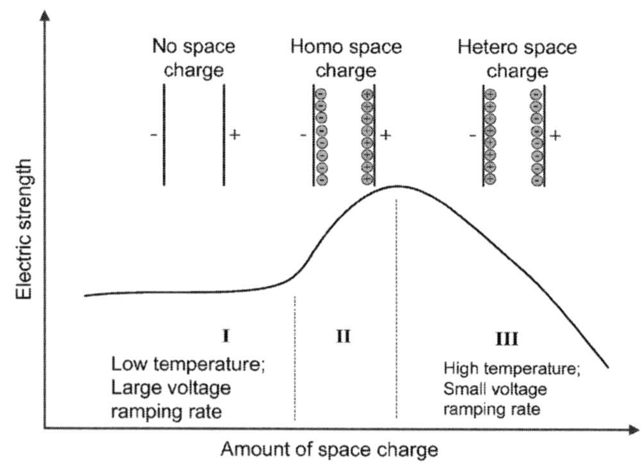

Figure 9.24 Relationship between space charge accumulation and electric strength [47].

It is well known that deep traps decrease both the carrier mobility and the density of free electrons, and lead to reduced conductivity [54]. Based on the free volume theory, the longest electron free path (l_e) and the barrier (E_μ) act as the key factors that determine the breakdown strength. Here, the deep traps increase the E_μ and thereby benefit the breakdown strength.

For nanocomposites, NPs would introduce more deep traps and similarly enhance breakdown strength.

Besides, space charge also affects the breakdown behavior of polymers [55–57]. The dynamics of space charge under a high voltage prior to breakdown is shown in Fig. 9.25 [55]. Breakdown occurs once the internal electric field, strongly influenced by the hole charge packet, exceeds a critical value, which is the intrinsic breakdown strength of the matrix. Here, the hole charge packet is significantly affected by the charge trapping and de-trapping processes. The bipolar charge model described in Fig. 9.25 can well explain the breakdown behavior in polymers. Chen et al. quantitatively studied the widely observed thickness dependence of the breakdown strength by calculating the inverse power index in LDPE [56]. Their calculation also verities that the breakdown strength would be enhanced on increasing the voltage rise rate.

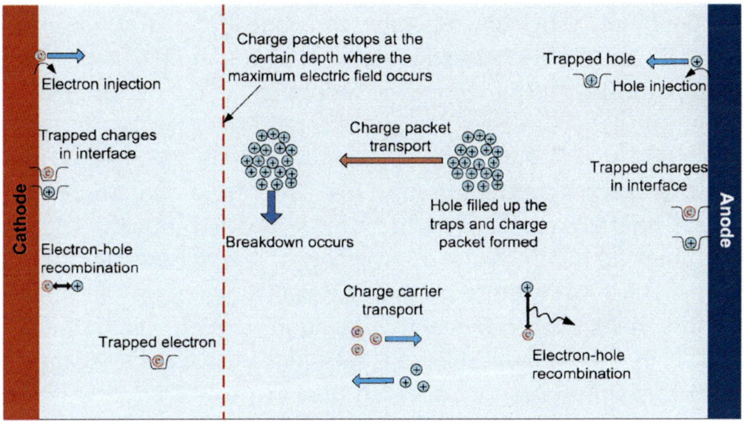

Figure 9.25 Schematic graph of relationship between space charge accumulation and breakdown in insulating material.

It can be concluded that space charge dynamics determine the maximum local electric field inside the sample [57]. Considering the dc condition, the maximum field would increase with the time of applied voltage and move from the anode towards the cathode. Hence, dc breakdown is always initiated from the bulk. Conversely, under ac conditions, the transport direction of the charge carriers keeps alternating; this causes significant charge accumulation near the electrodes due to local trapping. In this case, the electric field near to the electrodes is higher than that inside the bulk samples. Hence, breakdown tends to initiate at the interfaces between the electrode and the sample, and then quickly runs through the bulk. Macroscopically, the dc breakdown strength is relatively higher than that under ac conditions.

For nanocomposites, NPs lead to changes in the overall properties of polymer. Hereinto, the changes in the bulk properties contribute to the dc breakdown strength while that in the surface properties response to the ac breakdown strength.

9.3.3 Other Mechanisms

In addition to the mechanisms discussed above, additional interpretations have been proposed in order to understand the breakdown behavior of nanocomposites.

For metal NP-modified nanocomposites, e.g., EP/Ag, the ac breakdown strength was enhanced. This can be understood via "Coulomb Blockade Effects," introduced by the uniformly dispersed, isolated silver NPs acting as "Coulomb Islands." High loadings of silver NPs lead to a reduction in the ac breakdown strength [58].

A multi-core model was proposed to explain the enhancement of the breakdown strength by NPs [59]. Here, the inter-space between the NPs and the matrix can be divided into three layers from inner to outer, each of which has a different size. In the outmost third layer, there are many defects; electrons can interact with the defects to reduce the mean free path, leading to an increase in breakdown strength. In addition, it is also suggested that electrons lose energy because they are scattered or attracted by the Coulombic force when they move inside the Debye shielding length. In this case, electrons are decelerated and so increase the breakdown strength. However, such a model cannot apply in the case that excessive NPs decrease the breakdown strength.

9.4 Concluding Remarks

As demonstrated above, NPs do indeed affect the breakdown behavior of nanocomposites. Such modifications can be understood from the concept that additional "traps" are introduced by the NPs. Here it should be noted that the trap is crucial to the interpretation of the electrical properties of nanocomposites. Particularly considering the breakdown behavior discussed in this chapter, NPs can increase the density and the depth of the traps, especially the deep traps, which hinder the transport of charge carriers and subsequently change the dynamics of space charge and the distribution of the internal electric field. As a macroscopic phenomenon, the breakdown strength changes, either increasing or decreasing.

It should be noted that the above conclusions are based on experimental results from "FIRST" generation nanodielectric, as defined by Prof. Tanaka. Here, the first generation nanodielectrics are considered to exhibit less stability and reliability, which naturally leads to a poor understanding from physical or chemical points of view and, practically, leads to limited improvements in certain properties. To promote the development of nanodielectrics, great efforts should therefore be made to address the following aspects of the topic.

First, the fabrication of high quality nanocomposites with designed interfaces is the crucial task for "SECOND" generation nanodielectrics. Such nanodielectrics would provide enhanced breakdown strength as well as other physical properties, rather than at the expense of other properties, like the "FIRST" generation nanodielectrics do. Much attention should be addressed towards interfacial engineering, which plays the most important role in deciding the behavior of nanodielectrics. Here, two areas need to be considered; advanced fabrication techniques to achieve optimal dispersion of the NPs within the matrix and superior technologies to characterize the obtained interfaces and then feed back to the fabrication process. In the former area, living/controlled polymerization methods (including atom transfer radical polymerization (ATRP) and reversible addition-fragmentation chain transfer polymerization (RAFT)), tailoring synthesis methods ("Click chemistry," template polymerization, etc.) and in situ polymerization have demonstrated themselves to be very promising candidates to provide nanodielectrics of high quality. Considering characterization method, techniques focused on the interface (such as atomic force microscope (AFM) and positron annihilation spectroscopy) may provide new insights into relationships between morphology and properties. Such information is very important for improving processing to deliver certain properties. Through appropriate developments in fabrication, processing and characterizing techniques, systems defined as "SECOND" generation nanodielectrics would exhibit multifunctional superiorities and subsequently satisfy many real world applications.

Second, the time dependent evolution of physical properties from microscopic, mesoscopic to macroscopic scale (3M) is the key mission for nanocomposites, as pointed out by Lei QQ at the 354th Xiangshan scientific symposium [60]. It is well accepted that the physical properties of polymers exhibit a strong hierarchical nature in the space-time scale, since the basic morphology varies both temporally and spatially. With the incorporation of NPs, the morphology of the polymer would be greatly changed; considering how modifications through NP addition combine with this strong hierarchy in the space-time scale constitutes a major new challenge for nanocomposites. As discussed in this chapter, the interface plays the dominant role in nanocomposite, the dimensions of which fall within the mesoscopic scale (~10 nm). Hence, how to understand the effect of traps at this mesoscopic level and then link this to the 3M will be an essential element in developing the dielectric physics of nanocomposites. It should be pointed out that Fowler–Nordheim electron emission and quantum effect at the microscopic scale should be involved in the analysis of traps at the mesoscopic level and will eventually contribute to macroscopic properties. Practically for breakdown behavior, it is suggested that the thickness effect ($E_d \sim d^{-n}$) of short-term breakdown and the time dependence of discharges and electrical trees should be considered in terms of the traps introduced by the NPs. Besides, understand the "soft" or "hard" nature of the interface, which can be distinguished by the dimension of surface tension/elasticity modulus, is also an effective way to address breakdown problems in nanodielectrics.

Acknowledgments

The authors sincerely thank Prof. Qingquan Lei for helpful discussion and comments on the understanding of nanodielectrics. The authors thank Miss Shihu Yu for collecting the references.

The authors gratefully acknowledge financial supports from the National Natural science Foundation of China (51337008, 11275146, 51407140, 51161130524, 50625721, and 51221005).

References

1. Cao, Y., Irwin P. C., and Younsi, K. (2004). The future of nanodielectrics in the electrical power industry, *IEEE Trans. Dielectr. Electr. Insul.*, **11** pp. 797–807.
2. Li, S. T., Yin, G. L., Chen, G., et al. (2010). Short-term breakdown and long-term failure in nanodielectrics: A review, *IEEE Trans. Dielectr. Electr. Insul.*, **17**, pp. 1523–1535.
3. Murakami, Y., Nemoto, M., Okuzumi, S., et al. (2008). DC conduction and electrical breakdown of MgO/LDPE nanocomposite, *IEEE Trans. Dielectr. Electr. Insul.*, **15**, pp. 33–39.
4. Ma, D. L., Siegel, R. W., Hong, J. I., et al. (2004). Influence of nanoparticle surfaces on the electrical breakdown strength of nanoparticle-filled low-density polyethylene, *J. Mater. Res.*, **19**, pp. 857–863.
5. Tanaka, T., Montanari, G. C., and Mulhaupt, R. (2004). Polymer nanocomposites as dielectrics and electrical insulation-perspectives for processing technologies, material characterization and future applications, *IEEE Trans. Dielectr. Electr. Insul.*, **11**, pp. 763–784.
6. Nelson, J. K. (2007). Overview of nanodielectrics: Insulating materials of the future, *2007 Electr. Insul. Conf. Electr. Manufacturing Expo*, pp. 229–235.
7. Nelson, J. K., and Fothergill, J. C. (2004). Internal charge behaviour of nanocomposites, *Nanotechnology*, **15**, pp. 586–595.
8. Singha, S., and Thomas, M. J. (2008). Dielectric properties of epoxy nanocomposites, *IEEE Trans. Dielectr. Electr. Insul.*, **15**, pp. 12–23.
9. Lewis, T. J. (1994). Nanometric Dielectrics, *IEEE Trans. Dielectr. Electr. Insul.*, **1**, pp. 812–825.
10. Lewis, T. J. (2004). Interfaces are the dominant feature of dielectrics at the nanometric level, *IEEE Trans. Dielectr. Electr. Insul.*, **11**, pp. 739–753.
11. Roy, M., Nelson, J. K., MacCrone, R. K., et al. (2005). Polymer nanocomposite dielectrics: The role of the interface, *IEEE Trans. Dielectr. Electr. Insul.*, **12**, pp. 629–643.
12. Dissado, L. A., and Fothergill, J. C. (1992) *Electrical Degradation and Breakdown in Polymers*, IET Materials and Devices Series 9.

13. Roy, M., Nelson, J., MacCrone, R., et al. (2007). Candidate mechanisms controlling the electrical characteristics of silica/XLPE nanodielectrics, *J. Mater. Sci.*, **42**, pp. 3789–3799.

14. Tan, Q., Cao, Y., Irwin, P., et al. (2007). DC Breakdown in polyetherimide composites and implication for structural engineering, *IEEE Int. Conf. Solid Dielectr. (ICSD)*, pp. 411–414.

15. Tuncer, E., Polizos, G., Sauers, I., et al. (2012). Epoxy nanodielectrics fabricated with in situ and ex situ techniques, *J. Exp. Nanosci.*, **7**, pp. 274–281.

16. Hang, X. Y., Lei, F., and Jang, P. K. (2010). Effect of nanoparticle surface treatment on morphology, electrical and water treeing behavior of LLDPE composites, *IEEE Trans. Dielectr. Electr. Insul.*, **17**, pp. 1697–1704.

17. Kim, P., Jones, S. C., Hotchkiss, P. J., et al. (2007). Phosphonic acid-modified barium titanate polymer nanocomposites with high permittivity and dielectric strength, *Adv. Mater.*, **19**, pp. 1001–1005.

18. Li, S. T., Yin, G. L., Bai, S. N., et al. (2011). A new potential barrier model in epoxy resin nanodielectrics, *IEEE Trans. Dielectr. Electr Insul.*, **18**, pp. 1535–1543.

19. Li, S. T., Yin, G. L., and Li, J. Y. (2012). Breakdown performance of low density polyethylene nanocomposites, *2012 IEEE 10th Int. Conf. Proper. Appl. Dielectr. Mater. (ICPADM)*, pp. 1–4.

20. Wang, L., Xu, M., Feng, J. Q., et al. (2006). Study on AC breakdown property of Nano-Ag/epoxy resin composite, *Proc. 8th Int. Conf. Propert. Appl. Dielectr. Mater.*, pp. 163–166.

21. Kojima, Y., Usuki, A., Kawasumi, M., et al. (1993). One-pot synthesis of nylon-6 clay hybrid, *J. Polym. Sci. Pol. Chem.*, **31**, pp. 1755–1758.

22. Imai, T., Sawa, F., Ozaki, T., et al. (2006). Influence of temperature on mechanical and insulation properties of epoxy-layered silicate nanocomposite, *IEEE Trans. Dielectr. Electr. Insul.*, **13**, pp. 445–452.

23. Tuncer, E., Rondinone, A. J., Woodward, J., et al. (2009). Cobalt iron-oxide nanoparticle modified poly(methyl methacrylate) nanodielectrics, *Appl. Phys. A-Mater. Sci. Process.*, **94**, pp. 843–852.

24. Takala, M., Karttunen, M., Salovaara, P., et al. (2008). Dielectric properties of nanostructured polypropylene-polyhedral oligomeric silsesquioxane compounds, *IEEE Trans. Dielectr. Electr. Insul.*, **15**, pp. 40–51.

25. Chen, Y., Cheng, Y. H., Zhou, J. B., et al. (2008). Pulsed vacuum flashover of Al_2O_3/epoxy nanocomposites, *Int. Symposium Electr. Insul. Mater., 2008. (ISEIM)*, pp. 36–39.

26. Chen, Y., Cheng, Y. H., Wu, K., et al. (2009). Flashover characteristic of epoxy composites filled with different micro-inorganic oxide particles under nanosecond pulse in vacuum, *IEEE Trans. Plasma Sci.*, **37**, pp. 195–203.

27. Cheng, Y. H., Wang, Z. B., and Wu, K. (2012). Pulsed vacuum surface flashover characteristics of TiO_2/epoxy nano-micro composites, *IEEE Trans. Plasma Sci.*, **40**, pp. 68–77.

28. Kirkici, H., Serkan, M., and Koppisetty, K. (2007). Nano/micro dielectric surface flashover in partial vacuum, *IEEE Trans. Dielectr. Electr. Insul.*, **14**, pp. 790–795.

29. Fang, L., and Kirkici, H. (2010). Nanodielectric surface flashover studies under kHZ range pulsed fields in partial vacuum, *2010 IEEE Int. Power Modulator and High Voltage Conf. (IPMHVC)*, pp. 453–456.

30. Zhao W. B., Zhang G. J., Yang Y., et al. (2007). Correlation between trapping parameters and surface insulation strength of solid dielectric under pulse voltage in vacuum, *IEEE Trans. Dielectr. Electr. Insul.*, **14**, pp. 170–178.

31. Tanaka, T. (2005). Dielectric nanocomposites with insulating properties, *IEEE Trans. Dielectr. Electr. Insul.*, **12**, pp. 914–928.

32. Li, S. T., Wang, W. W., Ni, F. Y., et al. (2011). Surface flashover in vacuum and bulk breakdown in polystyrene nanocomposites, *2011 Int. Conf. Electr. Insul. Mater. (ISEIM 2011)*, pp. 486–490.

33. Wang, W. W., Li, S. T., Tang, F., et al. (2012). Characteristics on surface flashover of polyethylene nanocomposites film in vacuum, *2012 IEEE Int. Conf. Condition Monitoring and Diagnosis (CMD 2012). Proceedings*, pp. 1055–1058.

34. Sabuni, H., and Nelson, J. K. (1976). Factors determining the electric strength of polymeric dielectrics, *J. Mater. Sci.*, **11**, pp. 1574–1576.

35. Crine, J. P., and Vijh, A. K. (1988). On the relation between the cohesive energy density and the electric strength of solid polymer dielectrics, *Conf. Electr. Insul. Dielectr. Phenom. (CEIDP)*, pp. 424–429.

36. Ieda, M. (1980). Dielectric-breakdown process of polymers, *IEEE Trans. Electr. Insul.*, **15**, pp. 206–224.

37. Park, C. H., Okajima, K., Hara, M., et al. (1983). Effect of heat treatment on dielectric strength of polyethylene terephthalate under compressive stress, *IEEE Trans. Electr. Insul.*, **EI-18**, pp. 380–389.

38. Li, S. T., Wang, W. W., Yu, S. H., et al. (2014). Influence of hydrostatic pressure on dielectric properties of polyethylene/aluminum oxide nanocomposites, *IEEE Trans. Dielectr. Electr. Insul.*, **21**, pp. 519–528.

39. Crank, J., and Park, G. S. (1968) *Diffusion in Polymers*, Academic, London.

40. McCrum, N. G., Read, B. A., and Williams, G. (1967) *Anelastic and Dielectric Effects in Polymeric Solids*, London, Wiley, London.

41. Utracki, L. A., Simha, R., and Garcia-Rejon, A. (2003). Pressure-volume-temperature dependence of poly-epsilon-caprolactam/clay nanocomposites, *Macromolecules,* **36**, pp. 2114–2121.

42. Deng, Q., and Jean, Y. C. (1993). Free-volume distributions of an epoxy polymer probed by positron annihilation: pressure dependence, *Macromolecules*, **26**, pp. 30–34.

43. Nelson, J. K., Utracki, L. A., MacCrone, R. K., et al. (2004). Role of the interface in determining the dielectric properties of nanocomposites, *2004 Annual Report Conf. Electr. Insul. Dielectr. Phenom., (CEIDP)*, pp. 314–317.

44. Artbauer, J. (1996). Electric strength of polymers, *J. Phys. D: Appl. Phys.*, **29**, pp. 446–456.

45. Frechette, M. F., Larocque, R. Y., Trudeau, M., et al. (2008). Nanostructured polymer microcomposites: A distinct class of insulating materials, *IEEE Trans. Dielectr. Electr. Insul.,* **15**, pp. 90–105.

46. Frechette, M. F. (2010). Nanodielectrics: A panacea for solving all electrical insulation problems?, *Proc. 2010 IEEE Int. Conf. Solid Dielectr. (ICSD 2010)*, pp. 1–29.

47. Ieda, M., Nagao, M., and Hikita, M. (1994). High-field conduction and breakdown in insulating polymers. Present situation and future prospects, *IEEE Trans. Dielectr. Electr. Insul.,* **1**, pp. 934–945.

48. Kao, K. C. (2004) *Dielectric Phenomena in Solids*, Elsevier Academic Press.

49. Zeller, H. R., Pfluger, P., and Bernasconi, J. (1984). High-mobility states and dielectric breakdown in polymeric dielectrics, *IEEE Trans. Electr. Insul.,* **19**, pp. 200–204.

50. Jonscher, A. K., and Lacoste, R. (1984). On a cumulative model of dielectric breakdown in solids, *IEEE Trans. Electr. Insul.*, **EI-19**, pp. 567–577.

51. Muralidharan, M., Kumar, S. A., and Thomas, S. (2008). Morphology and transport characteristics of poly(ethylene-co-vinylacetate)/clay nanocomposites, *J. Membrane Sci.,* **315**, pp. 147–154.

52. Nelson, J. K., and Hu, Y. (2005). Nanocomposite dielectrics: Properties and implications, *J. Phys. D: Appl. Phys.,* **38**, pp. 213–222.

53. Nelson, J. K. (2010). *Dielectric Polymer Nanocomposites,* Springer, New York, Dordrecht, Heidelberg, London.

54. Frenkel, J. (1938). On pre-breakdown phenomena in insulators and electronic semi-conductors, *Phys. Rev.,* **54**, pp. 647–648.

55. Matsui, K., Tanaka, Y., Takada, T., et al. (2005). Space charge behavior in low-density polyethylene at pre-breakdown, *IEEE Trans. Dielectr. Electr. Insul.,* **12**, pp. 406–415.

56. Chen, G., Zhao, J., Li, S. T., et al. (2012). Origin of thickness dependent dc electrical breakdown in dielectrics, *Appl. Phys. Lett.,* **100**, pp. 222904.

57. Chen, G., and Zhao, J. W. (2011). Observation of negative differential mobility and charge packet in polyethylene, *J. Phys. D: Appl. Phys.,* **44**, pp. 212001.

58. Wang, L., Xu, M., Feng, J. Q., Cao, X. L. (2006). Study on AC Breakdown Property of NANO-Ag/EPOXY Resin Composite, *2006 IEEE 8th Int. Conf. Proper. Appl. Dielectr. Mater. (ICPADM)*, pp. 163–166.

59. Tanaka, T., Kozako, M., Fuse, N., et al. (2005). Proposal of a multi-core model for polymer nanocomposite dielectrics, *IEEE Trans. Dielectr. Electr. Insul.,* **12**, pp. 669–681.

60. Lei Q. Q. (2009). Multi-hierarchical structures of nanodielectrics and their macro-properties, *The 354th Xiangshan Academic Seminar*, pp. 7–19 (in Chinese).

Chapter 10

Suppression of Surface Erosion by Surface-Treated Fillers

Isaías Ramírez-Vázquez

Gerencia de Transmisión y Distribución, Instituto de Investigaciones Eléctricas, Reforma 113, Col. Palmira, Cuernavaca, Morelos 62490, México

iramirez@iie.org.mx

Outdoors, moisture as rain, drizzle or in another form in combination with environmental pollution causes electrical discharges on polymer insulation that result in degradation in the form of electrical tracking and erosion. Even indoors, it is possible that this degradation can occur, although the phenomena take a longer time to evolve. In such cases, degradation is detrimental to the life of the insulation. In order to make the insulation more resistant to these discharges, inorganic fillers are added to polymeric materials which, at the same time, reduces material costs. Indeed, there are more than twenty reasons for including fillers in a polymer matrix. Microfillers have been used widely to modify physical properties and, consequently, the effects of this on the properties of the resultant composite are well known. However, in recent times, nanofillers are being

Tailoring of Nanocomposite Dielectrics: From Fundamentals to Devices and Applications
Edited by Toshikatsu Tanaka and Alun S. Vaughan

ISBN 978-981-4669-80-1 (Hardcover), 978-981-4669-81-8 (eBook)
www.panstanford.com

used increasingly in some formulations, in order to improve both mechanical and electrical properties. The main problem in using nanofillers is that the nanoparticles agglomerate easily, because of their high surface energy, such that conventional mixing techniques are unable to achieve effective dispersion. A secondary problem is the incompatibility of the hydrophobic polymer with the hydrophilic nanoparticles, which results in poor interfacial interactions. Nevertheless, if the nanoparticles are well dispersed, the electrical properties of these materials will be significantly improved.

To improve nanoparticle dispersion, several techniques are available apart from mixing, which include surface modification of the nanoparticles by chemical and physical methods through the use of surfactants.

To evaluate the erosion resistance of nanocomposites, several test methods, including laser ablation, corona resistance, arcing resistance, salt fog, and inclined plane tests are used. An example included in this chapter confirms that combinations of microfiller and nanofiller with a surfactant results in composites with improved erosion resistance to dry band arcing. In order to understand this mechanism, fumed nanosilica, natural nanosilica, and nano alumina were introduced into a silicone rubber (SiR) matrix and the thermally decomposed silicone and the residual char that is formed during laser ablation tests were examined. The formation of a continuous layer on the surface behaves as a thermal insulator, protecting the material underneath from further decomposition.

10.1 The Surface Erosion Issue in Insulation Systems

In general, any conductor at a high voltage requires electrical insulation and in all such applications, insulating properties must be retained up to a certain level of voltage. For example, in electrical generators and motors, the conductor used for the winding has to be insulated; in power lines it is necessary to insulate the high voltage from the ground side and, consequently, appropriate insulators are needed. However, outdoor high voltage insulation subjected to environmental factors such as any type of moisture, UV, and pollution develops a small leakage current that can end in tracking and erosion of the surface. Usually, for high voltage silicone rubber insulators, the insulation thickness is 3 mm and over long times under polluted conditions, the leakage current can erode the complete thickness of the insulation cover (housing), as shown in Fig. 10.1. For windings in electrical machines, elevated temperatures, dust, and environmental moisture can similarly lead to erosion and tracking.

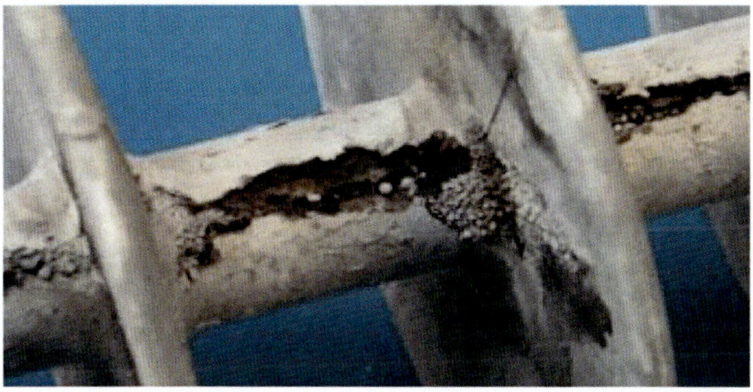

Figure 10.1 Tracking and erosion in silicone rubber insulators.

10.2 The Role of Fillers in Nanocomposites

Microfillers are commonly used to modify the physical properties of polymers and, consequently, the properties of the resultant composites are well known. Often, the amount of microfiller used in a composite is very high; conversely, in nanocomposites, small amounts of nanofiller are required but, despite this, their inclusion has been shown to lead to significant improvements in surface erosion resistance, mechanical properties, etc.

Some of the reported effects of nanofillers are

(a) Improved thermal conductivity of the compound, thereby improving heat dissipation
(b) Reduced organic material exposure to the heat caused by electrical surface discharges
(c) Improved dielectric strength, elongation, scratch resistance, hardness, and mechanical strength
(d) Longer time in lifetime tests
(e) Prevention of electrical treeing

For a given formulation, the quantity and type of nanofiller included is critical.

10.3 The Problem of Efficient Mixing

Mixing is a critical step in the preparation of polymeric nanocomposites. The dispersion rate seems to be linear for short mixing times but exponential over longer mixing intervals, depending on the type of mixer used. Another problem is the incompatibility of the hydrophilic nanoparticles with the hydrophobic polymer, resulting in poor interfacial interactions. The optimum mixing time and temperature are specific for each system and, currently, key parameters need to be established empirically for every formulation. Many different types of mixers have been used as shown in Table 10.1, but those that result in high shear forces are preferable, because they help to break apart agglomerates. However, these mixers do not universally give satisfactory results when dispersing nanoparticles into matrix materials [1]. Also ultrasonic homogenizers are helpful to achieve a more uniform dispersion of the nanofillers within the matrix.

Table 10. 1 Different type of mixers and their characteristics [2]

Mixer type	Blade type	Characteristics
Single stirrer mixer	(a) Anchor blades (b) Helical ribbons	Useful for viscous mixtures such as thick pastes or polymer solutions
Change can mixer	—	Allows the separation of the mixing blade(s) from the mixing vessel. They are available over a wide range of sizes from 1 L to 4 m^3
Double arm kneading mixer	Blade shapes available: (a) Sigma (b) Dispersion (c) Multiwing overlap (d) Double naben	Sigma-blade mixers range in size from 1 L to 5 m^3 The Banbury mixer is frequently used for very viscous materials such as rubber
Plow mixers, ribbon		They are used for free-flowing paste mixing applications, as

blenders, cone and screw mixers, and Mullers		well as just for solids blending purposes
Continuous mixers	Single-screw extruders	One or two rotors operate in an open or closed trough, which may be jacketed. The rotors may be cored to provide additional heat transfer area
Laboratory high shear force	Slotted head, disintegrating head, fine screen head, and axial flow head.	Used for homogenization, solubilization, emulsification, powder wet-out, grinding, and particle size reduction in liquids
Ultrasonic homogenizers	Probes: Microtips, tapered and flat tips for use in different volumes.	Emulsifying of hardly mixable liquids, deagglomeration of nanoparticles (nanostructured material), acceleration of chemical reactions, production of dispersions

Several high shear mixers are available commercially. For example, if a laboratory high shear force mixer with a blade diameter of 16 mm is used, it has been suggested that tip speeds of 8–10 m/s (peripheral velocity) should be used in order to obtain an adequate dispersion [3]. For such a mixer, the rpm is calculated as follows:

$$\frac{\text{rpm} = \text{peripheral velocity (m/s)} \cdot 60 \cdot 1000}{d\,(\text{mm}) \cdot \pi} \tag{10.1}$$

The wet-in time is defined as the time required for all the nanoparticles to be wetted by the liquid medium, which can be achieved using low shear mixing (e.g., <6,000 rpm). Once the nanoparticles are wetted, the mixing speed can then be increased to higher speed to begin dispersion.

Another critical aspect is the blade to vessel ratio. This ratio is chosen to be in the range 1:2 to 1:3, to produce a clear vortex during mixing.

In the case of ultrasonic stirrers, it is recommended that the required mixing time is determined by verifying the dispersion of the nanoparticles using an appropriate microscopy technique.

10.4 Dispersion of Nanoparticles

The feasibility of a mixing process for highly cohesive nanoparticles can be determined by calculating the forces generated during mixing and demonstrating that these forces are larger than the cohesive or adhesive forces acting on the powder particles. Interparticle forces can be of the van der Waals type (inversely proportional to the distance between the particles), electrostatic or magnetic attraction, chemical bonding, or through physical interactions like surface tension.

According to several researchers, the mixing of particles in the size range of a few nanometers is extremely difficult because, for such systems, interparticle forces are stronger than for particles with dimension of the order of micrometers.

In order to improve particle dispersion, diverse techniques are available apart from mixing [4, 5]. These techniques can be surface modification of the nanoparticles by physical and chemical methods. Also, calcination is another technique that can be used to improve particle dispersion.

10.4.1 Surface Modification of Nanofillers

Improved dispersion can be achieved through physical or chemical interactions between the filler and the modifier. When nanoparticles are modified by one of these methods, either the character of the surface is changed from hydrophilic to hydrophobic (and vice-versa), or specific groups are chemically bonded on their surface (change in functionality). Fillers are typically hydrophilic and do not disperse easily within most polymeric materials, which are usually hydrophobic. Such modification not only contributes to reinforcement, but also increases the interactions of the particles to impact rheological properties, prevents sedimentation and agglomeration, and aids dispersion. Also, improved dispersion of the nanoparticles can be achieved with solvents. These solvents can be hexane, methanol, ethanol, and toluene. Ultrasonic stirrers are used in combination with these solvents to improve nanofiller dispersion.

10.4.1.1 Physical methods

Surface modification by physical methods can be achieved by using a surfactant resulting in secondary (van der Waals, electrostatic, and/or hydrogen bonding) forces between the nanoparticles and the modifier. A surfactant (surface active agent) is a molecule which, when added to a liquid at low concentration, changes the properties of that liquid at a surface or interface [6]. The principle of surfactant treatment is the preferential adsorption, or tendency for a surfactant molecule to collect at the interface, due to polar groups on the surfactant molecules interacting favorably with the high energy surface of the filler. A surfactant is characterized by its tendency to adsorb at surfaces and interfaces. The surfactant concentration at the boundary depends on the surfactant structure and also on the nature of the two phases defining the interface. There is no effective surfactant for all uses; the choice depends on the application [7].

Surfactant adsorption is determined by two main factors: the interactions of the surfactant with the surface and the hydrophobicity of the surfactant. In addition, hydrophilic nanofillers stored under ordinary conditions (reagents room, laboratory, etc.) have their surface partly blocked by adsorbed species, and this condition lowers their effectiveness. To activate the surface of the nanofiller, thermal treatment may be required.

All surfactants consist of at least two parts, one that is soluble in a specific fluid, known as the lyophilic part (or hydrophilic part in aqueous systems) and one that is insoluble, known as the lyophobic (or hydrophobic) part. Thus, for example, when a surfactant is adsorbed from an aqueous solution onto a hydrophobic surface, the hydrophobic component becomes oriented towards the surface such that the polar group is exposed to the water. Since the surface then becomes more hydrophilic, the interfacial tension between the surface and water is reduced.

By 1993, around 2900 different types of surfactants were in use in industry [8]. Rahul et al. [9] developed a micro emulsion-based surface modification method for the introduction of different functional groups onto the surface of nanosilica particles, in order to keep the particles well dispersed. Tetraethyl orthosilicate (TEOS), Triton^{TM} X-100, and various organosilane reagents have also been used for this purpose. By using such a surface modification scheme, nanoparticle aggregation can be minimized.

Several mixing processes were used by Dongguang et al. [10], and their results indicate that solvent-based methods are the most effective (with ethanol and hexane). Thereafter, a mixing process consisting of stirring the nanoparticles in a heated and pressurized vessel with CO_2 was performed, which was shown to have a somewhat reduced efficiency. Nevertheless, this work also concluded that simple dry powder processing methods performed nearly as well as the solvent-based method in terms of producing a homogeneous mixture of nanoparticles.

Elsewhere, composites containing 5 wt% of calcined fumed nanosilica or 2.5 wt% of nano alumina in a silicone rubber matrix were prepared, with and without surfactant, to observe the dispersion of the nanoparticles in each case [11]. Resulting SEM images are shown in Fig. 10.2, where the left and right columns show specimens prepared without and with the surfactant, respectively. The composites prepared without the surfactant clearly contain larger agglomerates, which is in agreement with the findings of Rätzke et al. [12].

(a)

(b)

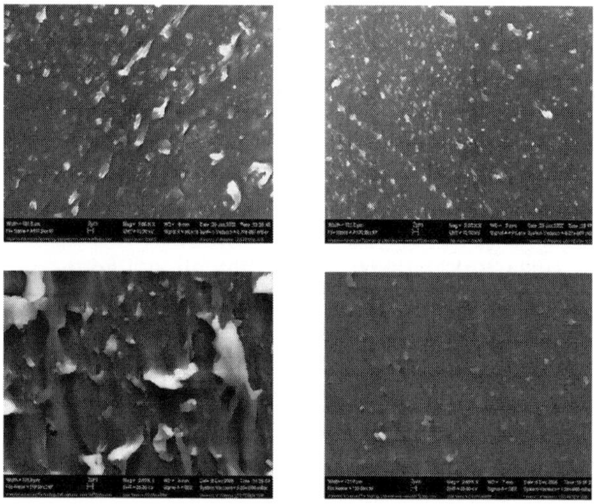

Figure 10.2 Effect of Triton on nanoparticle dispersion in composites containing nanofiller + silicone rubber (left column without surfactant; right column with surfactant) [11]. (a) Composite 5% by weight calcined fumed nanosilica. (b) Composite 2.5% by weight nano Al_2O_3.

The micrographs on the right hand side of Fig. 10.2 show that the surfactant helps to achieve a better dispersion of the nanofiller, although some agglomeration persists. It is clear that the formation of large agglomerates of calcined fumed nanosilica and nano alumina is favored in the silicone rubber matrix when no surfactant is used.

Other SEM tests have been conducted with TiO_2, ZnO, and $BaTiO_3$ nanofillers in silicone rubber matrix, to test the generality of the findings discussed above. For these nanofillers, the surfactant does not appear to improve the dispersion (Figs. 10.3–10.5).

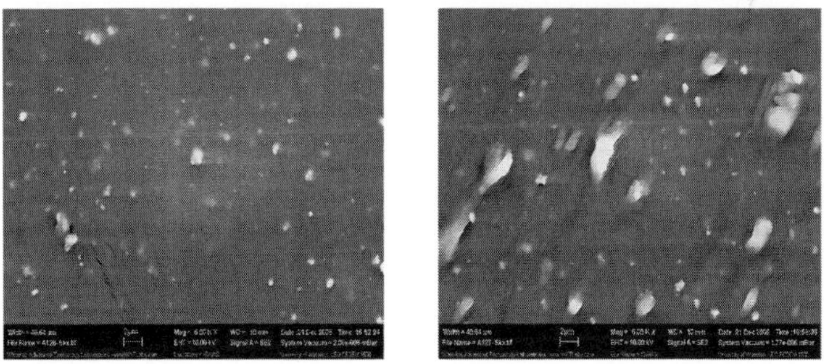

Figure 10.3 Effect of Triton in composites containing 2.5% by weight nano TiO_2 + 97.5% silicone rubber (left micrograph without surfactant; right micrograph with surfactant) [11].

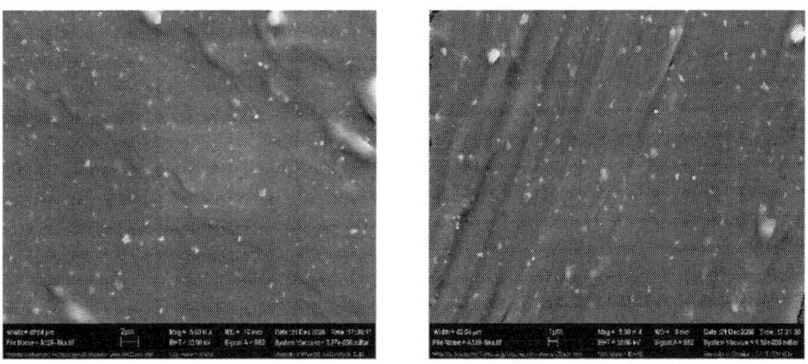

Figure 10.4 Effect of Triton in composites containing 2.5% by weight nano ZnO + 97.5% silicone rubber (left micrograph without surfactant; right micrograph with surfactant) [11].

In concert, the results presented above indicate that only certain surfactants are able to adsorb efficiently onto the surface of specific nanofillers. In this case, Triton appears to interact favorably with nanosilica, presumably through its hydrophilic poly(ethylene oxide) segment and with the silicone rubber matrix through its hydrophobic component. It does not, however, interact efficiently with other nanofillers.

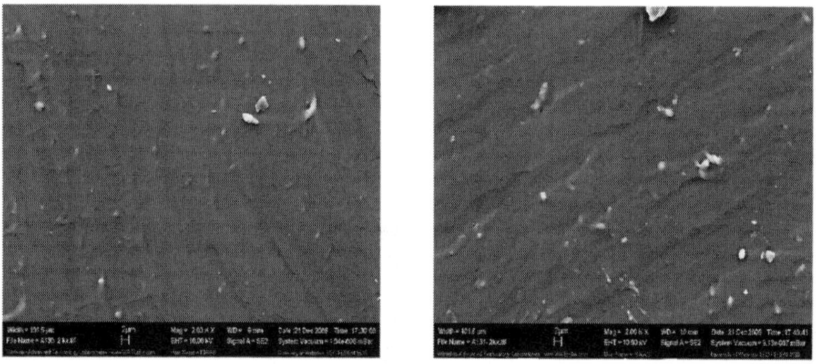

Figure 10.5 Effect of Triton in composites containing 2.5% by weight nano $BaTiO_3$ + 97.5% silicone rubber (left micrograph without surfactant; right micrograph with surfactant) [11].

10.4.1.2 Chemical methods

A variety of coupling agents including silanes, titanates, and zirconates have been used to improve the adhesion between inorganic fillers and organic matrices. For example, silanes can be used to reduce the hydrophilic properties of silica. However, uniform surface coverage by coupling agents is hard to achieve with nanoparticles by simple physical and mechanical means.

Stearic acid has been used with nano $CaCO_3$, with the result that virtually no nanoparticle agglomerates remain when an appropriate dosage of the dispersing agent is used [13]. Some results for the surface modification of SiO_2 nanoparticles with oleic acid obtained by Zongwei et al. [14] also confirmed that nanoparticle agglomeration decreased because of increased spacing (decreased attraction) between the nanoparticle cores due to the layer of surface modifier.

Another non-reactive modifier used by Seon et al. [15] to make the filler surface hydrophobic is stearic acid. The presence of adsorbed stearic acid on the surface of the silica nanoparticles reduced the interactions between the silica nanoparticles within agglomerates, which could consequently be broken down more easily.

10.4.1.3 Thermal treatment of nanoparticles (calcination)

Nanofillers stored under conditions normally found in reagent rooms or laboratories have their surface partially blocked by adsorbed species, mainly moisture. For silica, this condition translates into a lower effectiveness in dispersion. Calcination is one way to activate the silica surface, and an additional benefit of this process is the disintegration of the silica aggregates and pellets formed during storage [16].

Many of the properties of silica such as adsorption, adhesion, chemical, and catalytic properties depend on the chemistry and geometry of their surface. Silanol (–OH) groups on the silica surface are the main centers for the adsorption of water molecules [17, 18]. The concentration of silanol groups on the silica surface, α_{OH}, expressed as the number of –OH groups per square nanometer, is often called the silanol number. The numerical value, obtained by Zhuravlev, is α_{OH} = 4.6 (3.65 for fumed silica) [17].

Dehydration of the silica surface, i.e. the removal of physisorbed (adsorbed) water occurs at temperatures below 473 K. The concentration of silanol groups on the surface also decreases monotonically with increasing temperature when silica is heated under vacuum, according to Zhuravlev [17].

In the research of Wypych [19], most of the adsorbed water is removed at approximately 423 K. Internal silanol groups (structurally bound water inside the silica skeleton and very fine pores of diameter lower than 1 nm) begin to condense at about 873 to 1073 K, and in some cases at lower temperatures. At higher temperatures, from 1273 to 1373 K, only isolated silanol groups remain on the silica surface.

At a sufficient surface concentration, the –OH groups make the silica surface hydrophilic, whereas, when siloxane bridges predominate on the silica surface, the surface is hydrophobic. Figure 10.6 relates the concentration of surface hydroxyl groups to the temperature of thermal treatment in vacuum up to 1373 K [17].

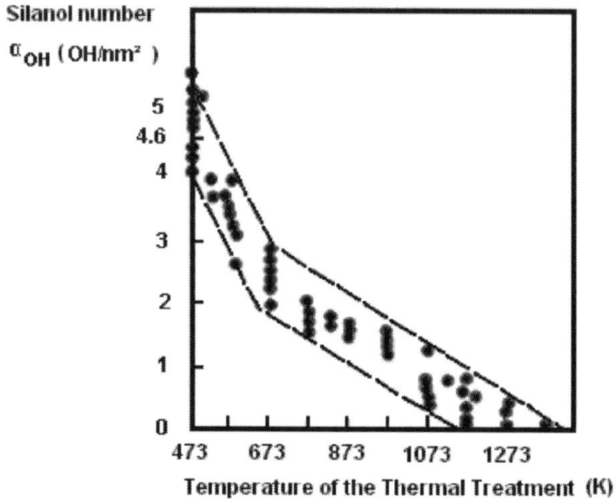

Figure 10.6 Silanol number as a function of the thermal treatment temperature of SiO_2 [17].

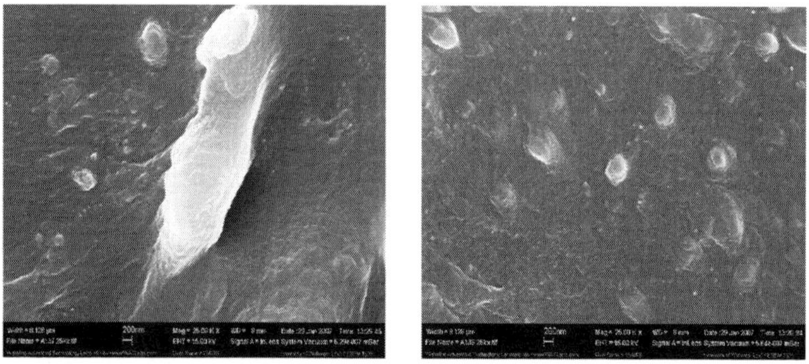

Figure 10.7 Effect of Triton in composites containing 5% by weight of calcined fumed nanosilica + 95% silicone rubber (left column without surfactant; right column with surfactant) [11].

For nanocomposites based on silicone rubber and 5% by weight of calcined fumed nanosilica, specimens have been prepared with and without added surfactant, to examine the effect of changing the surface chemistry of the nanoparticles on dispersion (see Fig. 10.7). The SEM images on the right show the system containing 33 pph of surfactant, while those on the left were obtained from the system without the surfactant. From these micrographs, it is clear that the presence of the surfactant improves the dispersion of the nanoparticles [11]. Even after calcination at 1273 K, some hydroxyl groups still remain on the surface of nanosilica. According to Zhuravlev et al. [17] $\alpha_{OH} < 1$, but these silanol groups nevertheless interact with the surfactant and facilitate dispersion.

10.5 Treated Nanofillers

In some areas, polymer insulators are exposed to a large amount of pollutants and moisture from the environment. Under these conditions, degradation of the housing takes place causing tracking and erosion. The capacity or ability of the polymer to resist electrical discharges or corona discharges can be improved by using nanofillers as reinforcement.

The surface of the filler is more favorable for some applications, depending of the treatment given to the nanofiller. Consequently, many different surface treatments have been employed, involving the application of silanes, siloxanes, and organic molecules to inorganic nanofillers. For example, dichlorodimethyl silane (DDS), hexamethyldisilazane (HMDS), octamethyl cyclotetrasiloxane (D4), dimethyldichlorosilane, and polydimethylsiloxanes have been widely used. The nanofillers are modified mainly by granulation processes, mechanical treatment, dispersion and densification, with the objective of modifying adhesion, rheological, mechanical, optical or electrical properties, antisetting, heat stability or pigmentation [20].

Many commercial materials, such as Aerosil R812, Aerosil R202, Aerosil R8200, Cabo-Sil TS-720, Aeroxide LE1, untreated fumed nanosilica (nfs); different surfactants such as Triton-100 red, Triton-114 red, Triton-405 red, and tetraethyl orthosilicate (TEOS), have been used in studies of the laser ablation behavior of silicone rubber-based nanocomposites. Evaluation of the results showed that surface treatment of the nanofiller has an influence in the erosion performance of the material. The nanocomposite that led to the lowest eroded mass corresponded to the 2.5% nfs + 20 pph of Triton X-100 surfactant, as shown in Fig. 10.8. It is believed that the improved performance of this system stems from a combination of improved dispersion of the nanofiller and the existence of favorable interactions between silanol groups on the silica and the silicone rubber matrix, in line with the images shown in Figs. 10.7 and 10.8.

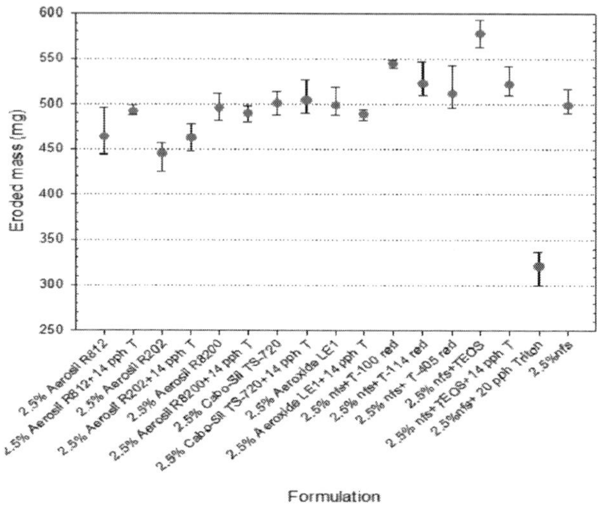

Figure 10.8 Eroded mass of composites with different surface treated nanoparticles.

10.6 Tests to Assess Erosion and Tracking Resistance

10.6.1 Inclined Plane Test

This technique is based on the ASTM D 2303 standard [21]. Five samples are tested; each sample is 50 mm wide × 130 mm long × 5 mm thick as shown in Fig. 10.9. For each formulation, the upper side and the mould side are tested. Specimens are initially energized at 3 kV, and then the voltage is increased by 0.25 kV every hour. The final step occurs at 3.75 kV, after 3 h; the duration for the complete test is 4 h. The flow rate of the NH_4Cl solution is 0.30 mL per minute and the concentration 1 g/dm^3 in deionized water.

After the test, the composites are inspected and cleaned. The eroded mass of the material is determined from the difference between the initial and final sample mass. The mean and the standard deviation of the eroded mass are finally calculated for each formulation.

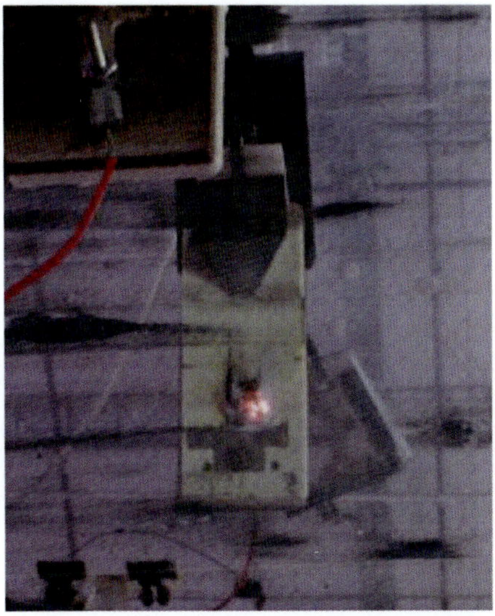

Figure 10.9 A sample undergoing the inclined plane test.

10.6.2 Laser Ablation Test

In this test, a pre-determined energy is deposited locally in each sample using an appropriate laser source (e.g., a Coherent infrared model FAP laser with an operating wavelength of 802 nm) and the resulting mass loss is determined. The absorbed energy produces a local increase in temperature (i.e. molecular vibrations), causing molecular bonds within the polymer to break [22].

Heat from dry band arcing is the main degradation factor in SiR used in outdoor insulation and, consequently, the degradation is thermal in nature. Laser ablation testing can therefore be used to simulate the effects of dry band arcing. Several tests are needed to adjust the laser, which in the example described below is operated in the continuous wave (CW) mode with a current of 17.5 A (power equivalent to 8.8 W) for 7 min, corresponding to a calculated total energy of 3700 J. The sample is located 50 mm from the laser source in all the tests as shown in Fig. 10.10.

In this laser test, the samples under test need to exhibit a dark color, in order to ensure uniform absorption of the laser radiation. This is achieved by including 2.5 wt% of iron oxide

(Fe_2O_3) in the mixture for all the composites [22]. Since Fe_2O_3 is stable at high temperatures, above the decomposition temperature of the SiR matrix, it may be considered otherwise inert [23].

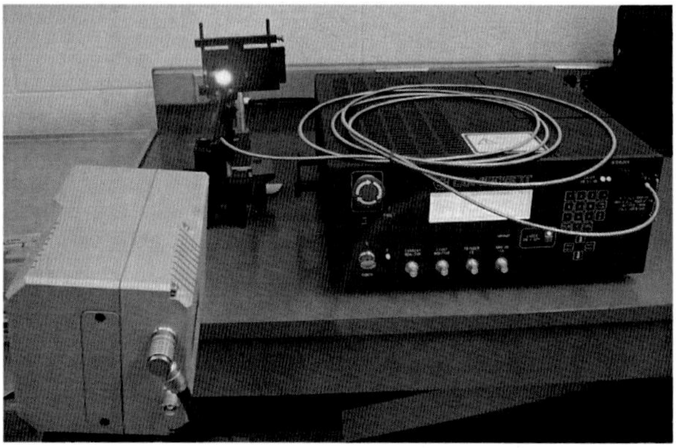

Figure 10.10 View of the laser ablation test.

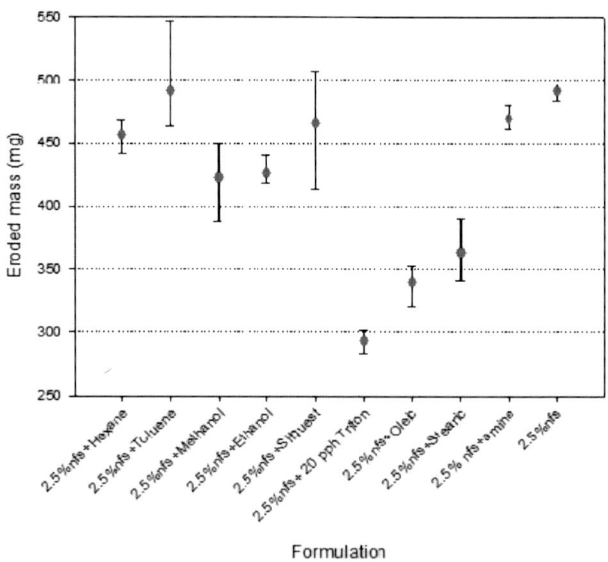

Figure 10.11 Eroded mass of composites with fumed nanosilica mixed with solvents or surfactants. The average eroded mass is represented by the dot; the 25th and 75th percentiles are represented by the horizontal lines of the bars.

Figure 10.11 contains data obtained in a laser ablation study of a range of different nanocomposite systems containing 2.5 wt% of fumed nanosilica processed using various solvents, a silane (silquest), a surfactant, stearic and oleic acid and a primary amine to aid dispersion. In Fig. 10.11, the average eroded mass obtained from each system is shown, from which it is evident that the lowest eroded mass is obtained from the formulation containing 2.5 wt% of nanofiller plus Triton, albeit that the performance of this system is comparable with that of composites processed with oleic and stearic acid. Thus, the surfactant consistently modifies the surface of the nanofiller and enhances erosion resistance.

10.6.3 Salt Fog Test

The procedure for this test is described in the IEC 62217 standard [24]. A salt fog chamber of maximum 15 m^3 is used and several nozzles are located on each side of the chamber. Salt water is prepared by dissolving 1 to 8 kg of NaCl per cubic meter of water. The selected concentration is maintained throughout the entire test and the flow rate should be 0.3 L/m^3 h. The applied voltage corresponds to a stress of 20 mm/kV, based on the creepage length of the insulator under test, and the maximum test duration is 1000 h. The test/fail criterion is that no tracking, puncture or erosion depth greater than 3 mm is permitted. A salt fog chamber is shown in Fig. 10.12.

Figure 10.12 View of a salt fog chamber.

10.6.4 Tracking Wheel Test

A typical setup for tracking wheel testing is shown in Fig. 10.13. In this procedure, insulators are tested to evaluate their tracking resistance, as described in IEC/TR 62730 [25]. Prior to testing on the tracking wheel, two insulators are cleaned with deionized water and then tested up to failure. In the test, each insulator remains stationary for 40 s in each of the four wheel positions and the travel time between positions is 8 s. The concentration of the saline solution is 1.4 kg of NaCl per cubic meter, which should be changed weekly, and the applied voltage is the creepage distance in mm divided by 28.6. In order to pass the test, the specimens should survive 30,000 cycles.

Figure 10.13 View of the tracking wheel test.

10.7 Representative Studies of Nanocomposite Systems: Erosion by Laser Ablation

10.7.1 Composites with Different Amounts of Surfactant and Filler

To illustrate the effect of nanofiller loading level and surface treatment, consider a series of silicone-based nanocomposites formulated with 2.5 or 5 wt% of nanofiller, together with various amounts of Triton as surface modifier [11]. Each formulation was subjected to localized erosion by laser ablation and the average eroded mass for fixed irradiation conditions was then derived from three equivalent specimens. In Figs. 10.14–10.19, the resulting data are displayed as the average eroded mass for each system as a function of the quantity of Triton used in the system, expressed in pph of nanofiller by weight. Linear regression analysis of these data up to 33 pph of surfactant was finally conducted.

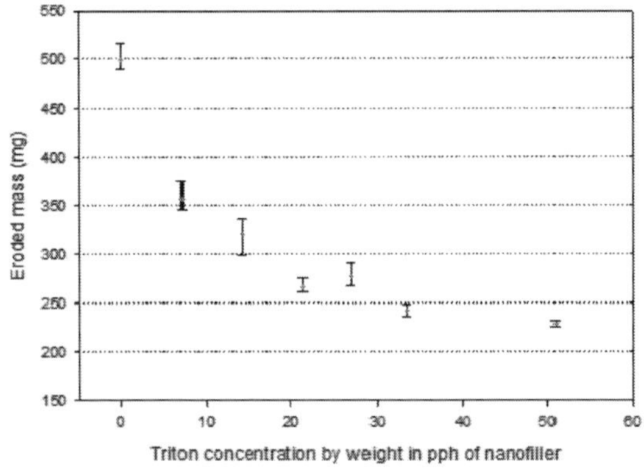

Figure 10.14 Eroded mass of silicone-based nanocomposites containing 2.5 wt% fumed nanosilica for various Triton concentrations (in pph of nanofiller). The average eroded mass derived from three repeat measurements is shown for each formulation.

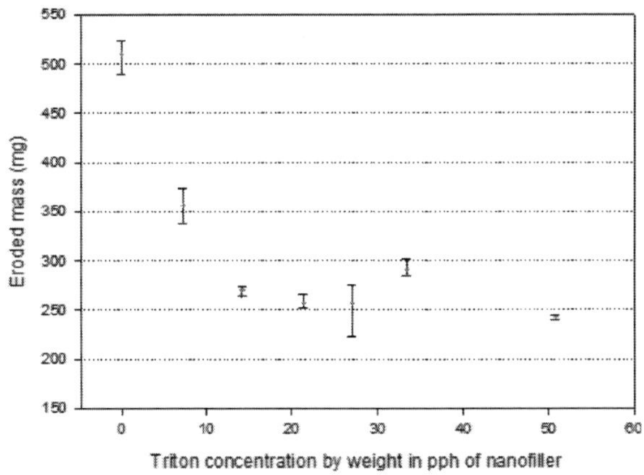

Figure 10.15 Eroded mass of silicone-based nanocomposites containing 5 wt% fumed nanosilica for various Triton concentrations (in pph of nanofiller). The average eroded mass derived from three repeat measurements is shown for each formulation.

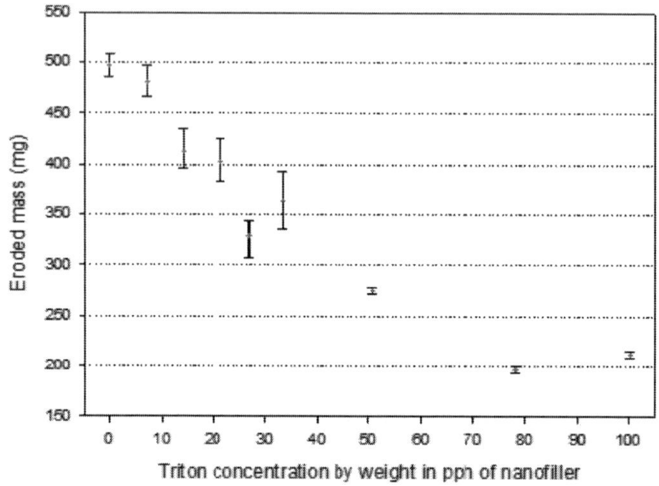

Figure 10.16 Eroded mass of silicone-based composites containing 2.5 wt% of natural nanosilica for various Triton concentrations (in pph of nanofiller). The average eroded mass derived from three repeat measurements is shown for each formulation.

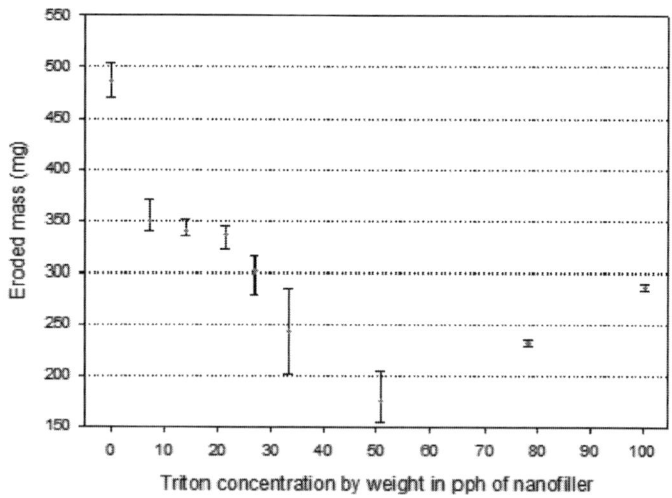

Figure 10.17 Eroded mass of silicone-based nano composites containing 5 wt% of natural nanosilica for various Triton concentrations (in pph of nanofiller). The average eroded mass derived from three repeat measurements is shown for each formulation.

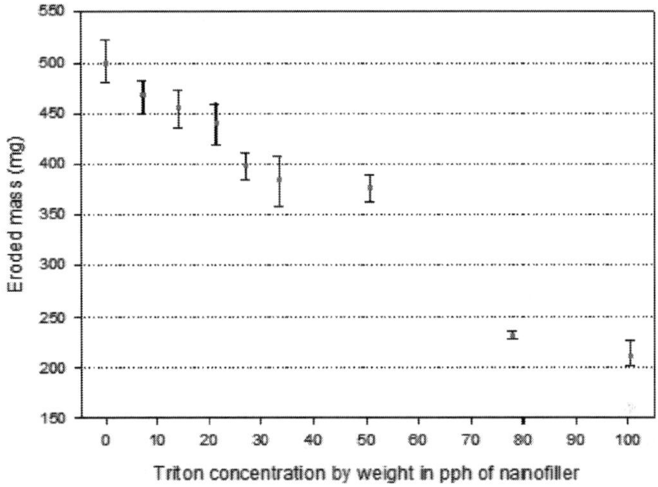

Figure 10.18 Eroded mass of silicone-based nano composites containing 2.5 wt% of nano alumina for various Triton concentrations (in pph of nanofiller). The average eroded mass derived from three repeat measurements is shown for each formulation.

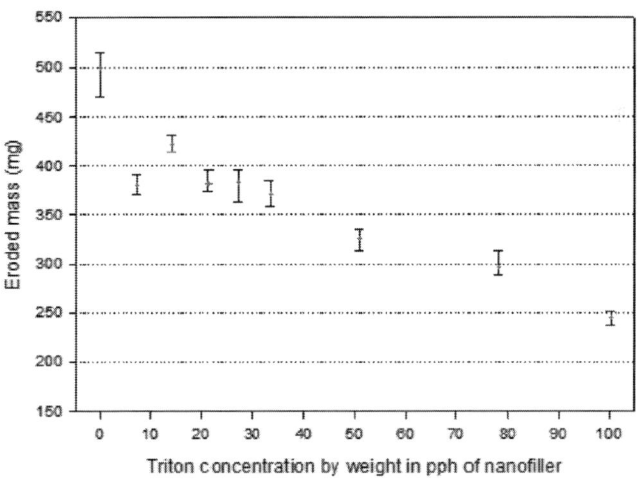

Figure 10.19 Eroded mass of silicone-based nano composites containing 5 wt% nano alumina for various Triton concentrations (in pph of nanofiller). The average eroded mass derived from three repeat measurements is shown for each formulation.

10.7.2 Nanocomposites Containing Fumed Nanosilica

Figures 10.14 and 10.15 show results obtained from systems containing 2.5 and 5 wt% of fumed nanosilica, respectively. In both cases, a marked decrease (about 50%) in the eroded mass is observed when approximately 20 pph of surfactant is included in the formulation at both loading levels. However, although a significant improvement in erosion resistance is observed with the addition of the surfactant, no significant difference in eroded mass is obtained on varying the quantity of nanofiller present in the system.

If it is assumed that Triton is adsorbed onto the filler particles in proportion to their BET surface area, as suggested by Seon et al. [15], then the following can be calculated. For a specific surface area of 350–430 m^2/g for fumed nanosilica and a wetting surface of 542 m^2/g for Triton, the maximum amount of surfactant that can be adsorbed per gram of nanofiller would be 0.65–0.8 g or 65–80 pph by weight. If it is further assumed that a homogeneous mixture is obtained and that this gives rise to uniform particle dispersion, then a linear relationship between the eroded mass and the amount of Triton added would be expected up to this maximum. Beyond this limit, the presence of excess Triton could affect the curing chemistry of the silicone.

10.7.3 Nanocomposites Containing Natural Nanosilica

Statistical analysis of the data shown below in Figs. 10.16 and 10.17 for systems containing 2.5 and 5 wt% of nanofiller, respectively, leads to first order linear regression coefficients of 5.1 and 5.9. Following the argument put forward for the fumed nanosilica, if the quantity of Triton taken up by the nanofiller is proportional to the BET surface area, then a maximum of 110–130 pph of Triton by weight of nanoparticles can be added. Consequently, it would be expected that a plot of eroded mass against Triton concentration should be linear up to this limit, as shown below. Once again, it appears that there is no additional reduction in eroded mass at higher concentrations of natural nanosilica. Furthermore, the effectiveness of natural nanosilica in reducing the mass loss is lower than for fumed nanosilica.

10.7.4 Nanocomposites Containing Nano Alumina

Erosion data obtained from these systems are presented in Figs. 10.18 and 10.19. Repeating the same linear regression process used above results in regression coefficients of 3.5 and 2.8 for nanocomposites containing 2.5 and 5 wt% of nano alumina, respectively. These coefficients are the lowest obtained for the three types of nanofillers analyzed. From the BET surface area, the optimum addition of Triton is more than 65 pph but less than 135 pph by weight for nano alumina.

10.8 Spectroscopic Analysis

FTIR spectroscopy has been used to determine the concentration of silanol groups on two type of fillers: fumed nanosilica and microsilica, as described by Kokai et al. [26]. In this method, solutions of organosilanols of known concentration in CCl_4 are used to calibrate the instrument, by integrating the absorption peaks in the spectral range 4300–4700 cm^{-1}. The concentration of silanol groups in the fillers was then determined by comparing the calibration curve with the equivalent peak area obtained from the samples of interest, dispersed in the same solvent.

FTIR spectra obtained at the highest concentrations of fumed nanosilica and microsilica are compared in Fig. 10.20. The ratio of the concentrations of silanol groups determined by FTIR for nano- and microsilica not subjected to calcination is 7.55. This is, per gram of filler, the concentration of silanol groups in the fumed nanosilica is about 7.55 times higher than in microsilica [27].

The large number of silanol groups on the fumed nanosilica originates from the manufacturing process used to obtain these materials and result in hydrogen bonds between the silica particles and the oxygen atoms of the polymer backbone in the silicone rubber matrix [17]. Because fumed nanosilica is characterized by a larger number of silanol groups than microsilica, it interacts more efficiently with the polymer matrix. Although this physical interaction is not as strong as covalent bonding, their large number, nevertheless, makes the interactions significant. This explains the higher residual weight and the higher degradation temperature observed in TGA for composites filled with 2.5% nfs as compared to 2.5% microsilica (Fig. 10.22).

Zhuravlev determined that the calcination of fumed nanosilica reduces the number of silanol groups present on the surface of the filler [17]. For fumed nanosilica calcined at 1173 K, the silanol absorption band is indeed significantly weaker than for the non-treated fumed nanosilica, as seen in Fig. 10.20 when comparing curves "c" and "a," respectively, obtained at the same filler concentration.

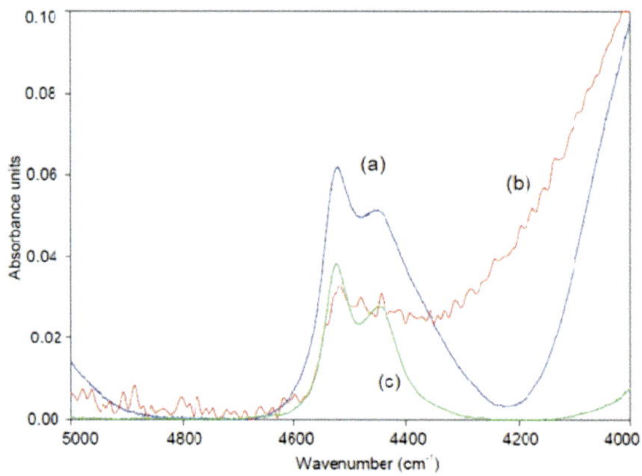

Figure 10.20 FTIR spectrum for dispersions of the following silicas in CCl_4: (a) fumed nanosilica (0.0394 g/mL); (b) micro silica (0.316 g/mL); (c) fumed nanosilica calcined at 1173 K (0.0394 g/mL).

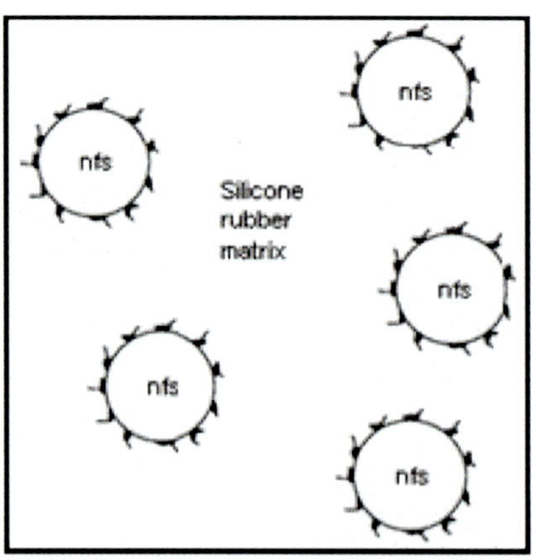

Figure 10.21 A simple adsorption model of non ionic surfactant on fumed nanosilica in a silicone rubber matrix.

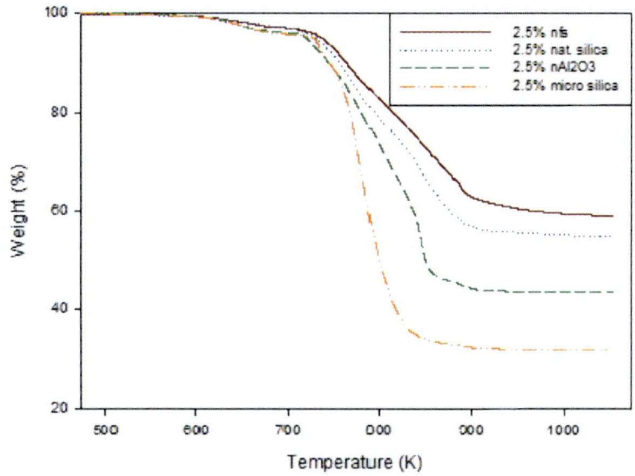

Figure 10.22 Thermal degradation of silicone rubber composites with different nano and micro fillers.

10.9 Interactions between Fumed Nanosilica and a Surfactant in a Polymer Matrix

The possible interaction mechanism leading to improved dispersion of fumed nanosilica in the presence of a surfactant is as follows. The fumed nanosilica is hydrophilic; this characteristic facilitates interactions with the hydrophilic poly(ethylene oxide) segment of the surfactant. When the surfactant is adsorbed, the hydrophobic group orients itself towards the silicone rubber matrix, exposing the hydrophilic group to the fumed nanosilica surface, as shown in Fig. 10.21. The surfactant consequently lowers the surface or interfacial tension between the nanofiller and the silicone matrix. This lowering of the interfacial tension makes it easier to separate the particles during mixing, thereby allowing improved dispersion of the nanofiller [27].

10.10 Thermal Gravimetric Analysis

Silicone rubber composites filled with fumed nanosilica, natural nanosilica, nano alumina, and microsilica have been subjected to thermogravimetric analysis, to assess the effect of the various fillers on decomposition behavior [27]. From this, it would appear that the fumed nanosilica imparts greater thermal stability to the silicone rubber matrix than the other fillers, as demonstrated by the lower slope and higher residual weight in the TGA curve shown in Fig. 10.22. Nevertheless, the onset of thermal degradation is similar for all composites. This result is consistent with the eroded mass behavior observed previously during the laser ablation tests (Figs. 10.14–10.19).

The higher residual weight observed for the composite containing 2.5 wt% fumed nanosilica than for the one with 2.5 wt% microsilica (and all the other fillers), as demonstrated in Fig. 10.22, is attributed to the large number of silanol groups in that material.

10.11 Thermal Conductivity of Composites

Thermal conductivity (TC) values have been obtained from heating and cooling profiles (decay profile) determined with an infrared camera, according to the methodology developed by Meyer et al. [28]. The principle is based on the known thermal conductivity of the unfilled polymer and on the assumption that heat transfer is mainly by conduction, such that heat transfer due to radiation and convection is negligible. The variation in thermal conductivity for all the evaluated composites is shown in Fig. 10.23; a baseline (indicated 100% SIR) is also plotted, to facilitate comparison of the composites and the unfilled silicone rubber [29].

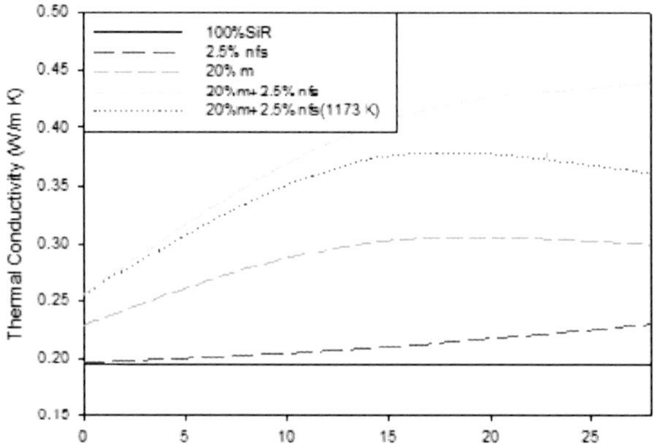

Figure 10.23 Variation in thermal conductivity with the amount of surfactant.

Table 10.2 compares TC values derived using the ASTM D5470 standard [30], theoretical models, proposed models, and results obtained from the decay profile. Theoretically, it is not possible to determine the effect on TC of adding the surfactant to each composite, since interactions between the filler, surfactant, and polymer matrix are not included in these models. However, with the decay profile methodology, it was possible to determine the TC of all of the composites.

Table 10.2 Thermal conductivity of composites containing different amounts of surfactant obtained using the approaches indicated

	Thermal conductivity (W/m K)						
	Calculated			Measured	Calculated from decay profile		
					Surfactant (pph)		
Formulation	Rule of mixtures	Lewis and Nielsen	Proposed model	ASTM D5470	0	14	28
2.5% nfs	0.217	0.202	0.202	0.220	0.197	0.209	0.229
20% m	0.309	0.234	0.233	0.280	0.229	0.301	0.300
20% m + 2.5% nfs	0.336	*	0.246	0.329	0.255	0.401	0.440
20% m + 2.5% nfs + 14 pph T	*	*	*	0.330	—	—	—
20% m + 2.5% nfs + 28 pph T	*	*	*	0.350	—	—	—
20% m + 2.5% nfs (1173 K)	*	*	*	**	0.254	0.373	0.361

*No calculation possible.
**Not determined.

In all the composites, the addition of the surfactant results in an increase in TC, as measured from the decay profiles. This increase in TC is attributed to better dispersion of the filler, as determined by SEM. As suggested in the literature, two different magnifications were used to examine the macroscopic and local filler distribution. The increase in TC is up to 72% when the surfactant is used in composites containing both micro- and nanofillers. If, however, the composite contains only nanofiller, then the addition of the surfactant results in an increase in TC of just 16%. The reduced effectiveness of the surfactant in composites containing only the nanofiller can ascribed to the weight percentage of nanofiller present in the formulation. Theoretically, in the absence of self-assembly effects, 10 to 30% by volume of filler is required to achieve percolation according to Rothon [31], albeit that this depends upon the type, particle size, shape, and the level of dispersion. This could be the reason why composites with micro- and nanofiller exhibit higher values of TC.

10.12 Mechanical Evaluation of Nanocomposites

The inclusion of a large quantity of surfactant may compromise interactions between silicone rubber chains and the nanofiller and, consequently, it is necessary to establish a balance between particle dispersibility and chain/nanoparticle interactions. Tensile strength, elongation at break, and hardness are bulk properties that are affected by the filler type and loading. The interactions (bonding) present within the polymer matrix will affect these properties and, since the addition of surfactant to disperse the filler may have adverse consequences for matrix/nanoparticle interactions, mechanical properties may be negatively impacted by excess surfactant.

To examine the influence of the surfactant on the mechanical properties of the composites, concentrations of 14, 28, and 100 pph were selected [27].

Figure 10.24 compares the ultimate tensile strength for a range of different systems, including a nanofiller and micro + nanofiller formulations, at several concentrations of surfactant. Numerical tensile strength values are summarized in Table 10.3, along with hardness and elongation at break data. While the average tensile strength is higher for the micro + nano-filled composites, compared with the nanofilled systems, it is evident that the addition of 14 pph surfactant has little effect on the tensile strength. At 28 pph surfactant an insignificant (approximately 4%) reduction in the average tensile strength is found with respect to equivalent composites without the surfactant. While 14–28 pph of surfactant sufficed to ensure proper dispersion of the nanofillers, these concentrations had little effect on tensile strength.

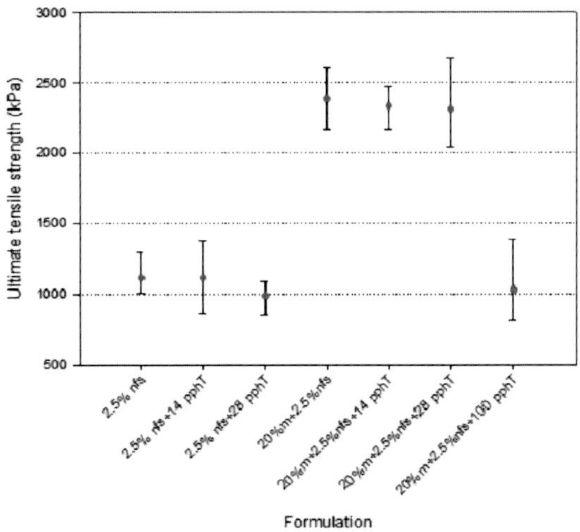

Figure 10.24 Ultimate tensile strength of nanofilled and micro + nanofilled composites. The average value is represented by the dot; the 25th and 75th percentiles are represented by the error bars.

Table 10.3 Summary of mechanical evaluation

Composite	Ultimate tensile strength (MPA)	Coefficient of variation, CV (%)	Elongation (%)	Hardness (type A)
2.5% nfs	1.11	15	238	52
2.5% nfs + 14 pph T	1.12	25	235	50
2.5% nfs + 28 pph T	0.99	14	219	50
20% m + 2.5% nfs	2.38	9	214	62
20% m + 2.5% nfs + 14 pph T	2.34	8	209	61
20% m + 2.5% nfs + 28 pph T	2.31	14.9	227	58
20% m + 2.5% nfs + 100 pph T	1.02	33.8	207	47

However, with 100 pph of surfactant, the tensile strength dropped by almost 60% as compared with the same formulation without the surfactant. A 25% drop in hardness is also observed for the same composite. Hence it can be stated that for

14–28 pph of surfactant (i.e. values that fall below the saturation limit of 65–80 pph for the nanofiller surface), neither tensile strength nor hardness is significantly affected. However, at 100 pph of surfactant, mechanical properties are negatively impacted.

10.13 Performance of Composites in Salt Fog Chamber, Inclined Plane, and Laser Ablation Tests

Figure 10.25 summarizes average eroded mass data obtained by (a) salt fog, (b) inclined plane and (c) laser tests [32]. As can be seen in this figure, the lowest eroded mass was obtained for the composition 20% m + 2.5% nfs + 14 pph Triton in the salt fog and inclined plane tests. For this composite in the inclined plane test, the eroded was only 5.2 mg.

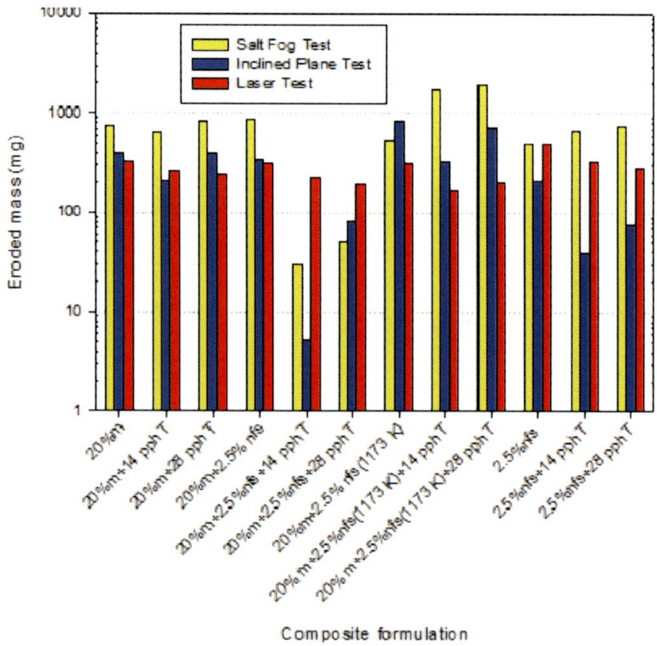

Figure 10.25 Effect of material formulation on average eroded mass (logarithmic scale), obtained using three different testing strategies.

At the other performance extreme, the formulation containing the calcined filler has the highest eroded mass in both the salt fog and inclined plane tests. Also, the addition of the surfactant is seen to improve performance in 18 out of 24 of the analyzed formulations. Specifically, the addition of 14 pph is beneficial in 10 out of 12 different cases, a result that is attributed to the surfactant improving the dispersion of the filler. The addition of 28 pph of surfactant results in an improvement in 8 out of 12 composites, relative to the equivalent formulation without surfactant. The addition of 28 pph of surfactant causes a slightly higher eroded mass in 9 out of 12 cases compared to 14 pph for the salt fog and inclined plane tests. However, in the case of laser ablation, the addition of 28 pph of surfactant results in lower eroded mass for 3 out of 4 formulations, compared with the systems containing just 14 pph.

10.14 Protective Mechanism of Nanofillers for Surface Erosion Resistance

During laser ablation testing of nanofilled composites with a silicone rubber matrix, a white deposit develops on the impinged surface of the composites at higher surfactant concentrations. In order to explore this protective mechanism, fumed nanosilica, natural nanosilica, and nano alumina were used as different fillers in a SiR matrix to produce the above mentioned surface deposit. As mentioned in Section 10.6.2, to ensure uniform absorption of the laser radiation 2.5 wt% of iron oxide was also included in all nanocomposites. The white deposit is shown in Fig. 10.26b. Each layer of white dust from the three nanofilled composites was then analyzed by SEM and energy dispersive X-ray spectroscopy (EDS) [29] and the results are presented in Figs. 10.27–10.29.

From the EDS spectra, the principal elements present in the dust obtained from both nanocomposites containing nanosilica were C, O, and Si. In the case of the nanosilica-filled systems and that containing nano alumina, the primary source of the Si is probably decomposition of the PDMS matrix. In the composite containing nano alumina, the elements found were C, O, and Si plus Al at less than 1% Al, which reflects the fact that the alumina is a minor component of the mixture. The Al distribution in the layer could also be inhomogeneous, e.g., depleted on the surface. It is important to mention that the white layer that was analyzed was located in the upper part of the formed char; the char itself was not analyzed by EDS for any all the three systems.

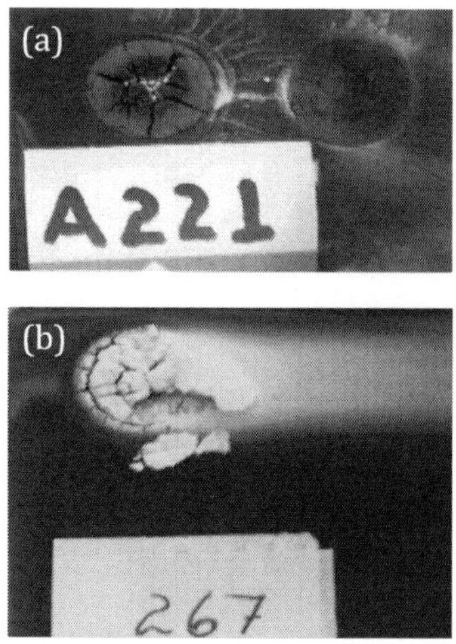

Figure 10.26 Eroded composites (a) with normal charred surface and after the char is removed (b) with the formation of a white deposit.

Figure 10.27 SEM image and EDS data obtained from the white dust layer obtained from a silicone rubber nanocomposite containing 2.5 wt% nano fumed silica + 51 pph Triton.

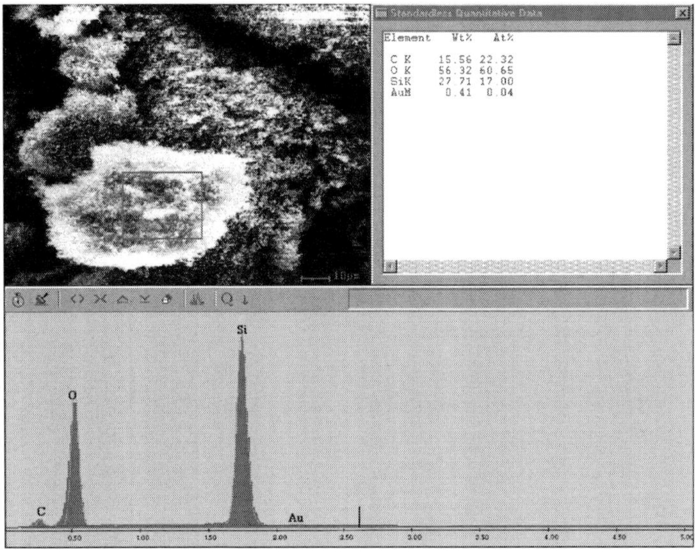

Figure 10.28 SEM image and EDS data obtained from the white dust layer obtained from a silicone rubber nanocomposite containing 5 wt% nano natural silica + 78 pph Triton.

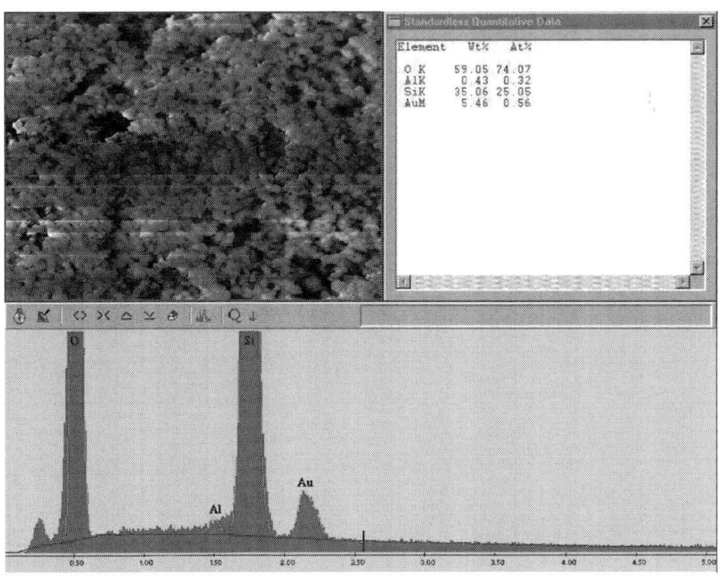

Figure 10.29 SEM image and EDS data obtained from the white dust layer obtained from a silicone rubber nanocomposite containing 5 wt% nano alumina + 156 pph Triton.

The likely protective mechanism against surface erosion for the nanocomposites based upon silicone rubber during laser ablation or dry band arcing is as follows. A local increase in temperature results in the formation of a char layer, which restricts mass and heat transfer between the most heated area and the polymer. This general process has much in common with the processes that occur during combustion of nanocomposites, a topic that is addressed in the following chapter.

10.15 Conclusions

The applications of nanofillers have expanded over the past few years in many research fields and it has been shown that some insulating materials have been reinforced by their inclusion. However, the use of nanofillers is not yet widespread. Considerable effort has been expended by many researchers to find a successful way to incorporate nanofillers into insulation applications and a main problem that has been encountered in mixing nanoparticle fillers with organic materials concerns their final dispersion state.

Surface treatment of nanofillers with a surfactant can be used to improve the dispersion of nanoparticles in a silicone rubber matrix. SEM analysis has shown how agglomeration of nanofillers is diminished, although it is not possible to eliminate all agglomerates.

The reinforcement of organic material with nano fumed silica in a microfilled composite yields a micro-nano composite with significantly improved resistance to surface erosion over compositions with only one or the other filler, particularly when the surfactant Triton™ X-100 is used to aid disperse the nanofiller. Evaluation of this micro-nano composite has been undertaken with one non-standard test and two standard tests. In the standard salt fog and inclined plane tests, electrical discharges cause thermal and chemical decomposition of the composite under test. In the non-standard test, a laser simulates the dry band arcing by heating the surface of the composite to around 1000 K. All three tests complement each other in evaluating such composites and demonstrate improved resistance to surface erosion originating from dry band arcing.

Fumed nanosilica has been shown to impart greater surface erosion resistance than either natural nanosilica or nano alumina. There is some difference in the erosion resistance of natural silica-or alumina-filled compositions. However, the selection of the nanofiller and the appropriate surface treatment method is important in terms of compatibility with the host material, in this case silicone rubber.

In summary, the surface treatment of nanofillers with a surfactant can be a very effective means of improving the dispersion of nanofillers in silicone composites, and the reinforcement of a microfilled composite with nanofillers has been shown to result in a composite that is more resistant to surface erosion from dry band arcing.

References

1. S. Singha and M. Joy Thomas (2006). Polymer composite/nanocomposite processing and its effect on the electrical properties, *IEEE Conference on Electrical Insulation and Dielectric Phenomena, CEIDP*, pp. 557–560.
2. E. L. Paul, V. A. Atiemo-Obeng, and S. M. Kresta, (2004). *Handbook of Industrial Mixing: Science and Practice* (John Wiley & Sons, Inc.) pp. 994–1008.
3. Degussa (2006). Successful use of Aerosil® fumed silica in liquid systems, *Technical Information No. 1279.*
4. D. Wei, R. Dave, and R. Pfeffer (2002). Mixing and characterization of nanosized powders: an assessment of different techniques, *Journal of Nanoparticle Research*, **4**, pp. 21–41.
5. M. Z. Rong, M. Q. Zhang, and W. H. Ruan (2006). Surface modification of nanoscale fillers for improving properties of polymer nanocomposites: a review, *Materials Science and Technology*, **22**(7), pp. 787–796.
6. M. J. Rosen (2004). *Surfactants and Interfacial Phenomena*, 3rd ed. (Wiley-Interscience).
7. K. Holmberg, B. Jönsson, B. Kronberg, and B. Lindman (2002). *Surfactants and Polymers in Aqueous Solution* (John Wiley & Sons, Ltd.).
8. E. W. Flick (1993). *Industrial Surfactants*, 2nd ed. (Noyes Publications).

9. R. P. Bagwe, L. R. Hilliard, and W. Tan (2006). Surface modification of silica nanoparticles to reduce aggregation and nonspecific binding, *American Chemical Society*, **22**(9), pp. 4357–4362.

10. D. Wei, R. Dave, and R. Pfeffer (2002). Mixing and characterization of nanosized powders: an assessment of different techniques, *Journal of Nanoparticle Research*, **4**, pp. 21–41.

11. I. Ramirez, E. A. Cherney, S. Jayaram, and M. Gauthier 2008). Nanofilled silicone dielectrics prepared with surfactant for outdoor insulation applications, *IEEE Transactions on Dielectrics and Electrical Insulation*, **15**(1), pp. 228–235.

12. S. Rätzke and J. Kindersberger (2005). Erosion behaviour of nanofilled silicone elastomers, *XIVth International Symposium on High Voltage Engineering*, Tsinghua University, Beijing, China, pp. C-09 1-6.

13. A. Lazzeri, S. M. Zebarjad, M. Pracella, K. Cavalier, and R. Rosa (2005). Filler toughening of plastics. Part 1: The effect of surface interactions on physico-mechanical properties and rheological behaviour of ultrafine $CaCO_3$/HDPE nanocomposites, *Polymer*, **46**, pp. 827–844.

14. Z. Li, and Y. Zhu (2003). Surface-modification of SiO_2 nanoparticles with oleic acid, *Applied Surface Science* **211**(1–4), pp. 315–320.

15. S. H. Ahn, S. H. Kim, and S. G. Lee (2004). Surface-modified silica nanoparticle-reinforced poly(ethylene 2,6-naphthalate), *Journal of Applied Polymer Science*, **94**, pp. 812–818.

16. W. W. Kubiak and E. Niewiara (2002). Influence of the electrolyte on Triton X-100 adsorption on fumed silica, *Electroanalysis*, **17**, pp. 1169–1175.

17. L. T. Zhuravlev (2000). The surface chemistry of amorphous silica. Zhuravlev model, *Colloids and Surfaces A: Physicochemical and Engineering Aspects*, **173**(1), pp. 1–38.

18. H. Barthel, L. Roesch, J. Weis (1994). Fumed silica: production, properties, and applications, in *Organosilicon Chemistry II: From Molecules to Materials*, pp. 761–778.

19. G. Wypych (2000). *Handbook of Fillers*, 2nd ed. (ChemTec Publishing).

20. Degussa (2006). Fine particles, *Technical Bulletin No. 11.*

21. ASTM D2303 (2013). Standard test methods for liquid-contaminant, inclined-plane tracking and erosion of insulating materials, *ASTM International*, West Conshohocken, PA.

22. L. H. Meyer, S. Jayaram, and E. A. Cherney (2005). A novel technique to evaluate the erosion resistance of silicone rubber composites for high voltage outdoor insulation using infrared laser erosion,

IEEE Transactions on Dielectrics and Electrical Insulation, **12**(6), pp. 1201–1208.

23. M. J. Billings, L. Warren, and R. Wilkins (1971). Thermal erosion of electrical insulating materials, *IEEE Transactions on Electrical Insulation*, **6**, pp. 82–90.
24. IEC 62217 (2012). *Polymeric HV Insulators for Indoor and Outdoor Use: General Definitions, Test Methods and Acceptance Criteria*, 2nd ed. (International Electrotechnical Commission).
25. IEC/TR 62730 (2012). *HV Polymeric Insulators for Indoor and Outdoor Use Tracking and Erosion Testing by Wheel Test and 5 000h Test*, 1st ed. (International Electrotechnical Commission).
26. Y. Kimata, S. Nakagawa and H. Kato, (2001). Measurement of concentration of silanol group, Japanese Patent No. 2001208683, pp. 1–5 (in Japanese).
27. I. Ramirez, S. Jayaram, E. A. Cherney, M. Gauthier, and L. Simon (2009). Erosion resistance and mechanical properties of silicone nanocomposite insulation, *IEEE Transactions on Dielectrics and Electrical Insulation*, **16**(1), pp. 52–59.
28. L. H. Meyer, S. Jayaram, and E. A. Cherney (2004). Thermal conductivity of filled silicone rubber and its relationship to erosion resistance in the inclined plane test, *IEEE Transactions on Dielectrics and Electrical Insulation*, **11**(4), pp. 620–630.
29. I. Ramirez, S. Jarayam, and E. A. Cherney (2010). Analysis of temperature profiles and protective mechanism against dry-band arcing in silicone rubber nanocomposites, *IEEE Transactions on Dielectrics and Electrical Insulation*, **17**(2), pp. 597–606.
30. ASTM D5470 (2006). *Standard Test Method for Thermal Transmission Properties of Thermally Conductive Electrical Insulation Materials* (ASTM International, West Conshohocken, PA).
31. R. N. Rothon (2003). *Particulate-Filled Polymer Composites*, 1st ed. (Rapra Technology Limited).
32. I. Ramirez, S. Jarayam, and E. A. Cherney (2010). Performance of silicone rubber nanocomposites in salt-fog, inclined plane, and laser ablation tests, *IEEE Transactions on Dielectrics and Electrical Insulation*, **17**(1), pp. 206–213.

Chapter 11

Degradation of Polymeric Micro- and Nanocomposites

Giulio Malucelli[a] and Alun S. Vaughan[b]

[a]*Politecnico di Torino, Department of Applied Science and Technology, Viale Teresa Michel 5, 15121 Alessandria, Italy*
[b]*ESC, Faculty of Physical Sciences and Engineering, University of Southampton, Southampton SO17 1BJ, UK*

giulio.malucelli@polito.it, asv@ecs.soton.ac.uk

This chapter reviews the main types, mechanisms, and effects of degradation phenomena that may occur during the life of material systems based upon thermoplastics or thermosetting polymers, which can affect their performances in service. Since degradation can be identified with any change in physical or chemical characteristics relative to initial or desirable properties, this generic term embraces any number of reactions that may occur in a polymer. The degradation of polymers may involve several physical and/or chemical processes (e.g., photodegradation, biodegradation, thermal and thermo-oxidative degradation, ultrasonic degradation, high-energy radiation-promoted degradation, chemical degradation, mechanical degradation/cracking, electrical aging, etching and corrosive degradation, abrasive

Tailoring of Nanocomposite Dielectrics: From Fundamentals to Devices and Applications
Edited by Toshikatsu Tanaka and Alun S. Vaughan

ISBN 978-981-4669-80-1 (Hardcover), 978-981-4669-81-8 (eBook)
www.panstanford.com

degradation), which promote changes that may lead to a significant deterioration of polymer quality and, hence, to the loss of its functional features. Particular attention here is focused on the role and effect of the presence of micro- and nanofillers on the degradation reactions and on the way by which such nanoscaled additives are able to modify and/or limit degradation phenomena.

11.1 Introduction

The typical properties of polymeric materials are closely related to their macromolecular structure and the interactions that take place between polymer chains. These factors are strictly related to the chemical nature of the repeating unit. Thus, any chemical reaction or physical process that promotes the modification of the polymeric molar mass or of the chemical structure of the constituent units can lead to irreversible changes in macroscopic characteristics. Indeed, even quantitatively limited modifications to the chemical properties of a polymer, which would be considered negligible for low-molar-mass compounds, can significantly affect the properties of a polymer. As an example, consider a linear polymer that possesses a polymerization degree of 1000, then scission of 1 in 1000 backbone bonds linking the repeating units (i.e., 0.1%) is enough to halve the polymer molar mass, thus lowering its viscosity in the molten state by about an order of magnitude [1, 2]. As a consequence, the word *degradation*, which is usually taken to indicate major changes in the molecular structure of any compound, has a different meaning where polymeric materials are considered. Indeed, for these systems, degradation can include processes that lead to limited chemical modifications, but which exert a substantial effect on the physico-mechanical features of the material.

Polymeric materials easily undergo degradation processes as a consequence of such physical agents as heat, light, high-energy radiations, mechanical deformation, as well as of chemical factors such as oxygen, ozone, water and atmospheric pollutants. From a philosophical point of view, *all* polymeric materials are exposed to degradation processes during two steps of their life. First, changes in molecular structure inevitably occur during initial processing (molding, extrusion, spinning, ...), due to the combined presence of mechanical stresses, heat and oxygen.

Second, physical and chemical changes will occur during use, as a consequence of the action exerted by light, oxygen, atmospheric pollutants, and possible specific factors that depend on the type of application. In the context of this review, applied electric fields are of particular relevance.

Usually, the degradation processes cause a gradual and irreversible deterioration of the thermal, mechanical, optical, electrical properties of the polymer, eventually rendering it unsuitable for the initially envisaged application. However, in some cases, degradation processes carried out under controlled conditions can be exploited to obtain materials with deliberately modified features according to a preset goal. Examples of this include the thermal degradation of polyacrylonitrile, as in the production of carbon fibers, or the molecular changes induced by UV or electron irradiation of resist materials, as used in microelectronic fabrication technologies.

The behavior of a polymeric material subjected to degradation processes depends on the type of degrading agent, as well as on the chemical structure and morphology of the polymer. Usually, polymeric materials have to be protected from relevant degradation processes by adding appropriate stabilizing systems. To this end, additives capable of slowing down or inhibiting the degradation phenomena are employed, such that the desirable characteristics of the material are retained, ideally, throughout the complete service life of the artifact. The choice of effective stabilizers is therefore strictly related to a detailed knowledge of the reaction mechanisms that have to be prevented.

Unfortunately, the study of degradation processes is, in general, extremely difficult, as a consequence of their complexity. This stems from the following: the presence of multiple mechanisms that, in general, act simultaneously; the occurrence of chemical reactions usually in the condensed phase and in heterogeneous systems; the lack or impossibility of using model compounds to simplify the study of the degradation reactions. In this final case, the difficulties stem from the presence of low concentrations of structural irregularities within the polymer chains, which play a dominant role in degradation. As is common in many areas of materials science, it is not the ideal bulk that dominates degradation/failure processes, but rather the occasional deviation from this. In the case of polymers, these defects may derive from the polymerization process itself (e.g., the presence of double bonds, branches, head-to-head defects, etc.) or from the exposure to atmospheric agents (e.g., oxidized or hydrolyzed structural units). These structural irregularities are not taken into proper consideration when the polymeric material is simply associated with the structure of the idealized repeating unit.

These irregularities are often unstable and represent the weakest point of the polymer chain, where subsequent degradation reactions start. As an example, when heated, polyethylene starts to decompose some 200 °C below its model compound, n-hexadecane. In many cases, the chemical nature of the defects cannot be identified, since their concentration is below the detection limits of characterization techniques that are currently used. However, the effect of the defects can be catastrophic because of the peculiar sensitivity of the polymer degradation phenomena to even very limited chemical modifications.

Historically, the aforementioned difficulties related to the investigation of the degradation mechanisms, combined with the imperative to solve practical problems that arise from degradation, have led to the adoption of stabilization methodologies that are largely based on empirical approaches. However, despite the undeniably good results that have been achieved, the necessity to have at our disposal more and more efficient stabilizers, as a consequence of the significant growth of the use of plastics in ever wider application fields, is forcing the development of new stabilization methods that are, rather, derived from a more scientific approach. This approach is becoming more and more realistic as the increasing availability of new, powerful analytical methods lead to increased knowledge and understanding of the fundamentals of degradation and stabilization mechanisms.

In parallel, the design and engineering of polymeric micro- and nanocomposites has driven the scientific community toward the study of the effect of micro- and nanofillers, having different morphologies, on the overall properties of different thermoplastics and thermosets. This has led to an enhanced knowledge of the effect of the dispersed phase on degradation and stabilization process in the polymer matrix.

This chapter will begin with a detailed description of physical and chemical degradation processes in polymers, of the mechanisms through which degradation takes place and of the overall final effects caused by degradation. Then, some examples concerning the influence of micro- or nanofillers on the degradation of different polymer matrices will be described, focusing on the effects that different experimental parameters (such as structure, concentration, morphology, and distribution of the filler within the polymer) exert on the degradation phenomena. Finally, we will consider how these general principles can be used to understand how electrical phenomena lead to the loss of desirable dielectric characteristics.

11.2 Types, Mechanisms and Effects of Degradation Phenomena on Polymeric Materials

Polymers are can be adversely affected by a wide range of degradation agents. Consequently, the following sections will be devoted to a summary of issues related to different types of degradation and the mechanisms involved. Furthermore, the main effects of the degradation reactions will be described.

11.2.1 Physical and/or Chemical Degradation Processes

11.2.1.1 Photodegradation

The energy associated with photons in the ultraviolet range (200–400 nm, 140–175 kcal/mol) of the electromagnetic spectrum is sufficient to lead to scission of chemical bonds located in the polymer chains (C–H 100 kcal/mol, C–C 80 kcal/mol, N–O 40 kcal/mol). However, photochemical reactions necessarily require the initial absorption of incident radiation and, therefore, photodegradation processes involve polymers that contain relevant chromophores. Chromophores can absorb radiation and move to an excited electronic state, which will subsequently relax and dissipate the excitation energy through a chemical reaction or through some alternative photophysical de-excitation process, such as energy transfer, internal conversion, fluorescence, phosphorescence, etc. If the photochemical process causes homolytic bond scission, the resulting radicals start a chain reaction that is analogously to radical degradation processes promoted by heat.

From an overall point of view, modifications induced by photodegradation are similar to those that occur during thermodegradation: both processes give rise to chain scission, cross-linking phenomena and to the formation of volatile species.

However, for a given polymer, the consequences of photodegradation processes can differ from those derived from thermodegradation, since (i) the electronic excited states can promote the occurrence of different reactions compared with those that take place in the ground state and (ii) the formation of radicals through photolysis requires lower temperatures compared to thermolysis [3–7].

As an example, polystyrene, which upon heating decomposes giving rise to styrene and oligomers, when exposed to UV radiation (235.7 nm) under vacuum at room temperature, starts to cross-link, releasing hydrogen and turning yellow, because of the following reactions [8]:

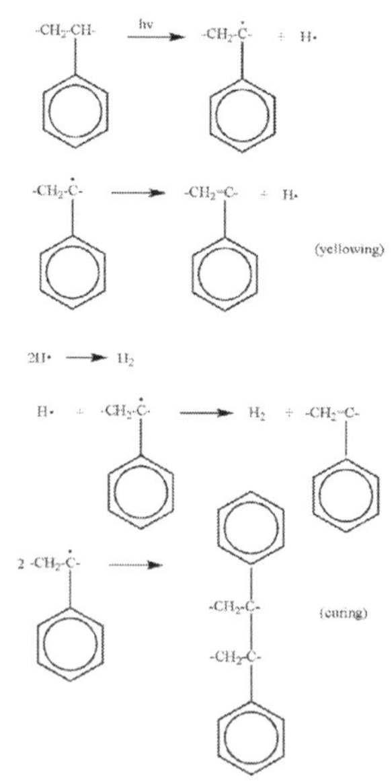

Figure 11.1 Reactions involved in the photodegradation of polystyrene.

The photochemical behavior of polymers bearing ketone groups is of particular interest, since these groups absorb in the UV range (>250 nm) and, from the excited state, undergo scission reactions (Norrish I, NI, and Norrish II, NII) which, apart from hydrogen abstractions, do not occur in the ground state. A typical example referring to ethylene copolymers containing carbonyl groups is shown schematically below:

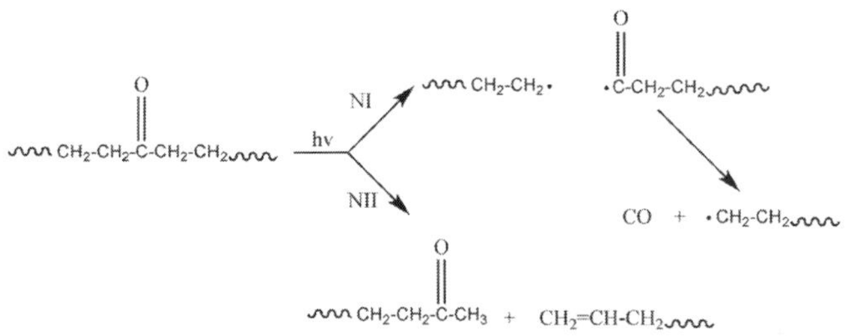

Figure 11.2 Scission reactions for ethylene copolymers containing carbonyl groups.

Although both of the reactions shown lead to scission of the polymer chain, Norrish I gives rise to the formation of two macroradicals plus carbon monoxide, while Norrish 2 generates two deactivated macromolecules.

Chain scission may also occur as a consequence of photolysis phenomena involving polymers that contain ketone functionalities in the side chain of the macromolecule. As an example, the behavior of poly(methyl-vinylketone) can be schematized as shown in Fig. 11.3. As depicted in this scheme, chain scission

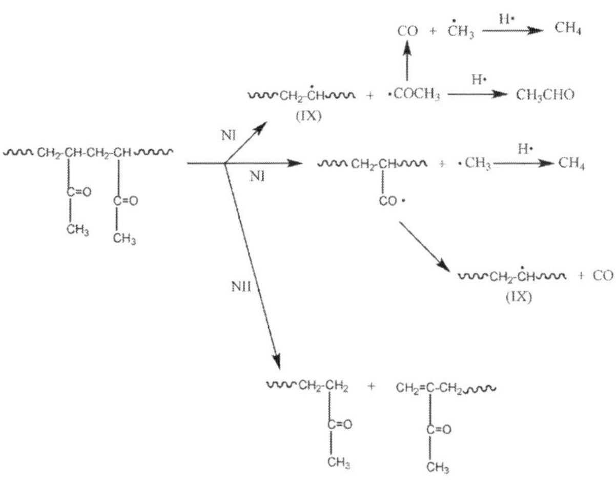

Figure 11.3 Photolysis of poly(methyl-vinylketone) leading to chain scission.

directly takes place as a consequence of Norrish II, while Norrish I causes the formation of low molecular weight compounds (i.e., methane and acetaldehyde) in addition to CO and the macroradical (IX), which can subsequently lead to additional chain scission or cross-linking reactions.

11.2.1.2 Biodegradation

Certain polymers are prone to biodegradation/biotic degradation [9]. That is, chemical degradation processes that are provoked by the action of microorganisms such as bacteria, fungi and algae. These microorganisms, together with their enzymes, are able to consume a polymer as a food source and, as a consequence, the original form of the substance is lost.

Usually, very small variations in the chemical structure of polymers can lead to large changes in biodegradability, which is markedly influenced by the molar mass, molecular form, porosity and crystallinity of the polymer. Generally speaking, biodegradability decreases with increasing molar mass; low-molar-mass moieties such as monomers, dimers and repeating units, degrade in an easy way. Biodegradation of polymers involves two main categories of enzymes, namely, extracellular and intracellular depolymerases, which are termed exo- and endoenzymes, respectively. Exoenzymes are responsible for depolymerization: these compounds are able to breakdown complex polymers into molecules that are small enough to diffuse through the cell walls to be utilized as carbon and energy sources. Mineralization occurs when the process gives rise to the formation of simple end products such as carbon dioxide, water, methane, nitrogen compounds, etc. Mineralization is complete when all the biodegradable materials or biomass is consumed and all the initial carbon is converted into carbon dioxide.

The biodegradation process of a polymer can be divided into aerobic and anaerobic degradation.

If oxygen is present, aerobic biodegradation occurs and carbon dioxide is produced. If there is no oxygen, anaerobic degradation occurs, leading to the production of methane instead of carbon dioxide.

Biodegradable polymers include inter alia, poly(lactic acid), polycaprolactone, polyhydroxyalkanoates, cellophane, polyethers, copolyesters, poly(vinyl alcohol), poly(vinyl chloride), acylated starch-plastic, modified starch, and vegetable oil.

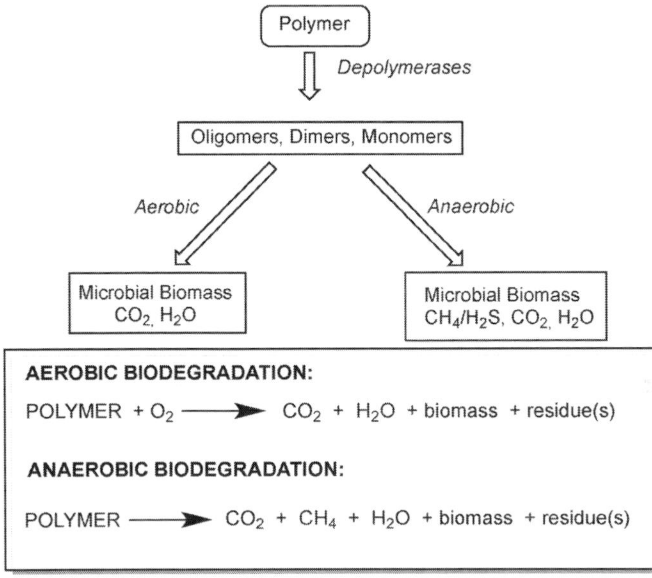

Figure 11.4 General overview of polymer biodegradation.

11.2.1.3 Thermal and thermo-oxidative degradation

Upon heating, the behavior of polymeric materials can be defined on the basis of two parameters indicated as "heat resistance" and "thermal resistance." The former refers to the maximum temperature at which the material can be heated without any significant variation of its mechanical properties [10–13]. Heat resistance therefore depends upon such characteristics as the glass transition and melting temperatures, which are strictly related to the physically reversible phenomena promoted by heat. Thermal resistance represents the maximum temperature to which a polymeric material can be heated without the occurrence of chemically irreversible modifications.

Different chemical bonds (e.g., C–C, C–H, C–N, C–O) are present in polymers and each of these will break, as a consequence of heating, at a temperature that depends on the bond dissociation energy. Usually, the associated temperatures fall in the range 150–500 °C. The primary thermolysis process is usually followed by a series of reactions (often chain-type reactions), which can progressively lead to polymer fragmentation and to the formation of low molecular weight and hence volatile products.

The thermal degradation processes that occur in polymeric materials can be broadly classified as follows:

(a) Reactions involving polymer chains. More specifically, scission involving the main polymer chain gives rise to rather volatile products, consisting of monomers and chain fragments having different molecular weight, where the polymer repeating units are still distinguishable:

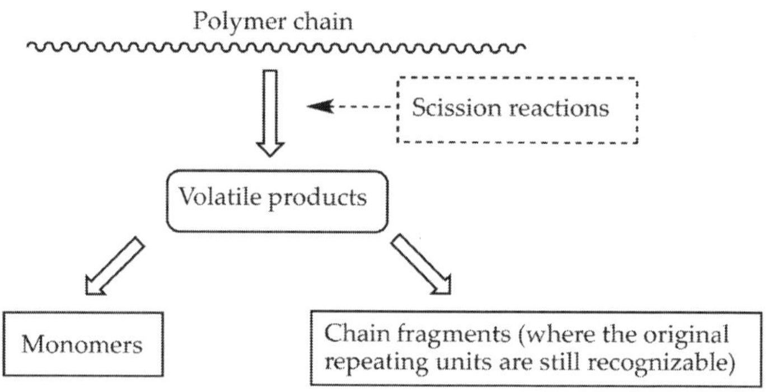

Figure 11.5 Scission reactions of polymer main chains.

(b) Reactions involving lateral substituents. In this case, the polymer retains a chain structure, but changes its chemical composition, because of the occurrence of reactions such as elimination, cyclization, etc.

In some cases, the effect of the heating process can lead to cross-linking phenomena, which can occur concurrently with backbone scission processes.

The type of reactions that occur as a consequence of heating depends not only on the chemical structure of the polymer but also on the imposed temperature. Thus, heating a polymeric material to increasing temperatures can promote consecutive degradation reactions, such as those involving substituent groups, followed at higher temperatures by chain scissions or cross-linking reactions.

11.2.1.3.1 Thermal degradation kinetics

A complete and exhaustive discussion of thermal degradation processes is very complicated, because of the large number of reactions involved. Also, standard kinetic laws, developed for reactions that take place in the gas phase or in dilute solutions, will not generally be suitable for investigating thermal degradation phenomena occurring in the condensed phase, where chemical processes will be affected by diffusion phenomena. Nevertheless, kinetic equations are greatly simplified when the thermal degradation process is dominated by a single reaction.

As an example, it can be useful to describe the kinetics of one situation, in which thermal degradation leads exclusively to the statistical scission of bonds located in the polymer backbone. That is, processes such as thermolysis or chemical degradation of polycondensation polymers (e.g., the hydrolysis of polyesters and polyamides) that lead to main-chain scission with negligible formation of low molecular weight products. This should be contrasted with the thermal depolymerization of vinyl polymers, which primarily involves macroradical intramolecular transfer reactions and the consequent formation of low-molar-mass products (i.e., volatiles at the degradation temperature).

The fundamental characteristic of this type of scission concerns the scission probability of the bonds linking the chain structural units, which is the same for any bond. The scission process of these bonds obeys first-order kinetics, the rate of which can be indicated as follows:

$$-\frac{dB_t}{dt} = kB_t \tag{11.1}$$

where B_t refers to the number of intact bonds present in the system at time t. The integration of Eq. (11.1) leads to

$$B_t = B_0 e^{-kt} \tag{11.2}$$

where B_0 is the total number of bonds at $t = 0$. Defining the fraction of bonds, α, that have undergone scission as

$$\alpha = \frac{B_0 - B_t}{B_0} = 1 - \frac{B_t}{B_0} \tag{11.3}$$

Equations (11.2) can be rewritten as

$$\text{In}(1 - \alpha) = -kt \tag{11.4}$$

Under the condition where $\alpha << 1$

$$\text{In}(1 - \alpha) \cong -\alpha \tag{11.5}$$

such that combining Eqs. (11.5) and (11.4) leads to

$$\alpha = kt \tag{11.6}$$

This result demonstrates that when the fraction of bonds broken is low, as will be the case for polymer degradation, this parameter varies linearly with reaction time. This relationship can be rewritten in terms of the degree of polymerization, where X_0 and X_t indicate the degree of polymerization at time 0 and time t respectively. If the initial number of polymer chains in the system is written N_0 and S scission reactions have occurred in time t, then

$$S = \frac{B_0 - B_t}{N_0} \tag{11.7}$$

Dividing Eq. (11.7) by Eq. (11.3) leads to

$$\frac{S}{\alpha} = \frac{B_0}{N_0} \tag{11.8}$$

and, since X_0 can be defined as

$$X_0 = \frac{B_0}{N_0} + 1 \tag{11.9}$$

then, by combining Eq. (11.9) and Eq. (11.8),

$$\frac{S}{\alpha} = X_0 - 1 \tag{11.10}$$

such that

$$\alpha = \frac{S}{X_0 - 1} \approx \frac{S}{X_0} \tag{11.11}$$

Starting from the definition of X_t

$$X_t = \frac{X_0}{S + 1} \tag{11.12}$$

Equations (11.11) can then be rewritten as

$$\alpha = \frac{X_0 - X_t}{X_0 X_t} = \frac{1}{X_t} - \frac{1}{X_0} \tag{11.13}$$

Finally, combining Eq. (11.13) with Eq. (11.6) gives the following result:

$$\frac{1}{X_t} - \frac{1}{X_0} = kt \tag{11.14}$$

These equations are not dependent on the molar mass distribution initially present in the polymer. As degradation proceeds, the distribution of molar masses present will change and it can be demonstrated that the random scission of a polymer leads to the "most probable" distribution.

11.2.1.4 Ultrasonic degradation

Unlike thermal degradation, where scission of the polymer backbone tends to take place at chain ends and/or randomly, in ultrasonic degradation, scission occurs preferentially at the midpoint. The driving force of the ultrasonic degradation mechanism is usually associated with the resulting hydrodynamic forces, which are attributed either to the increased frictional forces between ultrasonically accelerated faster moving solvent molecules and the less mobile polymer molecules or to the high pressure deriving from the collapse of bubbles [14, 15].

For the latter, ultrasonic effects are due to subsequent phenomena, i.e., to cavitation, growth, and rapid collapse of microscopic bubbles as the alternate rarefaction and compression phases of the longitudinal sound wave pass through the liquid medium. The growth and collapse of bubbles, which take place within microseconds, promote the movement of solvent molecules around the bubbles and create large shear fields. The polymer molecules near the shear field move faster than the polymer molecules far away from cavitation and, hence, cleavage of the polymer backbone is attributed to the relative motion of polymer chains and the consequent mechanical stresses which the polymer experiences. The cleavage confinement at the midpoint is a consequence of the force distribution induced in the polymer chains due to the relative motion of its constituent units: The maximum of this distribution corresponds to the center of the uncoiled chain.

Several parameters can affect the ultrasound degradation of polymer solutions. These include the solvent; the polymer concentration; the ultrasound intensity; the presence of dissolved gases; the system viscosity; the molar mass distribution. While the temperature does have an effect, this is rather limited; an increase in temperature usually results in decreased rate of degradation, thus confirming that the degradation has to be ascribed to hydro-mechanical phenomena. Ultrasonic degradation has been studied in many solutions of many different polymers, including polystyrene, polyvinyl acetate, polypropylene, polybutadiene, hydroxypropyl cellulose, carboxymethyl cellulose, polyacrylamide, poly(ε-caprolactone), poly(vinyl alcohol), poly(ethylene oxide), poly(vinyl pyrrolidone), dextran, and polyacrylates. One of the most significant outcomes from this type of degradation is an exponential decrease in molar mass toward some limiting value.

11.2.1.5 Degradation by high-energy radiation

The radiation chemistry of polymers is of great interest in that while irradiation can, on the one hand, promote degradation, it can also be used to modify materials beneficially. Indeed, exposure to high-energy radiation may result in structural changes, which can be exploited to improve stability or tailor the thermo-mechanical characteristics of the system [16].

High-energy radiation may induce many modifications in the structure of polymers, including cross-linking; bond scission in the molecular backbone and in the side groups; gas desorption; changes in crystallinity and unsaturation. Examples of industry sectors that exploit radiation-induced cross-linking are wire and cable, packing films and foams. Radiation curing is also used on a large scale in surface finishing coatings, lacquers and inks. In the case of backbone scission, this has considerable utility in connection with the processing of polytetrafluorethylene (PTFE), in order to reduce its molar mass and thereby improve its processability and in the so-called "visbreaking" of polyolefins [17]. Furthermore, high-energy radiation is used commercially to modify polymers deliberately so as to control molecular weight, cross-linking and long chain branching. Products from the high-energy irradiation of polymers currently include cross-linked or branched polymers to give modified grades from large-scale polymerization plants, fast curing coatings, hot water pipes, heat-shrinkable insulation, and non-melting insulation for high voltage circuits, as well as high-resolution lithography for the manufacture of micro-circuits and grafted monomers on polymer substrates. Finally, radiation-induced grafting is another powerful method for modifying existing polymers, creating an almost unlimited range of new materials exhibiting specific features.

The types of radiation that are commonly used in such applications are γ-rays, X-rays, and electron (e^-) beams; the use of heavy particles and neutrons is less common. Key parameters that need to be considered include the interaction cross section and penetration depth.

When a macromolecule (M) is exposed to high-energy irradiation, reactive species such as ions and free radicals are formed, as presented in reactions R1, R2, and R3 shown in Fig. 11.6.

$$\mathrm{M} \xrightarrow[\text{or e-beam}]{\gamma\,\text{radiation}} \mathrm{M^+} + e^- \quad \text{or} \quad \mathrm{M^\bullet} \qquad \text{(R1)}$$

$$\mathrm{M^+} + e^- \longrightarrow \mathrm{M^\bullet} \qquad \text{(R2)}$$

$$\mathrm{M^\bullet} \longrightarrow \mathrm{R_1^\bullet} + \mathrm{R_2^\bullet} \qquad \text{(R3)}$$

$$\mathrm{R_1^\bullet}\ (\text{or } \mathrm{R_2^\bullet}) + \mathrm{M} \longrightarrow \mathrm{R_1M^\bullet} \qquad \text{(R4)}$$

Figure 11.6 Reactions occurring upon exposure of a macromolecule to high-energy irradiation.

In the case of gamma radiation, the Compton effect results in the ionization of the macromolecule, which initially produces one electron (reaction R1 above). This electron may induce other secondary ionization events, form more electrons, or may combine with a molecular ion (M^+), converting it into a radical (reaction R2). This will generally exist in an excited state such that if the energy becomes localized, C–C bond scission occurs, giving rise to the formation of free macroradicals (reaction R3). In the case of polyethylene lamellae, chain scission occurs preferentially at fold surfaces, rather than within the crystal, as a consequence of such localization effects. Some of the radicals shown in R3 above will go on to react with the neighboring macromolecules (M), producing chain branching (reaction R4).

The protection of polymers against high-energy radiation requires efficient additives, which are capable of preventing and/or stopping the aforementioned chain reactions.

11.2.1.6 Chemical degradation

Many polymers are used in outdoor applications and, consequently, there is a great interest in the chemical degradation phenomena that may arise under conditions where products are exposed to rain, sunlight and temperature fluctuations and, also, to environmental bacteria. All these factors are able to drive the material toward oxidation and hydrolysis, while acidic and/or basic conditions can catalyze further chemical degradation reactions.

As a consequence of exposure to the environment, permeants may enter a polymer via permeation (i.e., through absorption, diffusion and sorption) through the polymer bulk. Once a permeant penetrates into the polymer, it may cause a variety of chemical and physical modifications within the material, such as plasticization, swelling, and the creation of a weak boundary layer in the case of adhesives bonded to a substrate.

Notwithstanding biodegradation, which can be considered as a chemical process, the most common chemical degradation activity is represented by hydrolytic degradation, which occurs not only under water but also under atmospheric conditions (indeed, atmospheric moisture can promote hydrolysis at high temperatures) [18, 19].

Polycondensation polymers such as polyamides and polyesters are particularly sensitive to this kind of chemical degradation, albeit with some differences. For example, hydrolysis of polyamides tends to be enhanced under acidic conditions and, therefore, the rate of degradation of such systems increases with temperature and acidity. However, the acid–amine pairs generated by hydrolysis events can recombine easily when both are associated with large, macromolecular fragments. Conversely, when one fragment is a monomer or a low molecular weight oligomer, it can escape easily from the system, whereupon the recombination yield decreases. It has been reported that the diffusion rate of acidic solutions is promoted by a decrease in molar mass, due to molecular chain scission by hydrolysis.

The hydrolysis of polyesters occurs under both acidic and basic conditions and, under each condition, the permeation of the solution from the polymer surface and diffusion of the solution within the bulk of the polymer determines the overall degradation rate.

Polymers used in outdoor applications can be also attacked by reactive gases, such as SO_x and NO_x, which are associated with acidic rain. As many kinds of polymer products suffer on exposure to these gases, this type of polymer degradation is not negligible.

11.2.1.7 Mechanical degradation

Mechanisms degradation affects impact polymers through their complete lifecycle, from initial manufacture, through service, to failure. As such, this topic is of critical importance in defining optimum processing conditions, designing materials with improved macroscopic macroscopic properties and lifetime prediction.

A mechano-chemical process is a multistage phenomenon involving at least three main steps [20]:

(a) mechano-activation
(b) mechano-cracking
(c) macroscopic cleavage

The first step involves the modification of bond lengths and bond angles coupled with chain conformational changes (chain extension/unfolding): all these phenomena promote an increase of potential energy. This mechano-excited state precedes the second stage, during which chemical reactions occur, often mediated by mechano-radicals caused by bond cleavages. The third step refers to the evolution of molecular damage to a macroscopic level, which leads to structural reorganization of the polymer and which, in turn, promotes some changes in the mechanical characteristics.

Chain scission usually occurs near the midpoint of the polymer backbone, where frictional forces between monomer units tend to concentrate. When a coiled polymer is exposed to elongational strain, at first, the chain starts to uncoil and consequently elongates; scission occurs when the energy bond limit is exceeded. The stretching force is zero at the chain ends and maximum in the middle and is usually characterized by a Gaussian distribution within the middle 15% of the chain. The probability for chain scission increases with increasing polymer molecular weight, because of strain accumulation at critical sites. As a consequence, the forces reach their maximum value where all of the small contributions deriving from the monomer units of the two halves work together.

Although this chain scission mechanism is described in terms of homolytic bond cleavage, with the consequent formation of a pair of neutral free radical intermediates, heterolytic bond scission in the presence of polar solvents cannot be excluded.

The relationship between average molar mass and shear stress has been thoroughly investigated using several different models. According to Nguyen and Khaush [21, 22], the equilibrium molecular weight limit for mechanical degradation follows a $1/M$ relationship with increasing shear stress. Smith and Chou [23] have, conversely, hypothesized a $1/M^2$ dependence. In a stretched polymer chain, bond cleavage will occur when thermal fluctuations overcome the bond length limit: in other words, mechanical energy has to "cooperate" with vibrational (thermal) energy by, effectively, decreasing the activation energy for bond cleavage. Zhurkov [24] proposed the concept of mechanical bond cleavage as a thermally activated process, using an Arrhenius type equation for describing the activation energy for bond rupture and its dependence on the applied tensile strength.

Since mechano-activation not only affects the chemical structure of the polymer by inducing chain scission and/or cross-linking but also increases sensitivity toward chemical reactions, it can consequently be exploited for initiating chemical processes of practical use, as in reactive processing.

11.2.1.8 Electrical degradation

Electrical degradation of polymers can be divided into two broad areas: surface effects and bulk effects. In the former, electrically induced degradation occurs as a consequence of the proximity of the polymer surface to energetic electrical events, such as discharges or corona. Under such circumstances, a multitude of mechanisms can be involved but, in general, these can be discussed in terms of the processes considered above. For example, consider dry band arcing.

In the majority of technologically relevant insulation systems, virgin polymer surfaces contain non-polar moieties, the surface is hydrophobic and, consequently, a continuous water film does not form. However, as described in Section 11.2.1.6, environmental exposure commonly results in the oxidation and hydrolysis of polymers as a consequence of their exposure to air, rain, sunlight, etc. As a result, as the surface degrades, so it becomes increasingly hydrophilic because of the inclusion of species such as carbonyl and hydroxyl groups. Particularly under polluted conditions, the subsequent formation of a continuous water film results in the flow of surface currents, which tend to heat the system non-uniformly until dry, high resistance regions are formed, which can be bridged by local discharges, i.e., dry band arcs. As a consequence of their energetic nature, the presence of discharges immediately adjacent to a polymer surface can then result in further degradation processes becoming active. Thermal degradation may occur or, alternatively, active species may be formed in the vicinity of the arc, which subsequently degrade the system.

Early attempts to develop all dielectric self-supporting (ADSS) fiber optic communications cables that would be suspended along overhead lines provide an excellent illustration of the way in which all these processes act in concert to cause failure of the system. Initially, changes in the surface chemistry of the polymer sheathing result in a transition from a hydrophobic to a hydrophilic character. Due to their proximity, capacitive coupling between the ADSS cable and the high-voltage conductors then results in the flow of currents along the cables when wet, the formation of dry bands and local heating of the sheath. Local degradation/melting/flow of the polymer sheath exposes the fiber reinforced composite core of the structure to aggressive species formed in the electrically active region, chemical degradation occurs and, finally, the cable fractures.

In the case of bulk electrical degradation, the position is much less clear in terms of the totality of the process. As such, it is convenient to divide electrical degradation into the growth of electrical trees, which occurs after some initiation stage. Electrical trees are fractal structures made up of connected gas-filled tubules that evolve through a dielectric as a consequence of partial discharge activity within the gas. Although many of the details of the process remain to be explained, the overall physical and chemical principles have been broadly understood for many years [25]. When a void of appropriate size (of the order of micrometers) exists within a polymer, discharge activity will occur. These discharges degrade the material through the impact of charged particles on the surfaces and absorption of photons of varying energies. The consequence is material damage, erosion and the conversion of the solid dielectric to gaseous decomposition products, which are then likely to undergo further gas phase chemical reactions. The deposited charges affect the local field and the generation of gaseous decomposition products increases the gas pressure in the void. Both of these factors initially inhibit further discharge activity. However, the space charge will decay and the initial gas pressure will reduce with time through diffusion into the solid and by condensation of reactants onto the wall of the void; as the direction of the electric field is reversed as a consequence of the alternating applied voltage, the processes is repeated.

In view of the significance of partial discharges within voids that is ascribed to the growth phase of electrical trees, it is perhaps natural that similar processes have been proposed to occur prior to the existence of a visible tree. That is, that bond scission occurs as a consequence of the presence of sufficiently energetic charge carriers within the material and, in some way, the resulting defects grow in size, coalesce and generate micro-cavities that can, subsequently, support the partial discharges that drive tree growth [26]. Conversely, other workers have proposed that electrically induced mechanical stresses are important [27], that stable space charge centers result in electromechanical stresses, which serve to cause local chemical changes and then micro-cracks [28], or that manufacturing defects are key factors [26]. However, compelling evidence to support such ideas of electrical aging is scarce and, indeed, convincing proof of electrical trees growing from isolated discharging voids is hard to find.

To conclude, our understanding of many aspects of electrical degradation is still poor, despite many decades of study. Nevertheless, the mechanisms that are believed to be involved include many of the processes previously described above.

11.3 Flammability and Combustion Processes of Polymers

Polymeric materials are able to start or propagate fires since, upon heating, they decompose and give rise to the formation of volatile and combustible products [29–32]. Although the precise details of a polymer combustion process are extremely complex, the schematic representation presented in Fig. 11.7 highlights the main features.

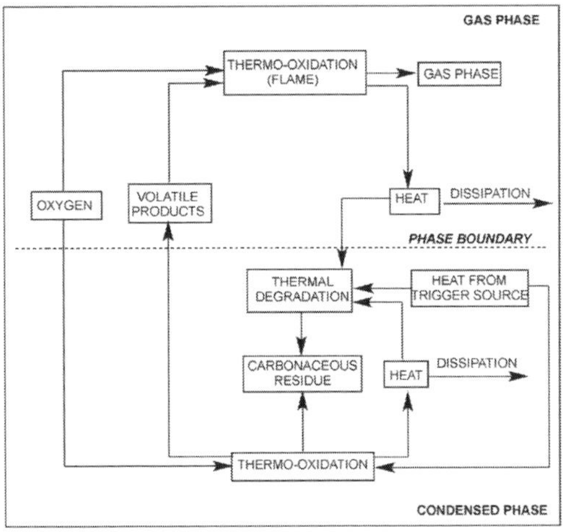

Figure 11.7 Schematic representation of the polymer combustion process.

Combustion starts when two conditions are met: First, the concentration of volatile products generated by the heat provided by the trigger source reaches the flammability range; second, the temperature exceeds that required for ignition. At this point, combustion occurs. If the heat input into the polymer is sufficient to sustain the above conditions, and thereby maintain the thermal degradation rate beyond the minimum value required for fuelling the flame itself, degradation will continue until the material is completely consumed. Otherwise, the flame extinguishes soon after ignition. When the heat provided by the trigger is negligible compared with that transferred to the polymer from combustion, combustion becomes self-sustainable only when the thermal requirements are fulfilled by the heat evolved during the thermo-oxidation reactions, which occur within the flame in the gas phase or in the burning material in the condensed phase.

Notwithstanding the general applicability of the general scheme shown in Fig. 11.7, some of the steps describing the combustion cycle (e.g., carbonization or thermo-oxidation in the condensed phase) can be missing or of little relevance, depending on the polymer of interest or the particular conditions under which combustion is taking place.

The flammability character of polymeric materials limits their applications in several fields, where fire risk (i.e., the probability that a fire occurs) is an important consideration. These include transportation, building, electrical and electronic sectors, etc. However, it is not possible to prepare an organic material that is completely non-combustible, while retaining desirable characteristics, such as ease of processing, low density, low cost, that are typical of many polymeric materials. Nevertheless, by using appropriate additives—so-called flame retardants—it is possible to limit the propensity to ignition and/or the combustion propagation rate, so that the time range required for intervention in order to extinguish the fire increases.

About half of the polymeric materials currently produced must, in some way, be protected from fire risks. Indeed, amongst polymer additives, flame retardants show the highest economic value and are continuously expanding because of more and more severe legislation on the use of materials for use in circumstances where fires may occur.

As far as flammability of polymers is concerned, this term does not have a specific scientific meaning, since it cannot be measured by assessing a single parameter, as in the case of gases, where flammability limits can be defined. Indeed, the propensity of polymeric materials toward combustion strictly depends on both intrinsic (e.g., chemical composition, morphology) and extrinsic factors (e.g., dimensions, shape, air speed, heat flux). Consequently, predicting the behavior of a polymeric material during a fire is difficult, due to the poor reproducibility that is typical of the fire itself.

Nevertheless, reliable methods for assessing the behavior of polymeric materials in relation to fire risk have been set, which substantially involve the measurement of such relevant parameters as ease of ignition, heat release rate, flame propagation rate, generation of toxic and darkening smokes, etc. These measurements need to be made under strictly defined conditions in order to provide a comparative classification of different materials' behavior. Large-scale tests allow real fire conditions to be simulated, but are very expensive and require the use of special facilities. For these reasons, it is usual to perform large-scale tests only on materials that have been identified on the basis of small, lab-scale experiments. Such measurements can be very useful for the development of new materials, as they require only small amounts of products; however, the adopted conditions are very far from real fire scenarios. Fortunately, empirical practice has demonstrated that the small-scale methodologies can provide reasonable conformity with the results deriving from more realistic, large-scale studies and, consequently, they do have considerable utility.

Amongst the range of fire tests that exist, limiting oxygen index (LOI) and cone calorimetry are the most commonly used for providing quantitative assessments of the combustion behavior of polymeric materials. In LOI evaluations, a bar specimen is vertically placed in a glass cylindrical channel and burns from the top to the bottom (c.f. a candle) within an ascending flow of a N_2/O_2 mixture. The specimen is initially ignited by a flame, which is subsequently removed, and by varying the composition of the gas, the minimum oxygen volume percentage at which self-sustained combustion occurs for 3 min can be determined. This represents the LOI of the material under test. Table 11.1 presents LOI values for a range of different polymeric materials.

Table 11.1 Limiting oxygen index values

Polymer	LOI (%)
Poly(urethane) foam	16.5
Cotton	16–17
Poly(styrene)	17.6–18.3
Ethylene vinyl acetate copolymer	19
Nylon 6	25–26
Poly(vinylchloride) (rigid)	45–49
Poly(tetrafluoroethylene)	95

Although any material characterized by an LOI value in excess of the oxygen content of the atmosphere should not give self-sustained combustion, because of the differences between lab-scale and fire test conditions, polymers are only considered flame retarded when the LOI is above 25.

Cone calorimetry tests exploit the empirical observation that heat released by burning materials is directly proportional to the quantity of oxygen consumed in the combustion process. In this approach, a square specimen is placed on a balance and exposed to the radiative heat flux provided by a cone heater; an electrical spark is used to ignite the volatile degradation products that are generated by the radiative heat flux. The combustion products (smokes) are then collected, where the concentrations of the unconsumed oxygen and of the evolved gases (namely CO and CO_2) are measured. An optical system also measures the obscuring power of the smoke. This apparatus provides a thorough description of the combustion process as a function of time, by assessing time to ignition, heat release (calculated on the basis of the consumed oxygen), sample weight loss, and opacity and composition of the smoke that is generated.

11.4 Effect of the Presence of Nanofillers on Degradation Phenomena

The motivation behind introducing nanoscale fillers into polymer matrices is the attainment of potentially unique properties, as a result of the nanometric dimensions of the dispersed phase. Indeed, polymer nanocomposites now represent an important class of nanostructured materials having wide applicability in a number of different industrial sectors.

Nanomaterials are usually classified into three categories, i.e., nanoparticles, nanotubes and nanolayers, depending on how many of the dimensions of the dispersed particles are nanometric. Inorganic nanofillers include layered silicates (such as montmorillonites, MMTs), nanotubes (mainly carbon nanotubes—CNTs), fullerenes, SiO_2, metal oxides (e.g., TiO_2, Fe_2O_3 Al_2O_3), metal nanoparticles (e.g., Au, Ag), polyhedral oligomeric silsesquioxanes (POSS), semiconductors (e.g., PbS, CdS), carbon black, nanodiamonds, etc. All these nanoparticles are employed in order to improve the physical, mainly mechanical, properties of the matrices within which they are homogeneously dispersed.

From a theoretical point of view, any kind of material can be processed to exist with nanometric dimensions. However, over the last few decades, none of the commercially available nanoparticles has attracted as much attention as clay-based systems (that is, layered silicates) and carbon nanotubes. Unlike conventional fillers, the effect of nanofillers on physical properties, including flammability, is achieved even when they are added to the polymer matrix in very low amounts (usually 3–5 wt.%, but not exceeding 10 wt.%), but this happens only when a good dispersion of the nanofiller in the polymer is attained [33–35]. Consequently, many different dispersion strategies (e.g., solution-adsorption, in situ intercalative polymerization, melt intercalation, in situ hydrothermal crystallization for nanoclays, direct melt mixing, solution mixing and in situ polymerization for carbon nanotubes) have been proposed and optimized.

From an applications perspective, the flammability of nanocomposites is an important topic, since these materials offer a number of potential benefits over traditional flame retardants such as aluminum hydroxide or halogenated compounds. First, although traditional flame retardants are known to be highly effective, it is generally necessary to add these at high loading levels to ensure an acceptable flame retardant performance, which often adversely affects mechanical properties or processability. It is also noteworthy that halogen-containing flame retardants are under pressure in many countries, because of environmental concerns. Furthermore, the addition of many flame retardants often favors the production of soot and carbon monoxide during combustion. Nanocomposites can offer advantages over traditional flame retardants in that processing of nanocomposites can be straightforward and, because of their halogen-free character, they may represent a valuable environmentally friendly alternative.

Many factors have been found to affect the thermal stability and flammability of nanocomposite materials. These include the intrinsic thermal resistance of the selected polymer matrix, the nanofiller content, the chemical structure of the organic modifier, the chemical character of polar compatibilizers and the access of oxygen to the nanocomposite material upon heating (i.e., barrier properties). However, for a given polymer/nanofiller system, enhancement of flame retardancy on addition of a nanofiller is generally ascribed to two different effects, one physical and one chemical.

The physical effect is related to the formation of a surface barrier layer due to local accumulation of the nanofiller. During combustion, the rapid ablation of the polymer allows the nanofiller to accumulate on the surface and this inorganic surface layer, which is thermally stable, acts as a barrier against the transfer of heat and oxygen from the flame region to the polymer and of degradation products from the polymer to the flame. The distribution of the nanofiller within the polymer determines the ease and rate of formation of this inorganic surface layer and, practically, a good dispersion of the nanofiller usually leads to the formation of a more effective barrier layer compared with nanocomposites that are characterized by significant agglomeration. This physical effect relating to surface accumulation of nanofiller has been accepted as the mechanism of improved flame retardancy for a range of different nanofillers with different dimensionalities, including isometric particles such as POSS, nanotubes and nanofibers, including CNTs and layered nanoclays.

The chemical effect is ascribed to the catalytic behavior of the nanofiller towards the formation of a carbonaceous residue (the so-called char). In the presence of nanofillers, the formation of char may be enhanced, which limits the production of the volatile species that are required to fuel the fire. Oxidative dehydrogenation is the mechanism proposed for the char formation in the presence of nanofillers, a process that can be catalyzed by the transition metals present within the structure of the nanofillers such as nanoclays.

In the following paragraphs, some examples regarding how the presence of nanofillers can improve polymer stability toward different types of degradation phenomena will be presented.

11.4.1 Thermal Degradation and Stabilization of Nanocomposites Based on Nanolayers

The incorporation of silicate nanolayers that are characterized by a high-aspect ratio (e.g., nanoclays) within different polymer matrices has been found to enhance thermal stability and flame retardancy [33]. The origins of this are associated with the nanofiller acting to (i) reduce heat transfer into the material, which improves the effective thermal stability of the polymer system, and (ii) enhance char formation. Also, since a general characteristic of nanocomposites including well-dispersed nanoclays is outstanding barrier properties, the presence of nanolayers will tend to inhibit the diffusion of volatile products generated during decomposition out of the solid phase into the flame. A key point here is that to achieve maximum benefit, the choice of surfactants used for modifying the nanoclay surface is critical. These are usually described as "onium" salts and ammonium salts such as tetra-alkyl ammonium–based cationic surfactants are the most commonly employed, in order to compatibilize the nanoclay with a given polymer system, thus favoring the obtainment of highly intercalated or exfoliated morphologies.

It has been demonstrated that in the case of nanocomposites, the temperature at which volatilization starts to occur significantly increases in comparison with microcomposite counterparts. Furthermore, the process of thermal oxidation is strongly delayed in nanocomposites that are characterized by a high-char yield: this is due to both the physical barrier effect, which is enhanced by ablative reassembling of the nanoclay, and to the chemical catalytic action associated with both the nanoclay itself and with the thermal decomposition of the nanoclay modifier, which creates strongly acid sites [35].

Within those nanocomposites that exhibit good fire retardant features, significant intermolecular reactions take place (e.g., inter-chain aminolysis/acidolysis, radical recombination and hydrogen abstraction). In the case of polymers in which a radical mechanism is involved in their degradation, the relative stability of the radical represents the key factor for predicting the effect that nanocomposite formation has on the reduction in the peak heat release rate. The higher the stability of the radical produced by the polymer, the better the fire retardancy of the nanocomposite, as measured by the reduction in the peak of the heat release rate.

11.4.2 Thermal Degradation and Stabilization of Nanocomposites Based on Nanotubes

The preparation of nanocomposites by the dispersion of carbon nanotubes within different polymers (e.g., polyolefins, polyesters, polyacrylates, polyamides, polyurethanes, etc.) has attracted considerable interest, since this type of nanofiller is able strongly to enhance flame retardancy, mechanical performance and the electrical properties of the host matrix [35]. Nevertheless, some significant limitations exist. For example, as the overall performance of CNT-based nanocomposites depends on the homogeneity of the dispersion of the nanofiller, functionalization techniques are usually required to improve polymer/CNT interactions and, thereby, enhance the level of dispersion that is achieved. Furthermore, although CNTs are known to be excellent conductive fillers for polymer matrices, their thermal stability and barrier properties are not as good as those exhibited by systems based upon nanoclays. Finally, while multi-functional high performance polymer composites are desperately needed in many different application areas, this is generally difficult to achieve through the incorporation of just a single type of nanofiller. For these reasons, multiphase polymer/CNT nanocomposites produced by adding combinations of nanofillers have been increasingly investigated. This approach has shown that positive synergistic effects between different nanofillers can occur. For example, CNTs are found to play an important role in reducing the heat release rate peak by generating a low permeability char containing graphitic carbon. Since the oxidation resistance of this char depends on the degree of graphitization, the overall performance of CNT/ethylene (vinyl acetate) (EVA) nanocomposites has been shown to be improved by adding a clay, since then, the tendency to form graphitic carbon is enhanced. In addition, the presence of nanotubes can also help to improve the mechanical integrity of any char that is formed; by reducing surface cracks of char, its barrier resistance to the evolution of flammable volatiles and the oxygen transfer to the condensed phase can be improved.

11.4.3 Electrical Surface Degradation of Nanocomposites

Although the effect of nanoparticles on surface erosion by electrical processes has been discussed in detail in the preceding chapter, it is, nevertheless worth briefly contrasting the work presented in Chapter 10 with the above discussion of the mechanisms by which nanoparticles affect thermal degradation. In both cases, improvements in performance primarily stem from nanofillers acting to promote the formation of a surface barrier layer. This occurs through accumulation of the nanofiller but also as a consequence of the nanofiller acting to promote the formation of char, an assertion that is consistent with excess residual masses observed in thermogravimetric data. In the case of flammability, the existence of a char *per se* may be beneficial but, in electrical applications, the formation of carbon-rich, electrically conducting residues may be less attractive. Nevertheless, this will depend upon the precise chemical pathways involved, which will, in turn, depend on the presence of appropriate catalyzing species (e.g., transition metal impurities in the case of nanoclays). The effectiveness of the surface barrier will then depend upon its physical properties (e.g., thermal conductivity, mechanical integrity, barrier characteristics, etc.) and, consequently, will depend upon the nature of the nanofiller, the quality of its dispersion, its surface chemistry, interactions with the matrix polymer, etc.

In the case of flammability, aspects of the chemical pathways that lead to char formation are understood (e.g., in polymers where degradation involves a radical mechanism, the greater the stability of the radical, the better the fire retardancy of the nanocomposite). In the electrical cases, pathways are likely to be more complex but by understanding these better and adopting a more active approach to design the material composition, it would appear that further improvements in material performance can be attained.

11.4.4 Effect of Nanoparticles on Degradation by Electrical Treeing

Many workers have explored the effect of the addition of nanoparticles on electrical treeing processes in polymers and, in some cases, dramatic improvements in performance have been reported. The effect of ZnO nanoparticles on electrical treeing in low density polyethylene (LDPE) was studied at 10 kV [36] and, in this work, the time to breakdown was found to increase monotonically from less than 2 h in the unfilled polymer to more than 100 h in the system containing 10% nano-ZnO. The effect of nanofiller aspect ratio in LDPE containing sheet-like sepiolite and MMT has been considered [37] and, again, significantly longer times to breakdown were reported at 10 kV; the system containing 5% MMT exhibiting the best performance out of the materials that were studied. Another comparative study of different nanofillers considered three systems that differ with respect to their dimensionality [38]: SiO_2 nanoparticles; a fibrous phyllosilicate; plate-like MMT. In this work, both tree growth rates and tree inception times were considered, as was the effect of the applied voltage. While the resulting data appear indicative of complex behavior, in general, tree inception times were found to be increased by including a nanofiller, particularly in systems containing MMT. Tree growth rates were, however, found to depend on the chosen voltage; at the higher voltages, where bush trees develop, tree growth rates were increased in the nanocomposites whereas, at the lower voltages where branch-type trees form, reduced tree growth rates were observed. Elsewhere, [39] the effect of nano-MgO on both tree inception and propagation rates in LDPE has been reported for systems containing up to 10 parts nanofiller per hundred parts resin (phr). Although no benefits of adding the nanofiller could be clearly inferred during the earliest stages of growth, "a slight positive effect on tree initiation is suspected and tree inception voltage might increase due to nanofillers." However, a reduction in tree growth rate was reported, even for the system containing just 1 phr of the nanofiller, such that the overall time to failure was found to increase monotonically with filler loading level.

While the fact that the addition of nanoparticles affects the development of electrical tree growth in polymers is well demonstrated, many explanations have been proposed to account for this. The simplest of these appears most applicable to the case of nanoclay-filled systems, where such structures may simply act as barriers that increase tortuosity. Many workers have suggested that the presence of nanoparticle/polymer interphase regions is critical, which may be either reinforcing or weakening, depending on their precise nature. Such effects have been attributed to the influence of the nanoparticles on the local distribution of trap sites and, in the work of Yamano and Iizuka [40], a nickel phthalocyanine was added specifically to vary this factor; increased tree inception voltages were reported. Finally, a key element in electrical tree growth concerns the degradation of the material through partial discharges within the tubules. As such, all the mechanisms discussed above in correction with surface erosion are also likely to come into play, with the implication that no single mechanism is likely to be responsible for the various effects that are observed.

11.5 Conclusions

Degradation of polymers can occur as a consequence of many different factors and this chapter has provided an overview of the topic. In many cases, the underlying processes have been thoroughly studied, precise mechanisms have been proposed and, as a consequence, strategies have been developed with a view to countering degradation. In the case of nanocomposites, significant parallels exist between topics such as flammability and electrical degradation. In both cases, the critical element concerns the optimization, during the primary degradation phase, of a protective surface layer that subsequently suppresses further degradation. This is not simply a case of the matrix polymer being removed, leaving the nanoparticles on the surface—clear evidence exists for the active role that nanoparticles play in char formation and, consequently, by optimizing nanofiller chemistry, nanocomposite composition and processing and matrix/nanoparticle interactions, superior systems may be actively designed.

References

1. Halim Hamid, S. (2000), *Handbook of Polymer Degradation*, 2nd Ed., Marcel Dekker, Inc., New. York.
2. Scott, G. (1990), *Mechanisms of Polymer Degradation and Stabilization*, Elsevier Applied Science, London.
3. Guillet, J. E. (1985), *Polymer Photophysics and Photochemistry*, Cambridge University Press, Cambridge.
4. McKellar, J. F., Allen, N. S. (1979), *Photochemistry of Man-Made Polymers*, Applied Science Publishers, London.
5. Ranby, U., Rabek, J. F. (1975), *Photodegradation, Photo-oxidation and Photostabilization of Polymers*, Wiley, London.
6. Yousif, E., Haddad, R. (2013), Photodegradation and photostabilization of polymers, especially polystyrene: Review, *SpringerPlus*, vol. 398, pp. 1–32.
7. Albertsson, A. C., karlsson, S. (1988), A the three stages in degradation of polymers-polyethylene as a model substance, *Journal of Applied Polymer Science*, vol. 35, pp. 1289–1302.

8. Massey, L. K. (2007), *The Effects of UV-Light and Weather on Plastics and Elastomers*, 2nd ed., William Andrew Publishing, Norwich.
9. Leja, K., Lewandowicz, G. (2010), Polymer biodegradation and biodegradable polymers: A review, *Polish Journal of Environmental Studies*, vol. 19, no. 2, pp. 255–266.
10. Pielichowski, K., Njuguna, J. (2005), *Thermal Degradation of Polymeric Materials*, Rapra Technology, Shawbury.
11. Crompton, T. R. (2010), Thermo-Oxidative Degradation of Polymers, iSmithers, Shawbury, UK.
12. Colin, X., Verdu, J. (2006), Polymer Degradation during Processing, *C. R. Chimie*, vol. 9, pp. 1380–1395.
13. Chattopadhyay, D. K., Webster, D. C. (2009), Thermal stability and flame retardancy of polyurethanes, *Progress in Polymer Science*, vol. 34, pp. 1068–1133.
14. Mohod, A. V., Gogate, P. R. (2011), Ultrasonic degradation of polymers: Effect of operating parameters and intensification using additives for carboxymethyl cellulose (CMC) and polyvinyl alcohol (PVA), *Ultrasonics Sonochemistry*, vol. 18, pp. 727–734.
15. Sathiskumar, P. S., Madras, G. (2012), Ultrasonic degradation of butadiene, styrene and their copolymers, *Ultrasonics Sonochemistry*, vol. 19, pp. 503–508.
16. Ciuprina, F., Zaharescu, T., Plesa, I. (2013), Effects of γ-radiation on dielectric properties of LDPE–Al_2O_3 nanocomposites, *Radiation Physics and Chemistry*, vol. 84, pp. 145–150.
17. Otaguro, H., de Lima, L. F. C. P., Parra, D. F., Lugao, A. B., Chinelatto, M. A., Canevarolo, S. V. (2008), High-energy radiation forming chain scission and branching in polypropylene, *Radiation Physics and Chemistry*, vol. 79, pp. 318–324.
18. Ito, M., Nagai, K. (2008), Degradation issues of polymer materials used in railway field, *Polymer Degradation and Stability*, vol. 93, pp. 1723–1735.
19. Gijpferich, A. (1996), Mechanisms of polymer degradation and erosion, *Biomaterials*, vol. 17, pp. 103–114.
20. Dondi, D., Zeffiro, A., Buttafava, A., Marciano, C., Bianchi, M., Faucitano, A. (2013), Mechanical degradation of elastomers in the presence of silica and inhibitors using a new design of mechano reactor, *Polymer Degradation and Stability*, vol. 98, pp. 392–407.
21. Nguyen, T. Q., Khausch, H. H. (1990) Effects of solvent viscosity on polystyrene degradation in transient elongational flow, *Macromolecules*, vol. 23, pp. 5137–5145.

22. Nguyen, T. Q. (2001), Dynamics of flexible polymer chains in elongational flow, *Chimia*, vol. 55, pp. 147–154.

23. Smith, D. E., Chu, S. (1998), Response of flexible polymers to a sudden elongational flow, *Science*, vol. 281, pp. 1335–1340.

24. Zhurkov, S. N., Korsukov, V. E. (1974), Atomic mechanism of fracture of solid polymers, *Journal Polymer Science Polymer Physics*, vol. 12, no. 2, pp. 385–398.

25. Dissado, L. A. (2002), Understanding electrical trees in solids: From experiment to theory, *IEEE Transactions on Dielectrics and Electrical Insulation*, vol. 9, pp. 483–497.

26. Montanari, G. C. (2013), Notes on theoretical and practical aspects of polymeric insulation aging, *IEEE Electrical Insulation Magazine*, vol. 29, no. 4, pp. 34–44.

27. Jones, J. P., Llewellyn, J. P., Lewis T. J. (2005) The contribution of field-induced morphological change to the electrical aging and breakdown of polyethylene, *IEEE Transactions on Dielectrics and Electrical Insulation*, vol. 12, pp. 951–966.

28. Mazzanti, G., Montanari, G. C., Dissado, L. A. (1999), A space-charge life model for a.c. electrical, aging of polymers, *IEEE Transactions on Dielectrics and Electrical Insulation*, vol. 6, pp. 864–875.

29. Horrocks, R., Price, D. (2001), *Fire Retardant Materials*, Woodhead Publishing Limited, Cambridge.

30. Morgan, A. B., Wilkie, C. A. (2007), *Flame Retardant Polymer Nanocomposites*, Wiley Interscience, New Jersey.

31. Horrocks, R., Price, D. (2008), *Advances in Fire Retardant Materials*, CRC Press, Cambridge.

32. Hull, T. R., Kandola, B. K. (2009), *Fire Retardancy of Polymers: New Strategies and Mechanisms*, RSC Publishing, Cambridge.

33. Chrissafis, K., Bikiaris, D. (2011), Can nanoparticles really enhance thermal stability of polymers? Part I: An overview on thermal decomposition of addition polymers, *Thermochimica Acta*, vol. 523, pp. 1– 24.

34. Bikiaris, D. (2011), Can nanoparticles really enhance thermal stability of polymers? Part II: An overview on thermal decomposition of polycondensation polymers, *Thermochimica Acta*, vol. 523, pp. 25–45.

35. Kumara, A. P., Depan, D., Tomer, N. S., Singh, R. P. (2009), Nanoscale particles for polymer degradation and stabilization: Trends and future perspectives, *Progress in Polymer Science*, vol. 34, pp. 479–515.

36. Tian, F., Lei, Q., Wang, X., Wang, Y. (2012), Investigation of electrical properties of LDPE/ZnO nanocomposite dielectrics, *IEEE Transactions on Dielectrics and Electrical Insulation*, vol. 19(3), pp. 763–769.

37. Guastavino, F., Dardano, A., Squarcia, S., Tiemblo, P., Guzman, J., Benito, E., Garcia, N. (2009), Electrical treeing in LDPE nanocomposite materials, *Annual Report—CEIDP*, pp. 697–700.

38. Tiemblo, P., Hoyos, M., Gomez-Elvira, J. M., Guzman, J., Garcia, N., Dardano, A., Guastavino, F. (2008) The development of electrical treeing in LDPE and its nanocomposites with spherical silica and fibrous and laminar silicates, *Journal of Physics D: Applied Physics*, vol. 41(12), pp. 125208–125215.

39. Tanaka, T., Iizuka, T., Sekiguchi, Y., Murata, Y., Ohki, Y. (2009) Tree initiation and growth in LDPE/MgO nanocomposites and roles of nano fillers, *Annual Report—CEIDP*, pp. 646–649.

40. Yamano, Y., Iizuka, M. (2011) Improvement of electrical tree resistance of LDPE by mixed addition of nanoparticles and phthalocyanine, *IEEE Transactions on Dielectrics and Electrical Insulation*, vol. 18(1), pp. 329–337.

Chapter 12

Permittivity Gradient Composite Material Structures

Naoki Hayakawa

Department of Electrical Engineering and Computer Science, Nagoya University, Furo-cho, Chikusa-ku, Nagoya 464-8603, Japan

nhayakaw@nuee.nagoya-u.ac.jp

This chapter introduces a novel technique for electric field grading and control in power apparatus with functionally graded material (FGM). FGM is characterized by the spatial distribution of dielectric permittivity and/or electrical conductivity in a solid insulator, which is expected to control the surrounding electric field distribution in power apparatus such as gas-insulated switchgears (GIS), epoxy molded transformers, and bushings.

12.1 Introduction

Innovation leading to higher performance and increased reliability of electrical power apparatus can be brought about through novel electrical insulating materials. For example, SF_6 gas-

Tailoring of Nanocomposite Dielectrics: From Fundamentals to Devices and Applications
Edited by Toshikatsu Tanaka and Alun S. Vaughan

ISBN 978-981-4669-80-1 (Hardcover), 978-981-4669-81-8 (eBook)
www.panstanford.com

insulated switchgear has contributed to the drastic reduction in size and footprint of such apparatus and, therefore, substations. However, because of the high sensitivity of SF_6 gas to electric field stress, the insulation design and configuration of gas-insulated power apparatus is complicated and costly to ensure appropriate electric field control, relaxation, and optimization. Thus, the development of novel and innovative approaches to electrical insulating materials to overcome the above issues is one of the most important tasks for future electrical power engineering.

From the above background, electric field grading techniques in power apparatus have been developed, using linear or non-linear semi-conductive materials such as carbon black, silicon carbide (SiC), zinc oxide (ZnO), and micro varistors as fillers in solid insulators for bushings, cable accessories, etc. [1, 2]. Functionally graded materials (FGM) with a spatial distribution of dielectric permittivity (ε) in solid insulators, i.e., ε-FGM, are being investigated for electric field grading and control of GIS spacer [1, 3–5]. In this chapter, different types and configurations of ε-FGM are introduced, including related concepts and results on electric field grading effect. This chapter also introduces another type of electric field grading through electrical conductivity (σ) on solid insulator surface, i.e., σ-FGM [6]. The effect on electric field grading and control for GIS spacer with σ-FGM is also discussed. Both ε-FGM and σ-FGM are expected to bring about the simplification of spacer configuration, increased compactness, energy saving, lower cost, etc., for power apparatus, compared with the conventional equivalents.

12.2 ε-FGM with Dielectric Permittivity Distribution

12.2.1 Concept of ε-FGM

In electrical power apparatus, electric, thermal, or mechanical stresses should be relaxed. For this purpose, FGM involving dielectric permittivity, electrical conductivity, thermal conductivity and thermal expansion coefficient, etc., can be considered. In this section, ε-FGM is introduced for the improvement of the electric field distribution in and around solid insulators.

The concept of ε-FGM is shown in Fig. 12.1. Compared with a uniform material shown in Fig. 12.1a, the ε-FGM in Fig. 12.1b has a spatial grading of filler content from material A to material B, which leads to the relaxation of the stress concentration at the interface. In a gas/solid insulation system, electric field stress under ac or impulse voltage application is generally intensified in the gas region, because the permittivity of gas is lower than that of solid materials. This stress intensification can be reduced by the use of an ε-FGM that is characterized by an appropriate ε-distribution inside the solid insulator. Consequently, a well-designed ε-FGM can drastically raise the electric field utilization factor in the insulating materials; then, the distribution of the electric field can be made more homogeneous in critical regions. Therefore, the key to the application of ε-FGM is to choose an appropriate ε-distribution such that the electric field utilization factor of the power apparatus be improved as much as possible.

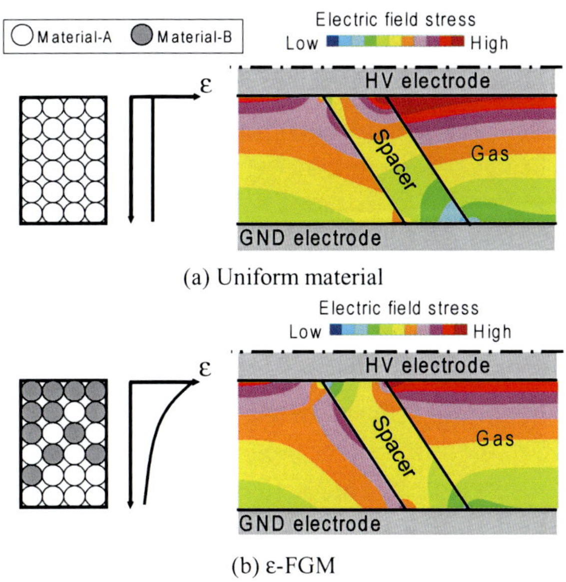

Figure 12.1 Concept of ε-FGM.

By the application of ε-FGM, the electric field distribution around the solid insulator can be improved while retaining the simple structure. According to the critical points of electrical insulation in GIS, the merits of the application of ε-FGM spacer can be listed as follows:

- Simplification of spacer and electrode configurations in gas/solid composite insulation systems
- Reduction of concentrated electric field stresses, such as at triple junctions
- Reduction of electric field stress at gas/solid insulator interfaces
- Reduction of the electric field stress at the embedded electrode within the solid insulator

12.2.2 Electric Field Grading Effect of ε-FGM

The ε-FGM in Fig. 12.1b is characterized by a progressive reduction in permittivity (grading to lower permittivity FGM [GLP-FGM]) from the high-voltage (HV) electrode to the grounded (GND) electrode of the GIS spacer. In this case, the maximum electric field stress that occurs around the triple junction of the cone-type GLP-FGM spacer was reduced by about 40%, compared with that for the conventional spacer manufactured with a uniform material shown in Fig. 12.1a [3]. Using a GLP-FGM spacer fabricated in the truncated cone-type configuration with a triple junction, the impulse breakdown voltage in SF_6 gas was verified to be improved by about 70%, compared with that of a conventional spacer [1]. Alternatively, Fig. 12.2 presents data for a post-type spacer in which the ε-FGM is characterized by with a U-shape ε-profile (U-FGM) [3]—this profile was obtained by the application of a centrifugal force of 3000 G for different times. The ε profile shown in Fig. 12.2a reduces the electric field stress at the interface between the spacer and HV or GND electrode, as shown in Fig. 12.2b. Based on the voltage–time (V–t) characteristics with the n-value of 10–20, the life time of the post-type U-FGM spacer is expected to be 2–6 times longer than that of the conventional spacer, as shown in Fig. 12.2c [3], for different centrifugal force and application time.

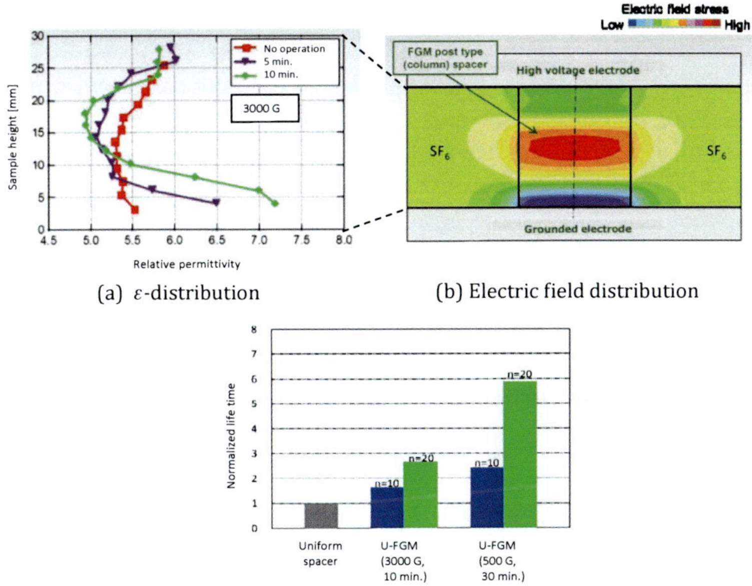

(a) ε-distribution (b) Electric field distribution

(c) Life time extension

Figure 12.2 ε-Distribution, electric field distribution, and life time extension of U-FGM.

12.2.3 Fabrication and Simulation Methods of ε-FGM

12.2.3.1 Fabrication methods of ε-FGM

Figure 12.3 shows different types of ε-FGM fabricated into post-type GIS spacers, where the permittivity profile is obtained by localizing different kinds of fillers, such as alumina (Al_2O_3), silica (SiO_2), and titania (TiO_2), in epoxy resin under a centrifugal force. The fabrication concepts and procedures are as follows [4]:

(i) GHP-FGM

Figure 12.3(i) shows the fabrication concept for GHP-FGM, where ε-grading to higher permittivity occurs along the direction of the centrifugal force. One type of filler with a permittivity higher than that of the epoxy resin is loaded into the epoxy, and the centrifugal force is applied. Then, a graded filler distribution along the centrifugal direction is obtained, as shown. When the high permittivity part of the GHP-FGM spacer is applied to the HV electrode of GIS, the electric field around the HV electrode can be reduced.

(ii) GLP-FGM

Figure 12.3(ii) shows the fabrication concept for GLP-FGM. Two kinds of fillers are first loaded in the epoxy resin and mixed to provide a uniform starting material. One filler is characterized by small diameter particles of high permittivity, while the other has large diameter particles of low permittivity. By the application of a centrifugal force, the larger filler particles (low permittivity) move in the centrifugal direction. Accordingly, the filler with small diameter (high permittivity) moves to the upper part of the sample instead. As a result, a GLP-FGM can be obtained, which can be applied to optimize the electric field distribution around disk-type spacers.

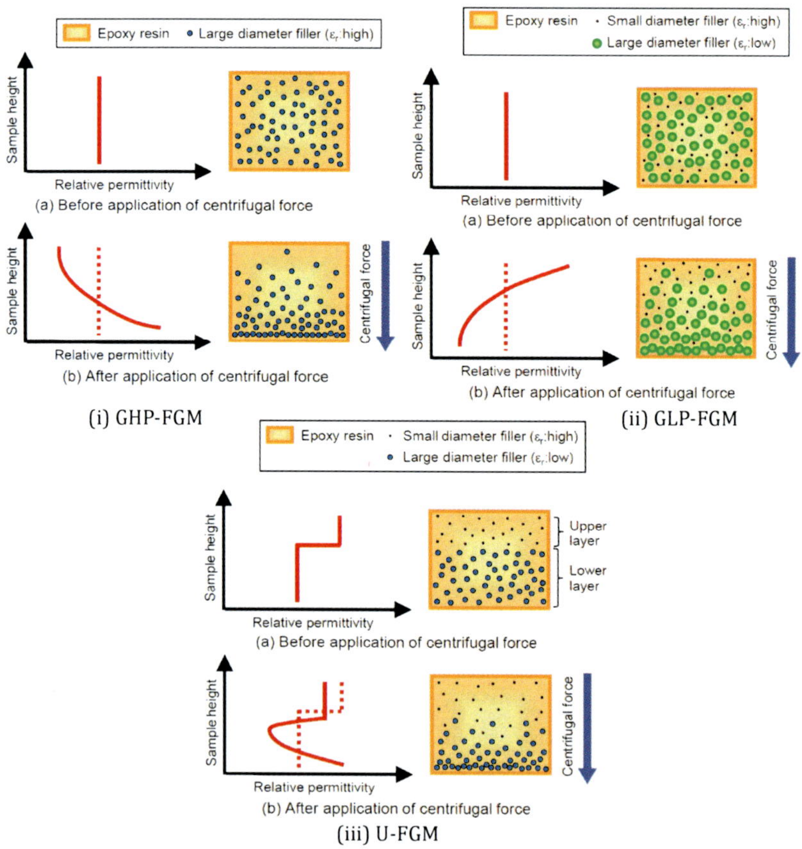

Figure 12.3 Fabrication concept of ε-FGM with different ε-grading for post-type configuration.

(iii) U-FGM

Figure 12.3(iii) shows the fabrication concept for U-FGM. Two kinds of fillers are loaded into the epoxy resin; one filler characterized by small diameter particles and a high permittivity is loaded into the upper layer of the sample, while the other consisting of large diameter particles of low permittivity is loaded into the lower layer. On the application of a centrifugal force, the large diameter filler moves along the centrifugal direction, whereas the small diameter filler does not move so much. Consequently, the permittivity is increased at the lower part of the sample where the density of large diameter filler particles is increased such that a U-shape permittivity distribution is obtained. U-FGM is effective at reducing the electric field on the HV and GND electrode surface in contact with post-type spacer, as shown in Fig. 12.2.

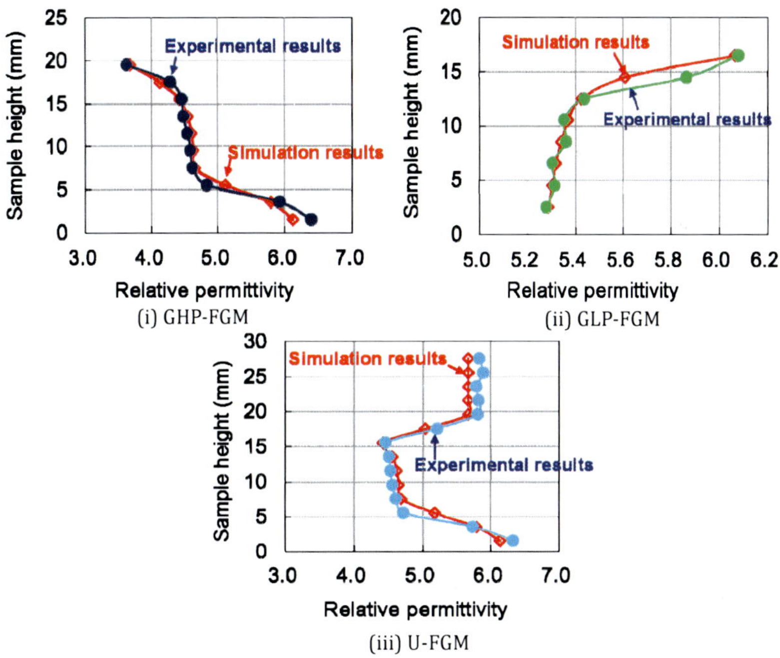

Figure 12.4 Fabrication concept of ε-FGM with different ε-grading for post-type configuration.

Permittivity profiles from differently fabricated ε-FGM systems are shown in Fig. 12.4: (i) GHP-FGM, (ii) GLP-FGM and

(iii) U-FGM [4]. Figure 12.4(i) shows that the GHP-FGM was fabricated with a relative permittivity (ε_r) that varied from $\varepsilon_r = 3.6$ to $\varepsilon_r = 6.4$ as a consequence of a graded distribution of Al_2O_3 particles. In the GLP-FGM shown in Fig. 12.4(ii), the permittivity varied from $\varepsilon_r = 6.1$ to $\varepsilon_r = 5.3$ as a consequence of the graded distribution of SiO_2 and TiO_2 particles. In the U-FGM case, the permittivity varied from $\varepsilon_r = 5.8$ to $\varepsilon_r = 6.3$ through $\varepsilon_r = 4.5$, as shown in Fig. 12.4(iii), as a consequence of the graded distribution of Al_2O_3 and TiO_2 particles. An arbitrary ε-distribution within post-type ε-FGM spacers can be obtained by optimizing the composition (material, size, density, relative permittivity, loading level, etc.) of the system and the centrifugal force conditions (acceleration, time, temperature and their combination pattern with time, etc.) used during manufacture.

12.2.3.2 Simulation methods of ε-FGM

Figure 12.5 shows the model used to simulate filler particle movement within a viscous fluid (epoxy resin mixed with hardener) under a centrifugal force [4]. Using such a one-dimensional model for a post-type ε-FGM spacer, the flow of particle density is calculated in each region of the sample under the centrifugal force. In the calculation, the balance of three forces on the filler particles is considered, centrifugal force, buoyancy and drag, which leads to the following equation:

$$M\frac{dv_r}{dt} = Mr\omega^2 - Mr\omega^2\frac{\rho_f}{\rho_p} - F_D \tag{12.1}$$

In this, M is mass of the filler particle, v_r is the relative velocity of the filler particle, r is the radius of rotation, ω is the angular speed, ρ_f and ρ_p represent the densities of the fluid and particle, respectively, and F_D is the viscous force.

Given $dv_r/dt = 0$, the terminal velocity (v_t) of filler particles is expressed as follows:

$$v_t = \frac{GD_p^2(\rho_p - \rho_f)}{18\eta}, \tag{12.2}$$

where G is the acceleration due to the centrifugal force, D_p is the diameter of the filler particle, η is the viscosity of the epoxy resin containing with filler particles. η is estimated and experimentally verified in consideration of the material and filler loading level, temperature, etc., and $F_D = 3\pi\eta D_p v_t$ according to Stokes' law.

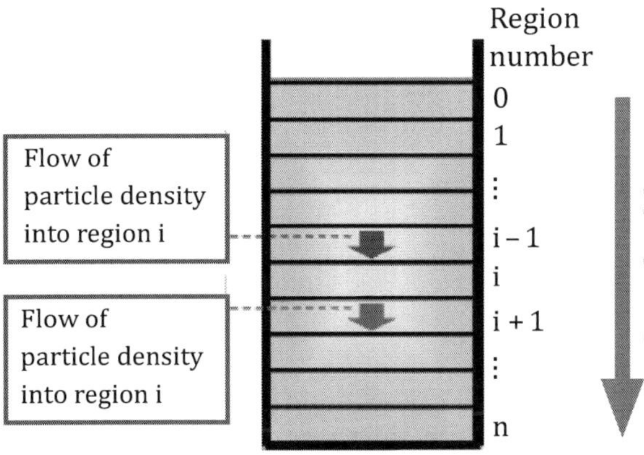

Figure 12.5 Calculation model of filler particle movement simulation for post-type ε-FGM.

In the case of the GHP-FGM, the movement of Al_2O_3 particles was simulated while, for the GLP-FGM, the movement of the larger diameter SiO_2 was considered, assuming that the local filler content (vs. epoxy) of the TiO_2 remained constant because these smaller particles do not move under the application of the centrifugal force. In the case of the U-FGM, the movement of both TiO_2 and Al_2O_3 particles was simulated. Finally, in order to estimate the local relative permittivity of the epoxy resin loaded with fillers, the extended Bruggeman equation was used.

Simulated ε-distributions results are shown in Figs. 12.4(i–iii) together with the equivalent experimental data. From each figure, it is evident that the simulation and experiment are in good agreement.

12.2.3.3 Extension to ε-FGM coaxial disk-type spacer

The fabrication and simulation techniques described above for post-type ε-FGM spacers have also been extended to coaxial disk-type ε-FGM spacers. In this sub-section, the fabrication of a GLP-FGM coaxial disk-type configuration is described and the resulting ε-distribution is verified by both measurement and through particle movement simulation.

Table 12.1 includes parameter data pertaining to the fillers used to fabricate a GLP-FGM coaxial disk-type spacer. In this case, TiO_2 was taken as the small diameter (high ε_r) filler and SiO_2 was taken as the large diameter (low ε_r) filler; Fig. 12.6 shows the diameter distribution for each filler. The loading level for each filler and the centrifuging conditions were as specified. A centrifugal separator with a rotational rotor system was used to produce ε-FGM coaxial disk-type spacers, whereas the revolution system was used for the post-type ε-FGM spacer discussed in the previous sub-section. The calculation model for the particle movement in the coaxial disk-type ε-FGM was modified from the rectangular-shape shown in Fig. 12.5 to a fan-shape.

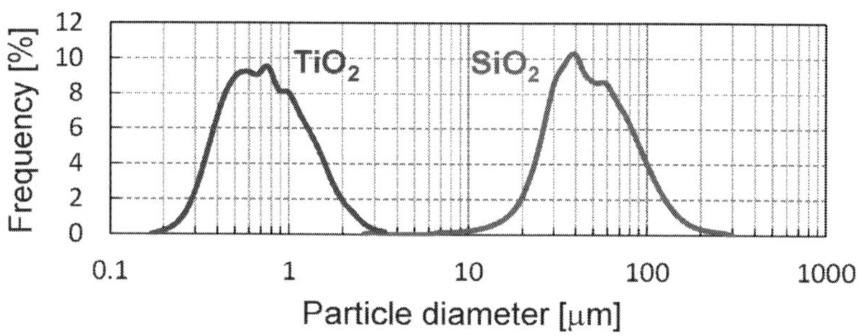

Figure 12.6 Particle diameter distribution.

Table 12.1 Specifications of filler particles

	Relative permittivity (@1 kHz)	Specific gravity [g/ml]	Mean diameter [μm]	Loading [vol%] (vs. total)	Centrifuging condition	
					Force [G]	Time [min]
TiO_2	114	4.2	0.672	15	4,000	30
SiO_2	3.7	2.21	43.1	40		

Figure 12.7 shows the ε-distribution across a fabricated coaxial disk-type GLP-FGM spacer with an inner radius of 10 mm and an outer radius of 30 mm. The GLP-FGM exhibited an ε-distribution varied from $\varepsilon_r = 8.2$ at the HV-electrode/spacer interface to $\varepsilon_r = 5.3$ at the GND-electrode/spacer interface. The experimental results of ε-distribution agreed well with the simulation results.

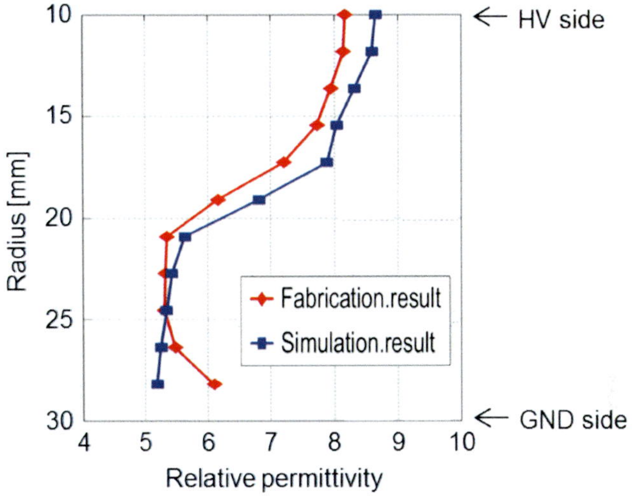

Figure 12.7 ε-Distribution of fabricated coaxial disk-type GLP-FGM.

The electric field distribution around the above coaxial disk-type GLP-FGM GIS spacer was determined and compared with a conventional spacer with a uniform $\varepsilon_r = 6$; Figs. 12.8 and 12.9 provide different representation of the electric field distribution in both cases. From these, the maximum electric field at the HV-electrode/spacer interface was reduced by 12% for the coaxial disk-type GLP-FGM spacer, compared with the conventional design, so demonstrating that the use of GLP-FGM leads to a reduction in electric field for GIS spacers.

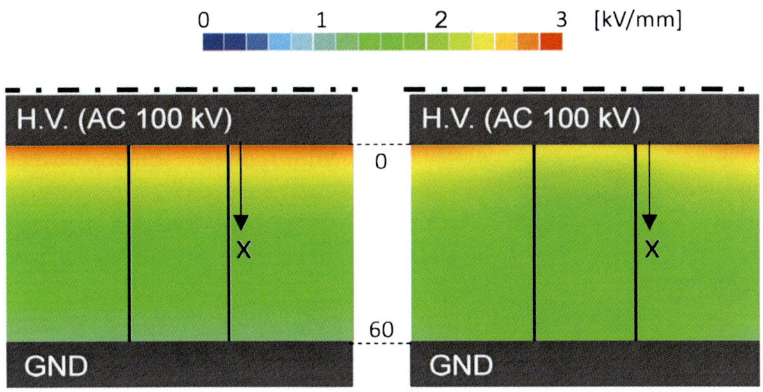

Figure 12.8 Electric field distribution around spacer with ε-grading.

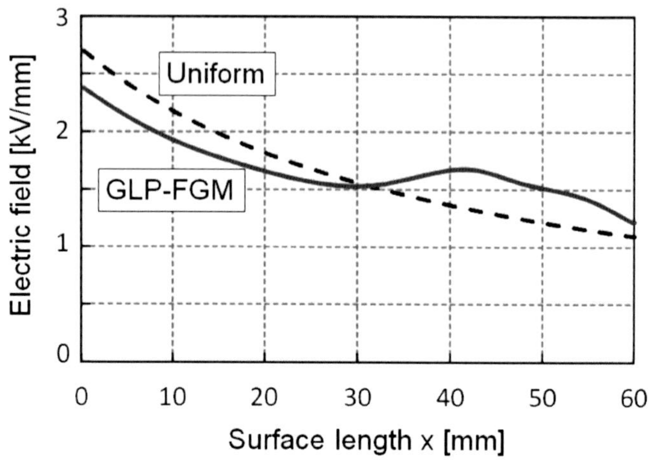

Figure 12.9 Electric field distribution on spacer surface with ε-grading.

ε-FGM can be fabricated by not only the centrifugal force application as has been described above, but also using continuous high-speed casting and so on for the actual GIS spacer and other power apparatus. The maximum electric field experienced within power apparatus is expected to be reduced by more than 20% through the use of ε-FGM with optimized ε-grading.

12.3 σ-FGM with Electrical Conductivity Distribution

12.3.1 Concept of σ-FGM

The complex permittivity ε^* of a solid insulator is given by $\varepsilon^* = \varepsilon - j\sigma/(2\pi f)$, where σ is the electrical conductivity and f is frequency. Thus, the electric field within power apparatus can be controlled through both dielectric permittivity ε and electrical conductivity σ. In this section, electric field grading through the control of σ is discussed, in the context of the surface of GIS spacers [6].

Figure 12.10 shows a schematic model of a GIS spacer with a cone-type configuration, in which a moderately conductive layer is superposed on the spacer surface. When the conductivity of the thin layer is graded, it is referred to as σ-FGM. In this model, the conductivity of the gas and the solid spacer, except within the conductive layer, was assumed to be zero and, for simplicity, an ac high voltage with a frequency of 1 kHz was applied to the inner electrode. The relative permittivity ε_r of the solid spacer was set at 6.0. In order to examine the influence of the σ-FGM layer on the electric field grading of GIS spacer, the electric field E_A at the triple junction (point A) and E_B at the tip of the thin layer (point B) were considered for different thicknesses, lengths, conductivities and its spatial distributions of the thin layer of σ-FGM.

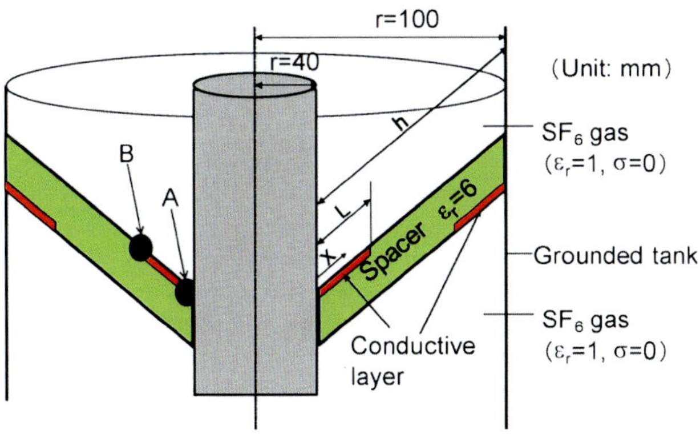

Figure 12.10 Calculation model of GIS spacer with a cone-type configuration.

12.3.2 Electric Field Grading Effect of σ-FGM

The electric field E_A and E_B is shown in Fig. 12.11 as a function of the uniform σ-value of the conductive layer with a fixed thickness of 0.2 mm and the length ratio L/h = 0.25 (these parameters are defined in Fig. 12.10). Figure 12.11 indicates that the maximum electric field appears near the point A, when the conductivity is low. However, as the conductivity increases, E_A is reduced while E_B increases and then exceeds E_A. The maximum electric field, i.e., the higher of E_A and E_B, was reduced to 4.8 kV/mm ($E_A = E_B$) at the moderate conductivity $\sigma = 10^{-5}$ [S/m], corresponding to 79% of 6.1 kV/mm at the low conductivity $\sigma = 10^{-8}$ [S/m], where the electric field was dominated by the permittivity ratio between the solid spacer and the insulating gas [6].

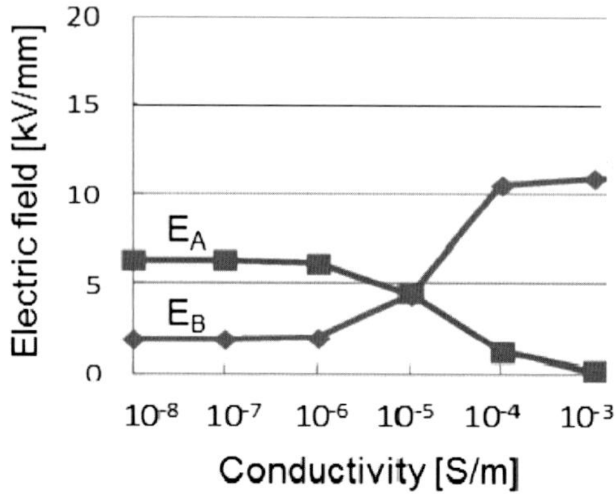

Figure 12.11 Electric field as a function of conductivity.

Figure 12.12 shows different σ-grading profiles as a function of x/L on the σ-FGM layer, while Fig. 12.13 shows a typical electric field distribution around the spacer for the situation corresponding to $L/h = 0.3$. The distortion or concentration of the electric field in the vicinity of the triple junction can, from Fig. 12.13b, be seen to be diminished by the appropriate σ-grading pattern. Figure 12.14 shows the electric field as a function of L/h for different σ-grading patterns. The electric field was reduced to 3.1 kV/mm for exponential σ-grading from 5×10^{-5} [S/m] to 5×10^{-6} [S/m] at L/h higher than 0.3, which corresponds to 50% of the previous value of 6.1 kV/mm at $\sigma = 10^{-8}$ [S/m] [6].

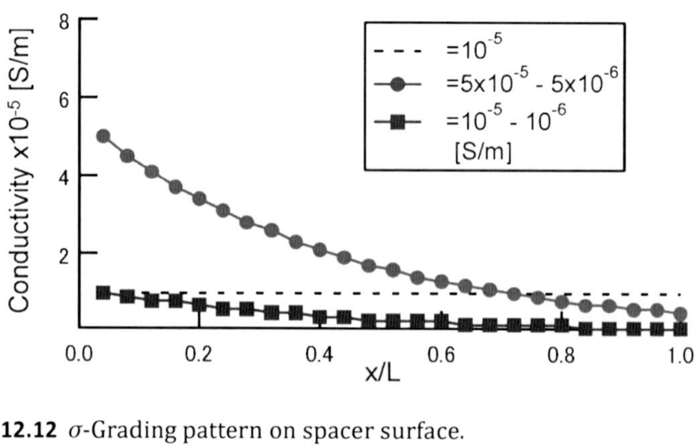

Figure 12.12 σ-Grading pattern on spacer surface.

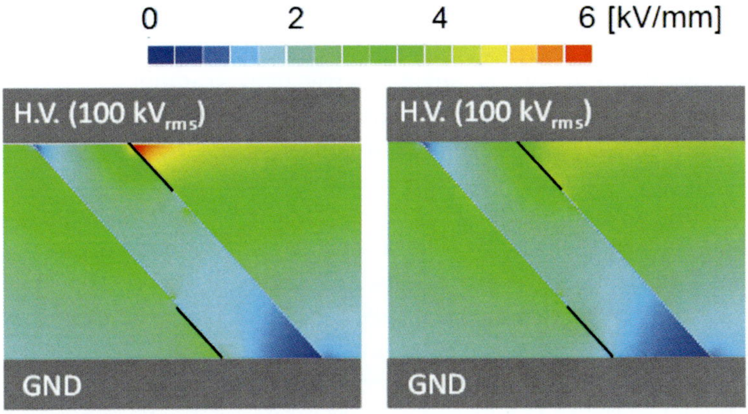

Figure 12.13 Electric field distribution around spacer with σ-grading.

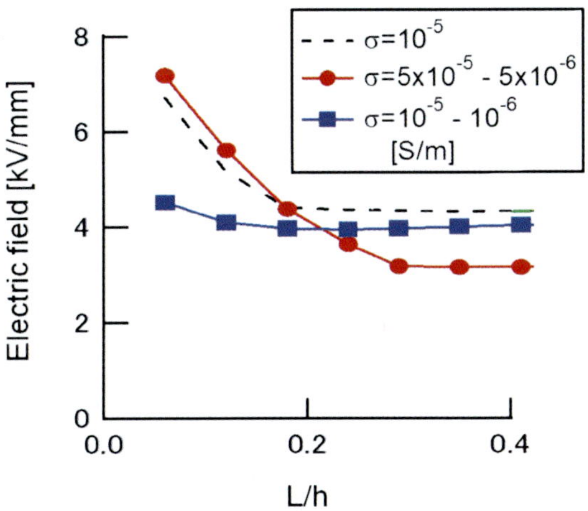

Figure 12.14 Electric field for different σ-grading.

As described above, by the appropriate choice of conductivity distribution of σ-FGM, lowered electric field stress values and more uniform electric field distributions along the GIS spacer surfaces can be obtained. Because of the local and graded conductivity, there are no problems related increased surface currents and temperatures of GIS spacer with σ-FGM. Determining the optimum conductivity distribution around the highly stressed region on GIS spacer surfaces would seem to be an effective means for electric field control.

12.4 Conclusions

This chapter introduced novel electric field grading and control techniques using functionally graded materials, namely, ε-FGM and σ-FGM, where the spatial distribution of dielectric permittivity and electrical conductivity are respectively optimized in connection with their use in solid insulators. FGM will contribute to the simplification of spacer configuration and thereby lead to increased compactness, energy saving, lower cost, etc., of power apparatus, compared with conventional electric field control techniques, such as embedded electrodes and shape control. The practical applicability of FGM within power apparatus can be expected through further R&D, experimental verification and optimization.

References

1. K. Kato, M. Kurimoto, H. Shumiya, H. Adachi, S. Sakuma, H. Okubo, Application of functionally graded material for solid insulator in gaseous insulation system, *IEEE Trans. Dielectr. Electr. Insul.*, vol. 13, no. 1, pp. 362–372, 2006.
2. L. Donzel, F. Greuter, T. Christen, Nonlinear resistive electric field grading. Part 2: Materials and applications, *IEEE Electr. Insul. Mag.*, vol. 27, no. 2, pp. 18–29, 2011.
3. M. Kurimoto, K. Kato, M. Hanai, Y. Hoshina, M. Takei, H. Okubo, Application of functionally graded material for reducing electric field on electrode and spacer interface, *IEEE Trans. Dielectr. Electr. Insul.*, vol. 17, no. 1, pp. 256–263, 2010.
4. H. Okubo, J. Shimomura, Y. Fujii, N. Hayakawa, M. Hanai, K. Kato, Fabrication and simulation techniques of permittivity graded materials for gas insulated power equipment, *International Symposium on High Voltage Engineering (ISH)*, E-095, 2011.
5. N. Hayakawa, J. Shimomura, T. Nakano, M. Hanai, K. Kato, H. Okubo, Fabrication technique of permittivity graded materials (FGM) for disk-type solid insulator, *Conf. Electr. Insul. Dielectr. Phenomena (CEIDP)*, pp. 32–35, 2012.
6. K. Kato, J. Ishiguro, N. Hayakawa, H. Okubo, Electric field relaxation by adding moderate conductivity layer on GIS spacer surface, *International Symposium on High Voltage Engineering (ISH)*, PE-21, 2013.

Chapter 13

Permeability Control by Nano-Magnetic Fillers: Case Study

Yoshimichi Ohki

Department of Electrical Engineering and Bioscience and Research Institute of Materials Science and Technology, Waseda University, Tokyo, 169-8555, Japan

yohki@waseda.jp

This chapter describes an exemplar case study describing research conducted with the aim of developing a new electrical insulating substance that possesses the properties of low electrical conductivity, high dielectric permittivity and high magnetic permeability, by dispersing ferromagnetic iron nanoparticles within an epoxy resin. Since ferromagnetic particles are agglomerative, surface treatment is especially important. In this case, the surface of iron nanoparticles was coated with oleylamine, which consists of amino and alkyl groups. As a result, composites with low electrical conductivity, high ferromagnetic resonance frequency, and low dielectric loss were obtained.

13.1 Introduction

As a unique example of the formulation of polymer nanocomposites, trials to develop an electrically insulating polymer that possesses

Tailoring of Nanocomposite Dielectrics: From Fundamentals to Devices and Applications
Edited by Toshikatsu Tanaka and Alun S. Vaughan

ISBN 978-981-4669-80-1 (Hardcover), 978-981-4669-81-8 (eBook)
www.panstanford.com

both high dielectric permittivity and magnetic permeability have been carried out. The objectives of making such composites can mostly be divided into two categories; one concerns pure science while the other is practical. A typical case of the former is a study of the minimum size necessary for ferromagnetic materials to exhibit ferromagnetic multi-domain structures; it has been reported that ferromagnetic nanoparticles smaller than a critical size possess a single domain structure, leading to magnetic properties superior to those of bulk materials.

From a practical perspective, industrial applications of such systems include their use in electronic devices for information and communication technologies where high magnetic permeability is required. For example, electromagnetic wave absorbers need high dielectric permittivity in addition to high magnetic permeability, as will be mentioned in the next session. Furthermore, the length of an electromagnetic wave in an insulating material is inversely proportional to the square root of both the permittivity and the permeability of the material. Therefore, if the electrically insulating material that is surrounding the antenna of a telecommunication device such as a cellular phone has a high permittivity and a high permeability, then the length of the antenna can be shortened, which makes it possible to make the device more compact. In this chapter, several examples of research done from this latter viewpoint are briefly described, while the work conducted by the author and his co-workers will be described in more detail.

13.2 Several Examples of Research on Magneto-Dielectric Nanocomposites

For the purpose of studying the effect of using magneto-dielectric nanocomposite materials on the miniaturization of the physical dimension of an active antenna, Borah and Bhattacharyya of the Microwave Research Laboratory in India incorporated nano-sized $CoFe_2O_4$ into low-density polyethylene (LDPE). Although details are not disclosed in literature [1], they seem to have incorporated the desired amount of the nano-sized ferrite particles into the LDPE using a co-precipitation process after nano-sized ferrite had been obtained by a heat treatment of a precipitated nitrate precursor. They reported that the size of a microstrip antenna can be miniaturized if the nano-cobalt ferrite magneto-dielectric composite is used as a substrate for the antenna, without suffering an appreciable increase in losses.

Y. J. An and his collaborators of National Defense Academy of Japan made composites of iron (Fe), titanium dioxide (TiO_2), and an ethylene propylene diene copolymer (EPDM) with an intention of developing a microwave absorber capable of absorbing power from electromagnetic waves within a specific frequency range [2]. The idea behind this work is as follows. High electric conductivity and high magnetic permeability are two key characteristics of electromagnetic wave absorbers. However, good electric conductors would absorb all the electromagnetic waves from very low to very high frequencies and, therefore, if the desired outcome is the absorption of electromagnetic waves with specific frequencies, a moderate value of electric conductivity is needed. Furthermore, if the frequency of the electromagnetic wave is high, a capacitive current will flow. This means that dielectric substances that exhibit a high permittivity can be a good absorber of electromagnetic waves with high frequencies, such as microwaves. An and his collaborators used iron particles with an average diameter of 4.0 μm and TiO_2 particles with an average diameter of 0.98 μm, and they changed the volume ratios of the two particles, keeping the sum of the two volume ratios nearly constant at about 60%. As a result, they found that a composite consisting of Fe (26 vol%), TiO_2 (36.5 vol%), and EPDM (37.5 vol%) satisfies the criterion of a reflection loss of −20 dB at 5.8 GHz.

Ta-I Yang of University of Maryland, US, and his co-workers produced composites consisting of surfactant-modified iron oxide (Fe_3O_4) nanoparticles of various sizes incorporated within a styrene-b-ethylene/butylene-b-styrene (SEBS) block copolymer [3]. According to their report, the bandwidth for antenna applications can be improved significantly at an appropriate ratio of dielectric permittivity and magnetic permeability, while keeping the product of the two at a constant value. They first made Fe_3O_4 nanoparticles with a size of 11 nm, which were then used as growth seeds to obtain larger particles. By changing the amount of iron-oleate complex, oleic acid, and seeds systematically, they studied their effects on the size of the nanoparticles.

13.3 Development of Low Loss Magneto-Dielectric Nanocomposites

In this section, a trial to develop low loss magneto-dielectric nanocomposite samples consisting of epoxy resin and iron nanoparticles is described in detail, which was carried out by the author and his co-workers [4].

13.3.1 Objective of the Development

All the television broadcast systems in Japan, both terrestrial and satellite, were moved to digital systems in July 2011. This accelerated even more a recent trend in society toward the installation of digital broadcast receivers within portable information devices. The length of an antenna to receive the television broadcasts should correspond to the broadcast using the lowest frequency, which is 470 MHz for current Japanese television broadcasts. The wavelength of a 470 MHz wave in vacuum is about 640 mm, and its quarter-wavelength, 160 mm, is longer than a half-length of ordinary cellular phones. Since the wavelength of an electromagnetic wave depends on the dielectric permittivity (ε) and magnetic permeability (μ) of the insulating medium surrounding the antenna, according to $\lambda \propto (\varepsilon\mu)^{-1/2}$, materials with high permittivity have been mainly used to make antennas short.

However, attempts to use materials with high permeability, such as magneto-dielectric composites, have recently been attracting attention, as outlined above. Although ferrite is a famous magnetic material, its permeability is low at GHz frequencies because of its small magnetic moment. On the other hand, bulk iron (Fe) cannot be used as an insulating substrate, although it has a large magnetic moment. Taking these facts into account, the author of this chapter and his co-workers tried to make a new composite consisting of nm-sized Fe particles and a polymer, aiming to realize simultaneously low eddy current losses and low dielectric losses.

13.3.2 Preparation of Samples

Diglycidyl ether of bisphenol A (816B, Mitsubishi Chemical) was used as the base epoxy resin and alicyclic polyamine (113, Mitsubishi Chemical) was used as the curing agent. Furthermore, alpha Fe particles with a quoted diameter of 70 nm were used as the nano-filler to be added to the epoxy resin.

Completely uniform dispersion of nanofillers is most essential for any composite. Regarding this, ferromagnetic Fe particles tend to agglomerate. Therefore, oleylamine ($C_{18}H_{35}NH_2$) was used as a surfactant in the present research, since it contains both an amino group ($-NH_2$), which tends to be adsorbed easily onto the surface of nanoparticles, and an alkyl group ($-C_{18}H_{35}$), which is easily dispersed in an organic solvent.

First, the Fe particles, oleylamine as the surfactant and zirconia beads for mixing were put into a container and the container was rotated in a planetary centrifugal mixer (P-5, Fritsch Japan). Here, beads with a diameter of 10 mm were used for the first 12 h, and those with a diameter of 1 mm were used for the next 12 h. In this way, agglomerates of Fe particles were broken up and the surface of each particle was coated with oleylamine. Next, the resulting nm-sized Fe particles were added into the uncured epoxy resin. Here, the designated values of Fe volume content (p) were set between 0 to 40%. The mixture was mixed three times for about 30 s at room temperature in an ultrasonic homogenizer (450D, Branson), which utilizes mixing forces induced by the collapse of ultrasonically created bubbles. After that, within 60 min, the liquid nanocomposite was injected into a mold, and cured for 3 h at 70°C followed by 3 h at 120°C. It should be noted that a nanomizer with very fine nozzles or a mixer with rotating blades is unsuitable for ferromagnetic Fe particles, because the particles tend to get stuck in the nozzles or adhere to the blades.

13.3.3 Experimental Methods

The Fe nanoparticles coated with oleylamine were observed by transmission electron microscopy (TEM), focusing on their surfaces. Furthermore, the composite sample was observed by scanning electron microscopy (SEM) in order to examine the nanofiller dispersion. For SEM observation, internal surfaces were prepared using cryo-fracture such that contamination was avoided. Prior to examination, the fracture surfaces were cleaned in an ultrasonic cleaner containing ethyl alcohol.

In order to determine the electrical conductivity, α, gold electrodes with a diameter of 20 mm were vacuum evaporated onto each side of a sheet sample with dimensions 50 × 50 × 0.5 mm^3 and the current was measured for 1200 s under an applied electric field of 5 kV/mm. This was conducted in N_2 at a pressure of one atmosphere and at a temperature of 60°C. The value of σ was calculated from the average steady state current value in a period from 1100 to 1200 s after the start of voltage application. Furthermore, the relative permittivity, ε_r, was measured for similar sheet samples with a main electrode (20 mmϕ) and a guard electrode (inner diameter 30 mm and outer diameter 40 mm) on one side and a lower electrode (40 mmϕ) on the other side over the frequency range from 10^{-1} to 10^5 Hz with an impedance analyzer (SI126096W, Solartron) at 60°C. In addition, the complex relative permeability, $\mu_r' - i\mu_r''$, was measured by the S parameter method over a frequency range from 0.2 to 6.0 GHz at room temperature. Here, the sample used for the permeability measurement is of a coaxial cylindrical shape with an outer and an inner radius of 3.5 and 1.5 mm, respectively, and a height of 10 mm.

13.3.4 Experimental Results

From the TEM image shown in Fig. 13.1, the surface of the Fe nanoparticles appears to be coated with oleylamine, which should prevent the particles from agglomerating with each other. However, the SEM images shown in Fig. 13.2 indicate that those samples containing 20 and 40 vol% of nanoparticles contain large agglomerations; agglomeration is, however, not evident when the filler content is 6 vol%.

Figure 13.3 shows the dependence of the electrical conductivity on the particle content, p. At $p < 2.6$ vol% or in region I, σ hardly increases, while at $2.6 < p < 30$ vol% in region II, σ shows a very rapid increase with increasing p. Finally, in region III, where $p > 30$ vol%, σ increases rather slowly. At the maximum content p = 40 vol%, σ is 1.2×10^{-5} S/m, which is lower than the conductivity of ferrite ($\geq 10^{-4}$ S/m) [5].

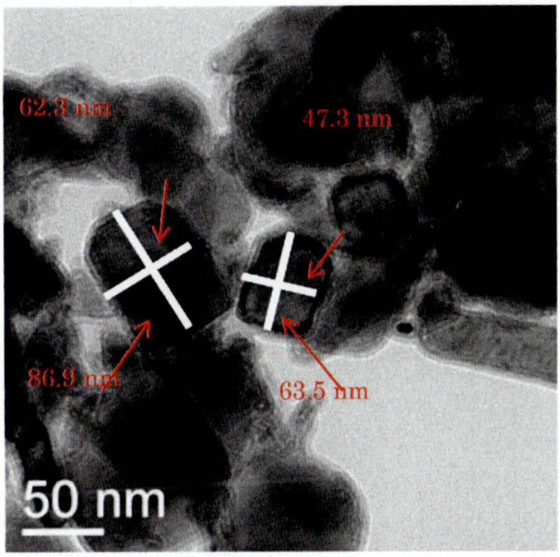

Figure 13.1 TEM image of Fe nanoparticles surface-treated by oleylamine.

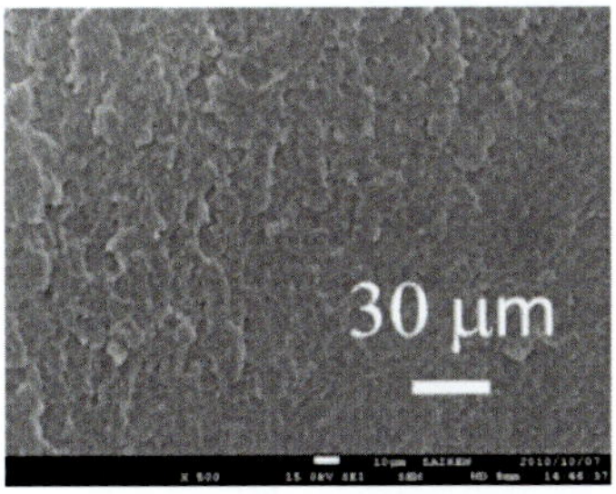

(a) 6 vol%

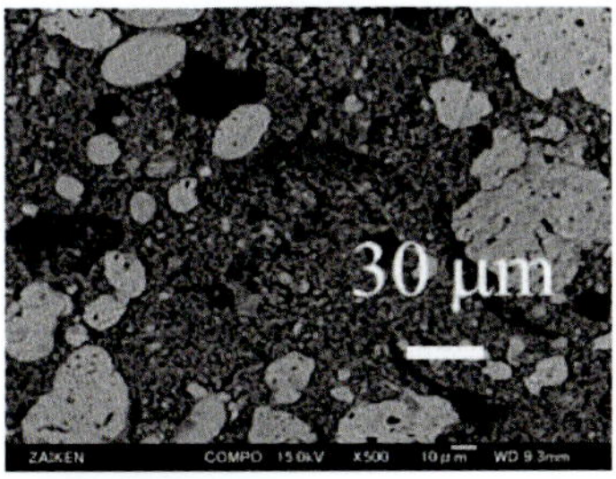

(a) 20 vol%

(a) 40 vol%

Figure 13.2 Cross-sectional SEM images observed for the samples with particle contents of 6, 20, and 40 vol%.

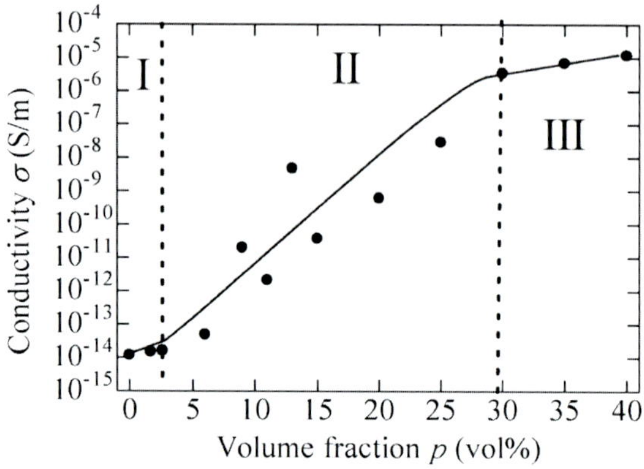

Figure 13.3 Conductivity σ as a function of particle content. Roman numerals I, II, and III indicate three regions where σ shows different dependences. —: curve fitted to data.

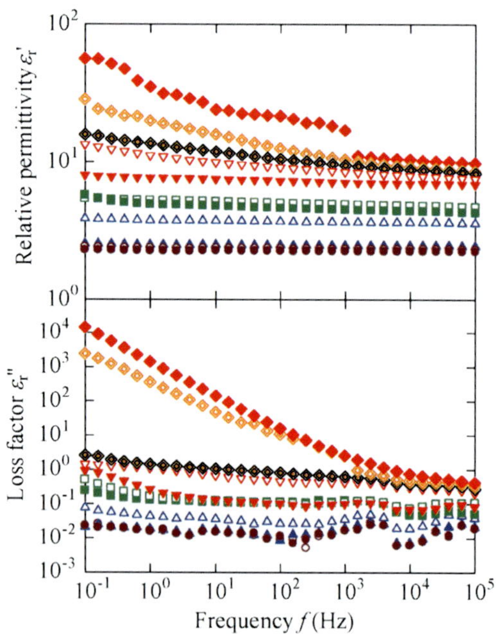

Figure 13.4 Relations between complex relative permittivity and frequency observed for each sample. ●: 0 vol%, ○: 0.8 vol%, ▲: 1.7 vol%, △: 2.6 vol%, ■: 9.0 vol%, □: 13 vol%, ▼: 20 vol%, ▽: 25 vol%, ◆: 30 vol%, ◈: 35 vol%, ◆: 40 vol%.

Figure 13.4 shows the frequency dependence of the real part of the relative permittivity, ε_r', and the imaginary part of relative permittivity, ε_r'', for samples containing various loading levels of nanoparticles. In the samples with $p = 35$ and $p = 40$ vol%, a very rapid increase both in ε_r' and ε_r'' is evident, especially at low frequencies. It is clearly shown in Fig. 13.5 that both ε_r' and ε_r'' at 10^5 Hz increase with increasing p. The sample with a filler content of 40 vol% shows ε_r' of ~9.8, which is higher than ε_r' for the epoxy resin (~3.9 [6]).

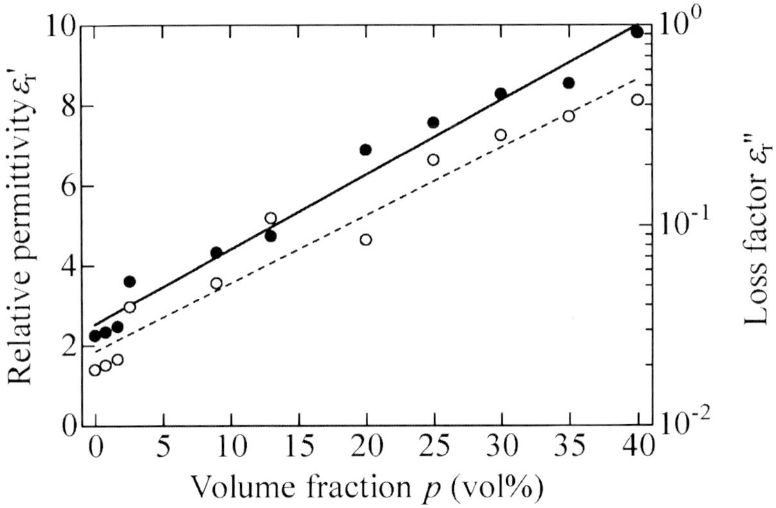

Figure 13.5 Relations between complex relative permittivity (ε_r': ●, ε_r'': ○) and particle content, measured at 10^5 Hz for each sample. —: fitted for ε_r', ---: fitted for ε_r''.

Figure 13.6 shows the frequency dependences of the complex relative permeability, μ_r' and μ_r'', measured for the samples with nanofiller contents of $20 < p < 40$ vol%. With the exception that μ_r'' is higher at 35 vol% than at 40 vol%, both μ_r' and μ_r'' increase monotonically with increasing p. Note: the measured μ_r' was much lower than the correct values as will be mentioned later. While μ_r' decreases with increasing p in the frequency range from 1.4 to 3.3 GHz, μ_r'' shows a peak at 2.3 GHz.

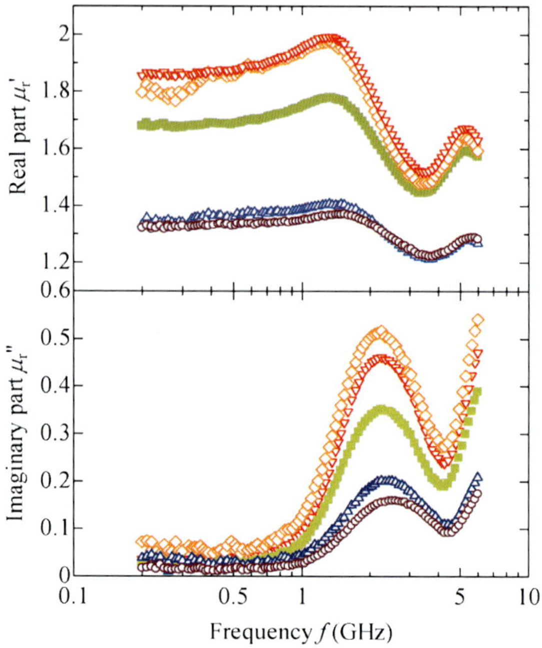

Figure 13.6 Complex permeability as a function of frequency measured for five kinds of samples with relatively high particle contents. ○: 20 vol%, △: 25 vol%, ■: 30 vol%, ◇: 35 vol%, ▽:40 vol%.

13.3.5 Discussion

As shown in Fig. 13.3, the dependence of σ on the particle content, p, changes at p = 2.6 and 30 vol%. At p < 2.6 vol%, the electrical conductivity of the epoxy resin dominates since, in this range, few conducting nanoparticles are present. For 2.6 < p < 30 vol%, the progressive linkage of Fe particles in contact with each other along the direction parallel to the electric field seems to occur. For p > 30 vol%, the effect of the increase in the total number of particles becomes relatively high compared to that of the increase in the number of percolating pathways aligned along the direction of electric field.

The electrical conductivity σ at p = 40 vol%, 1.2 $\times 10^{-5}$ S/m, is less than 1/10 of that of nickel-lead-ferrite, which is of the order of 10^{-4} S/m [5] and is within the permissible range for an electrically insulating substrate.

As shown in Figs. 13.4 and 13.5, both ε_r' and ε_r'' increase with increasing p. Since the permittivity of a conductor is infinite if the frequency is low, it is reasonable that the permittivity increases with the increase in the content of Fe nanoparticles. The permittivity ε_r' is 9.8 for the sample with $p = 40$ vol% at 10^5 Hz, which is fairly high for an electrical insulating material. Therefore, the attempt to raise ε_r' was successful. On the other hand, since conduction currents result in Joule heat, high electrical conductivity increases ε_r''. Although ε_r'' increases with increasing p, the values of ε_r'' in the important high frequency range remains at 0.42 even at 40 vol% as shown in Fig. 13.5. This is an acceptably low level for electronic devices.

As shown in Fig. 13.6, both μ_r' and μ_r'' increase with increasing p, although the measured value of μ_r' is much lower than the true value, as a consequence of the employed measurement method using a cylindrical sample. The wavelength of an electromagnetic wave with the highest measurement frequency of 6 GHz is 5 cm in vacuum. Although the wavelength becomes shorter in a magneto-dielectric material characterized by high ε_r' and high μ_r', it is still much larger than the size of the employed nanoparticles, i.e. 65 ± 20 nm. This indicates that eddy currents are not induced unless many Fe nanoparticles align with each other. However, the nanoparticle surface coating is effective in preventing eddy currents; eddy currents generally induce a new magnetic field that would prevent the change in the magnetic field, which decreases μ_r'. Therefore, the suppression of eddy currents is important for the development of a low loss magneto-dielectric materials.

The peak of μ_r'' at around 2.25 GHz that is seen in Fig. 13.6 is due to ferromagnetic resonance. This frequency is much higher than the theoretical value of the ferromagnetic resonance frequency of ordinary metallic bulk Fe (i.e. 1.54 GHz); the ferromagnetic resonance frequency is, however, not observable in metallic bulk Fe, due to the presence of eddy current losses. In the present composites, a good dispersion of oleylamine-coated Fe nanoparticles into epoxy resin successfully leads to the suppression of eddy current losses, which results in an increase of the ferromagnetic resonance frequency to a value that is higher than the theoretical value for bulk Fe.

13.4 Conclusion

Polymer nanocomposites with both high dielectric permittivity and high permeability have potentially scientifically interesting and industrially valuable features. In this chapter, several examples of research done from the above-mentioned viewpoint were reviewed as a case study. One investigation carried out by the author's group was described in detail. In this, composites of epoxy resin and Fe nanoparticles were generated for the purpose of developing a new insulating substrate characterized by small eddy-current losses, high permittivity, and high permeability. This work showed that coating of the Fe nanoparticles with oleylamine is important to prevent the nanoparticles from contacting mutually. This helps the electrical conductivity to remain low and the eddy-current losses to be negligibly small. The ferromagnetic resonance frequency of the composite was higher than the value of bulk Fe and the dielectric loss was low enough to be acceptable for electronic devices. Therefore, the composite has the potential to lead to more compact electromagnetic parts, such as antennas, for high-frequency use.

References

1. Borah, K., and Bhattacharyya, N. S., Magneto-dielectric material with nano ferrite inclusion for microstrip antennas: dielectric characterization, *IEEE Trans. Dielectr. Electr. Insul.*, vol. 17, no. 6, pp. 1676–1681 (2010).
2. An, Y., Nishida, K., Yamamoto, T., Ueda, S., and Deguchi, T., Microwave absorber properties of magnetic and dielectric composite materials, *IEEJ Trans. FM*, vol. 128, no. 6, pp. 441–448 (2008) (in Japanese).
3. Yang, T.-I., Brown, R. N. C., Kempel, L. C., and Kofinas, P., Magneto-dielectric properties of polymer–Fe_3O_4 nano-composites, *J. Magn. Magn. Mater.*, vol. 320, no. 21, pp. 2714–2720 (2008).
4. Hirose, Y., Hasegawa, D., Ohki, Y., Development of low loss magnetodielectric nanocomposites of epoxy resin and iron nanoparticles, *Electrical Engineering in Japan*, vol. 190, no. 2 (2015) (translated from *IEEJ Trans. FM*, vol. 133-A, no. 12, 2013, pp. 668–673).

5. Satyanarayana, R., Murthy, S. R., Rao, T. S., and Rao, S. M. D., Electrical conductivity of Ni-Zn ferrites, *J. Less-Common Met.*, vol. 90, no. 2, pp. 243–250 (1983).
6. Katayama, J., Ohki, Y., Fuse, N., Kozako, M., and Tanaka, T., Effects of nanofiller materials on the dielectric properties of epoxy nanocomposites, *IEEE Trans. Dielectr. Electr. Insul.*, vol. 20, no. 1, pp. 157–165 (2013).

Part 4

Devices and Applications

Chapter 14

High-Density Mounted Components for Electronic Devices

Takashi Ohta

Panasonic Corporation,
1048 Kadoma, Kadoma City, Osaka 571-8686, Japan

ohta.takashi@jp.panasonic.com

As the development of compact, high-performance electronic devices progresses, there is an increasing demand for high-density mounting of electronic components. Consequently, insulation with excellent performance characteristics is needed to satisfy the following requirements. The distance between the conductors is shortened. Die bonding material to bond the encapsulation material is employed to protect the connection of semiconductor elements and the package. Support structures between a semiconductor element and the substrate are devised. Ball grid array (BGA) and chip-scale package (CSP) underfill material mounting is sometimes used to keep the interlayer material of the printed circuit board in sound conditions. Moreover, high heat radiation (high thermal conductivity), heat resistance, and electrochemical migration resistance are also target for intensive research. Polymer nanocomposites are now recognized

Tailoring of Nanocomposite Dielectrics: From Fundamentals to Devices and Applications
Edited by Toshikatsu Tanaka and Alun S. Vaughan

ISBN 978-981-4669-80-1 (Hardcover), 978-981-4669-81-8 (eBook)
www.panstanford.com

as novel materials with the potential to solve all the requirements indicated above.

14.1 Compact High-Performance, High-Density Mounting of Electronic Equipment Parts

In high-performance electronic devices, high-density mounting of components is adopted. In order to cope with the ultra-high frequency band above 10 GHz, three-dimensional mounting technologies are progressing, to avoid the signal delay caused by adding extra wiring routes. In order to meet the need to shorten the distance between conductors and between conductors and the build-up substrate, substrates composed of ceramic and resin have been utilized. Also, the environments around electronic devices are becoming increasingly severe and to address such changes in requirements, the following measures have been taken. First of all, high-quality insulation is now under development, since the insulation of multi-layered structure is subjected to high electric stress. Interlayer interfaces are a particular focal point for research. Second, compact design creates severe thermal conditions; high heat dissipation necessitates the use of materials that are characterized by high thermal conductivity. As shown in Fig. 14.1, to take advantage of the feature that the size of the chip component is small, design considering the footprint of the gap (interval) and the printed circuit board between the components, requires only manufacturing techniques can respond to it.

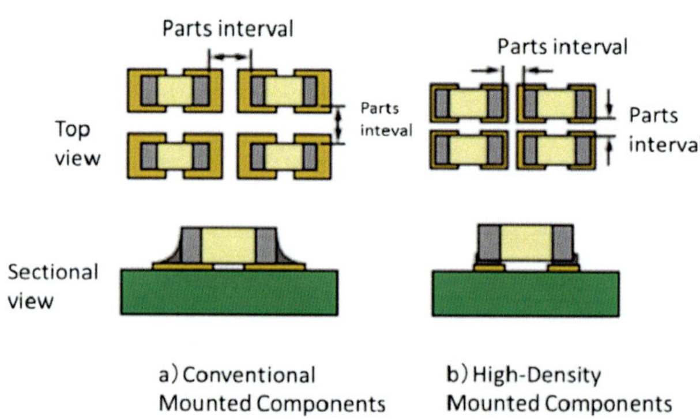

Figure 14.1 High-density mounted components for electronic devices.

To cope with the requirements stated above, polymer nanocomposites, or polymer nano-microcomposites are considered to be a good choice for use in connection with high-density electronics devices. It should be mentioned that materials and insulating substrates to be used for electronics devices are usually subjected to high temperatures during their component mounting processes and, for this reason, nano-microcomposites are considered to be better than mere nanocomposites. Many different materials are used in the production of electronic devices and include component materials used at the time of mounting die-bonding sealing resin, semiconductor elements and base board, underfill materials and internal capacitor materials. The sealing resin is utilized to protect the mechanical or thermal stress to the connecting portion of the semiconductor device with the package. The underfill is used when the BGA and CSP are mounted.

As for the insulating substrate, high-thermal-conductivity materials are available for substrate or resist. These materials should be close in thermal expansion coefficient to other components in order to prevent mechanical malfunction. Nano-microcomposites fortunately meet such requirement. Furthermore, since the distance between neighboring conductors is extremely short, say, from several micrometers to several tens of micrometers, electrochemical migration (ECM) and electrical treeing are likely to take place. Appropriate measures against such detrimental phenomena must be investigated for nano-microcomposites.

Figure 14.1 shows a comparison of conventiomnal and high-density mounted components for electronic devices.

14.2 Validity of Nanocomposites as a Sealing Resin for Electronic Components

14.2.1 Sealing Resins for Semiconductor Devices and Packaging

For nanocomposites to function with high reliability as sealing resins, they are required to have ample heat resistance. Conventionally, microfillers of the order of micrometers in size (silica fillers) have long been used to fill thermosetting resins, in order to convey a high heat resistance and low thermal expansion coefficient on the final composite system. However, microfilled thermosetting resins can suffer from the disadvantage that the specific surface area of the microfiller is small and interfacial adhesion between the resin is relatively weak, such that they are prone to moisture penetration and thereby material degradation. Therefore, since the fabrication of thermosetting resins with low thermal expansion coefficient is considered to be possible through the addition of a small amount of an appropriate nanofiller, such approaches are under R&D.

Sealing resins with low thermal expansion coefficient can be broadly divided into epoxy resin and silicone gel, depending on voltage class. Epoxy sealing resins are used in the voltage range up to 1 kV, while silicone gels are applicable at higher voltages.

Nanocomposites for use as sealing materials can be formulated by nanohybridization in epoxy resins or in silicone gels; outline details are provided below. First, epoxy resin insulation may contain complex structures, including the base materials for electronic devices and various composite parts. As a result of the presence of interfaces between dissimilar materials, the assembly is not stable to temperature change and can be prone to mechanical cracking. Consequently, to avoid this, epoxy resins are filled with inorganic fillers. Furthermore, to make high-density mounting a reality, a key research challenge is to enhance the thermal conductivity of sealing resins. In the case of thermosetting epoxy resins, adding organically modified layered silicate clays can result in improved mechanical properties, when properly processed. When homogeneously dispersed, the spacing between the silicate clay layers is of the order of nanometers and, under such circumstances, it has been shown that the resulting composite material has excellent thermal properties [1].

Layered silicate clays can be modified through the use of short-length guanidine urea alkyl chain (GU) to bring multi-functionality. Composite resins filled with the resultant modified clays are expected to exhibit lowered coefficients of thermal expansion (CTE) and elevated glass transition temperatures (T_g) compared to pure epoxy resins. However, some decrease in T_g has been observed in systems containing GU modified clays.

Since the GU includes high heat resistance amino group, it might be possible to increase or at least maintain T_g. Therefore, attempts have been made to lower the CTE, while maintaining the T_g, by dispersing the clays using the melt-kneading method. Active hydrogen associated with the amino group of guanylurea acts as a polyfunctional curing agent, and contributes to clay cleavage resulting in the increase in surface area. Size effects also influence flame retardancy and, in the case of fused silica loaded epoxy resins, the flame retardancy has been shown to increase with the increase of filler size.

Finally, silicone resin with a nanohybrid structure are widely used in this application [2]. Such systems are highly flexible after curing and exhibit excellent shock absorption, vibration damping property and crack resistance even under conditions of rapid temperature cycling. By exploiting such characteristics, integrally molded switching elements rated at 4.5 kV and 120. A have been developed. Pulse width modulation (PWM) three-phase inverters rated 110 kV with SiC commuted gate turn-off thyristor (SIGT) are imbedded in the resin. High temperature composite sealing resins able to withstand temperatures up to 250 °C can be fabricated using polysiloxane functional inorganic fine fillers by ultrasonic hybrid technology [3, 4].

14.2.2 Underfill Materials Used to Fabricate Semiconductor Flip Chips

Here, a film-like adhesive is used to fabricate the semiconductor package, including silicon semiconductor chips and substrates, support structures such as the lead frame, etc. This material contains curing agents and epoxy resins and acrylic polymer with cross-linkable copolymerized functional groups. Although epoxy resin is uniformly dispersed in the acrylic polymer in the B-stage prior to the curing of the epoxy resin with a curing agent, the composite might exhibit clear phase separation structure in the C stage, after curing. As a result, the storage modulus decreases significantly over a wide temperature range, while the degradation of the acrylic polymer phase is observed at around 40 °C [5]. Nevertheless, it is then possible to improve the heat resistance of the resin by incorporating nanofillers; the addition of ~5 vol% of a nanofiller also induces an increase (~0.02 MPa) in tensile elastic modulus at high temperatures while the tear strength of the composites is also improved relative to that of the unfilled resin. The next issue concerns underfill materials, which are used to mount the flip-chip components to make structural reinforcement by filling the gap between the chip and the substrate. Although the resin material is required rapidly to impregnate a narrow gap, void formation and filler sedimentation must not occur and excellent adhesion to various surfaces is expected.

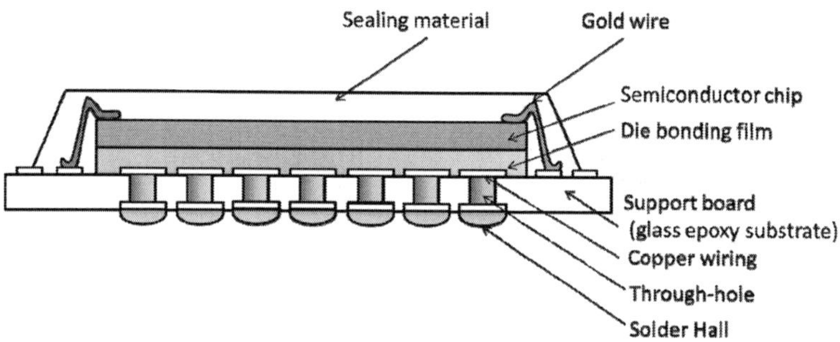

Figure 14.2 An example of a semiconductor package structure.

In order to make the high-density mounting of flip-chips feasible, modular techniques have been adopted in certain packages, such as digital cards. Different kinds of materials are used for substrates in this case and the underfill material used for this purpose usually need to meet design requirements relating to the thermal expansion coefficient and elastic modulus, such that low thermal stresses result. In addition, adhesion between different surfaces and retardation of moisture ingress are additional areas of great concern, especially when the materials are subjected to heat cycles. Good underfill materials have been obtained by combining epoxy resins and nanosilica, where the average size of the silica filler is ~120 nm and the filler particles are uniformly dispersed at a loading level of 50 wt% [6]. The resulting material is characterized by a viscosity of 8000 cps, a T_g value of 143 °C, and a CTE of 37 ppm/K. Figure 14.2 shows an example of a semiconductor package structure. A film-like adhesive is used to fabricate the semiconductor package and to bond silicon semiconductor chips and substrates and support structures such as a lead frame and to conduct in the die bonding processes [5].

14.2.3 High-Dielectric-Constant Capacitor Materials

Large-scale integration (LSI) involves different elements, such as resistors and capacitors, being built into the printed circuit substrate and, to minimize signal delay times, the shortest possible wiring lengths are required. Capacitor dielectrics may be produced using a high-dielectric constant filler dispersed in a resin matrix and coated on copper foils, in which case, methods of dispersion and surface treatment need to be improved. An example of such a system for commercial use is an epoxy resin filled with barium titanate ($BaTiO_3$) filler, where the filler loading is about 60 wt% and the resulting dielectric constant is 40 to 50. The dielectric loss tangent is, however, rather high, ranging from 0.02 to 0.03. New composite materials are also being developed for capacitor use, which are composed of $BaTiO_3$ in an alkali developable photosensitive positive or negative polyimide resin. The relative dielectric constant of these materials is around 50, when the filler loading level is about 50 wt%. Also, since the base resin is a polyimide, these materials are able to tolerate exposure to high temperatures: up to 700 °C and 400 °C for the positive and negative type, respectively. Other resins with similar performance have also been investigated, such as systems based upon silsesquioxane and cycloolefin resins filled with barium titanate nanofillers or alumina nanofillers. The materials such as those described above, the size and packing density of fillers are important factors and, in the case of $BaTiO_3$ nanofillers of 29 nm in size, it has been shown that the dielectric constant of composite films increases with filler loading level, but only up to a maximum value of 30.

However, an alternative approach using an electrophoretic deposition (EPD) method indicates that it is possible to fabricate thin films of $BaTiO_3$ nanofillers with the dielectric constant 152. This value is more than twice that exhibited by composite resins prepared by conventional spin coating methods. Formation of ceramic films on resin substrates is another option to fabricate large dielectric constant materials. This should take place at as low temperature as possible. Using an aerosol deposition method, ceramic powders are carried in a stream of gas which impinges at high speed upon the substrate. In this way, by overcoming the difficult adhesion of ceramics to substrates, it is possible to form films at room temperature. The reported dielectric constant is around 400 [7].

14.3 High-Heat-Dissipation, High-Thermal-Conductivity Material as Insulating Substrates for Electronic Equipment

Heat dissipation is an issue in many areas of electronics and in certain applications. This can be improved by using nano-microcomposites as insulating film substrates.

14.3.1 High-Heat-Dissipation, High-Thermal-Conductivity Insulating Substrates

Component mounting densities increase year-on-year to deliver electronic devices with reduced size and high-performance, automotive electronic devices, lighting equipment such as LEDs and power devices. In such applications, there is a great need for high-heat-dissipation, high-thermal-conductivity substrate materials. Optimum thermal performance requires the selection of suitable materials but and optimization of system structures. Thermal stress is a major concern and, to reduce it, buffer layers, called thermal stress relaxation layers, are usually provided on the internal components of substrates. Padding with high thermal conductivity, or a conventional large thermal "vias" (filled with thermally conductive material), is used for high-power IC chips to radiate heat away toward the outside. Heat transferred to the inner wiring parts from high-power IC chips is radiated to the lower substrate transmitted by the thermal vias. Heat dissipation measures may be modified when thick copper foils are used.

Nanocomposites or nano-microcomposites are a possible route to high-thermal-conductivity insulating substrates that can provide enhanced heat dissipation. Such systems can be formulated using high-thermal-conductivity inorganic fillers dispersed within a suitable polymer resin. First, resin materials that incorporate self-assembling elements, such as monomesogen-type biphenyl groups, exhibit higher order structures. As a consequence of this ordering, phonon scattering is reduced, leading to an increase in the thermal conductivity (0.96 W/m·K compared to a value of

0.19 W/m·K for conventional general-purpose epoxy resins). Second, the thermal conductivity of conventional resins can be increased by the inclusion of high-thermal-conductivity inorganic fillers. Such materials can be used with an alumina substrate with a copper foil on the other side, to form a metal foil laminated ceramic substrate. Similar substrates for LED lighting devices have been developed, using crystalline epoxy resins obtained by introducing a mesogen skeleton [8]. Furthermore, high-thermal-conductivity glass composite substrate materials have been investigated, including woven glass/epoxy resin copper-clad laminates and nonwoven glass/epoxy resin copper-clad laminates. Laminates filled with fillers of different sizes have also been produced, which are characterized by low drilling wear rate, high thermal conductivity, and good solder heat resistance. Such materials are employed as LED mounting substrates.

14.3.2 Solder Resist Films

A solder resist conductor package can be fabricated by covering predetermined portions other than copper wiring regions. For that purpose, it is necessary to ensure the quality of the insulation between wiring circuits and the solder, even during solder reflow processes [9]. Solder resist materials are therefore required to have high T_g, low thermal expansion coefficient, and excellent mechanical properties (such as heat cycle resistance). Furthermore, ECM resistance is also required for substrate materials containing embedded lead wires and/or inter-terminal regions with a fine pitch. In order to obtain the performance characteristics described above, epoxy resin–based, Cu-dispersed solder resist materials filled with nanofillers and microfillers have been developed [9]. It is has been shown that the EMC resistance of epoxy resins can be improved by adding nanofillers [10].

Favorable data have been obtained for solder resist films as follows: a glass transition temperature in the range 145~155 °C; thermal expansion coefficient of 20~25 ppm/K. Synergy effects associated with the presence of both nano- and microfillers are considered to lead to the improved performances of materials used for electronic devices. Co-filling with both types of filler provides a tool to optimize key material characteristics and thereby minimize the size and weight of package and assembly, such that high component densities can be achieved. Characteristics to be improved include insulation degradation, dielectric constant, heat resistance, and thermal conductivity. It is anticipated that nano-microcomposites will provide attractive solutions that will accommodate the ever-increasing demands for enhanced performance of materials to be used in the electronics sector.

References

1. R. Ikezawa, T. Ishiguro, T. Hayashi, S. Akagi (2006) Flame retardant-free system halogen-free sealing material, Hitachi Chemical technical report, 46, pp. 43–48 (in Japanese).
2. Asahi Denka Co., Ltd., Kansai Electric Power Co., Ltd. (2005) 400 to withstand °C ~ to open the way for ~ SiC inverter capacity for the development of nanotech resin KA-100, Asahi Denka news-scan release (in Japanese).
3. Nippon Shokubai (2012) SiC power applications development nanotechnology a semiconductor for high-heat-resistant sealant, Nippon Shokubai news lease (in Japanese).
4. Toray Dow Corning Co., Ltd. (2010) Joint disclosure and NEDO in next generation power semiconductor for new technology development nano tech 2010, Toray News Release (in Japanese).
5. T. Inada (2009) Development of Die-bonding film for semiconductor packages by applying reaction-induced phase separation: Pursuing soft, endurable and controllable materials, Hitachi Chemical technical report, 52, pp. 7–12 (in Japanese).
6. K. Gross, S. Hackett, W. Schultz, W. Thompson (2003). Nanocomposite underfills for flip-chip application, *2003 Electronic Components and Technology Conference*, p. 951.
7. M. Arimura, A. Makino, K. Fujiyoshi, M. Kuwabara (2009) Development of a printed wiring high-capacity thin film capacitor for board built-in: Preparation of barium titanate nanoparticles thin film by electrophoretic deposition method, Fukuoka Industrial Technology Center research report, 19, pp. 29–32 (in Japanese).
8. D. Baba, T. Sawada (2011) Technological trends of high heat dissipation circuit board materials, Panasonic Electric Works technical report, 59, 1, pp. 17–23 (in Japanese).

9. T. Yoshino, M. Joumen, H. Katagi (2006) Advanced photo-definable solder mask for high-performance semiconductor packages, Hitachi Chemical technical report, 46, pp. 29–34 (in Japanese).
10. Y. Ohki, Y. Horose, G. Wada, H. Asakawa, T. Maeno, K. Okamoto (2012) Two methods for improving electrochemical migration resistance of printed wiring boards, *Proceedings of the 2012 International Conference on High Voltage Engineering and Application*, pp. 687–691.

Chapter 15

Power Applications

Toshikatsu Tanaka[a] **and Takahiro Imai**[b]

[a]*IPS Research Center, Waseda University,*
2-7 Hibikino, Wakamatsu-ku, Kitakyushu-shi 808-0135, Japan
[b]*Power and Industrial Systems R&D Center, Toshiba Corporation,*
1 Toshiba-cho, Fuchu-shi, Tokyo 183-8511, Japan

t-tanaka@waseda.jp, takahiro2.imai@toshiba.co.jp

This chapter describes the application of newly emerging nanocomposite dielectrics to the electrical insulation of power cables and apparatus, where it has been found that nanocomposite dielectrics are advantageous with respect to a variety of properties compared with the base resins. Amongst these are PD resistance, treeing resistance, short time breakdown strength, space charge free properties, and small arc resistance. They are advanced materials but based on conventional insulating materials such as epoxy resins, polyethylene, polypropylene, polyimide, and silicone rubbers. In that sense, the continuity of technology is smooth and advantageous. Therefore, although nanocomposite dielectrics may be considered to be easily acceptable for replacing conventional insulation, only a few products are on the market. Let us explain the possibility and probability of applying these materials to the power sector.

Tailoring of Nanocomposite Dielectrics: From Fundamentals to Devices and Applications
Edited by Toshikatsu Tanaka and Alun S. Vaughan

ISBN 978-981-4669-80-1 (Hardcover), 978-981-4669-81-8 (eBook)
www.panstanford.com

15.1 Introduction

This section gives a simplified picture of an electric power network from a power generating station down to consumers, as shown in Fig. 15.1. Electric power generators rated at 20 kV class are installed in generator stations to supply electricity downstream through power lines. Power lines consist of overhead power transmission lines (275–500 kV class) and underground cable power transmission and distribution lines. Overhead lines need insulators. Porcelain and glass insulators are in service, but the use of light-weight and mechanically strong polymer insulators is increasing. Extra-high-voltage (150–275 kV class) cable transmission lines are in service in urban areas, while high-voltage (20–30 kV class) and low-voltage (6 kV class) distribution lines operate in residential areas. In some cases, gas insulated power lines (GIL) are utilized. DC submarine cables are sometimes preferred between islands and continents. Power network systems include several stages of substations from high to low voltages. Such substations are driven by switchgear such as gas-insulated switchgear (GIS) and power capacitors. At the consumer level, electric motors are needed for power drive, many of which are driven by inverters. In the near future, smart grid and/or microgrids will emerge with coexisting downstream and upstream power flows. Mixed AC and DC power flows are also envisaged, which requires more reliable electrical insulation materials, such as nanocomposite dielectrics.

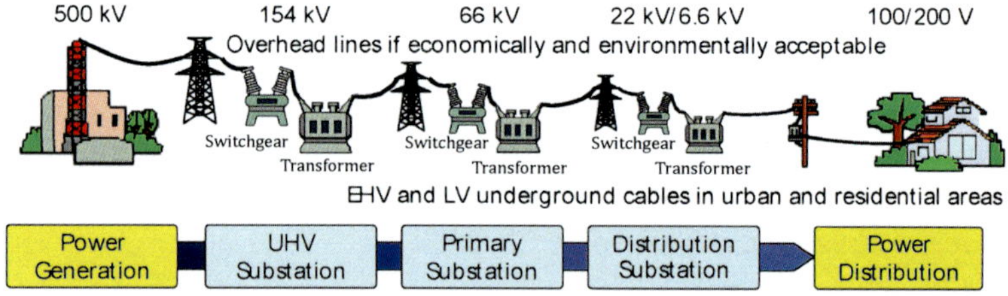

Figure 15.1 A simplified picture of electric power transmission systems.

Nanocomposites offer potential advantages over their base resins and, consequently, many applications are subject to R&D [1]. Table 15.1 shows potential applications of nanocomposite dielectrics in the power sector, which utilize some of their specific favorable characteristics. These are explained in subsequent subsections.

Table 15.1 Possible applications of nanocomposite dielectrics to power apparatus and cables by utilizing specific characteristics

Applications	Preferred characteristics
Motor windings (magnet wires) and power generators	High PD resistance
EHV DC cables	Space charge free characteristics
AC cables and capacitors	Increased dielectric breakdown strength
Cable accessories and end winding of power generators	Nonlinear *V–I* characteristics
GIS and GIL spacers; DC GIS spacers	Functional grading characteristics of permittivity and conductivity
Polymer insulators	Small arc and pollution resistance

15.2 Improved Insulation for Motors and Power Generators

Nanocomposite dielectrics are more resistant to partial discharges (PD) than their base resins and two exemplars are provided here. The first relates to inverter-fed induction motor insulation for industrial purposes, while the second concerns power generator winding insulation for electric utilities. In both cases, insulation systems with increased thermal and PD resistance would be highly beneficial, in order to make the whole motor or generator more compact.

15.2.1 Inverter-Fed Motors

As for inverter-fed motors, medium-voltage motors with form-wound stator windings, and low-voltage (below 690 V) motors with random-wound stator windings are commercially available. Magnet wires are used for the windings insulated with heat-resistant enamels such as polyester-imide, polyimide, and polyamide-imide. Unlike the former, the latter is not designed to accept a certain level of PD. Motors of this kind are controlled through pulse width modulation (PWM), which inevitably produces voltage spikes and circulating currents at higher harmonic frequencies. In turn, such effects may result in PD and excess heating, which eventually leads to insulation failure. In particular, PD is critical for conventional enamel coated conductor wires and, consequently, PD-resistant enamel coating materials have been developed to cope with this situation.

Illustrations of the slot profile in a random-wound stator and the wave shape of inverter-fed voltage pulses applied between inter turn wires are shown in Fig. 15.2. Enamel wires for this application are usually evaluated by PD resistance test using a twisted pair of enameled wires. Of course, enameled wires should fulfill specifications other than PD resistance; necessary characteristics include flexibility, resistance to abrasion, coefficient of static friction, breakdown voltage, cut-through temperature, heat shock resistance, varnish bonding strength, and thermal index.

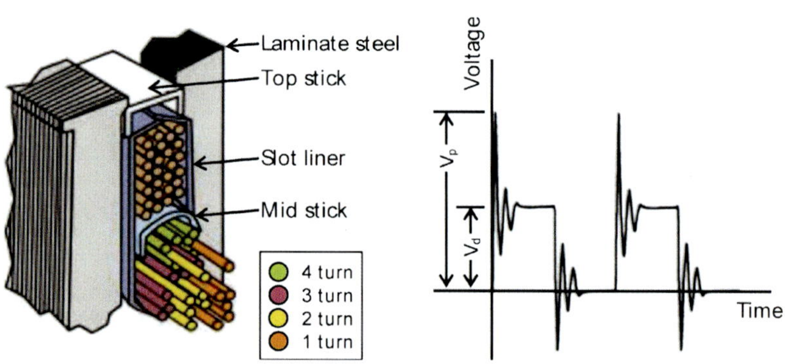

Figure 15.2 Slot profile of a random-wound stator and a wave shape of inverter-fed voltage pulses applied between inter-turn wires.

Inverter surge–resistant enameled wires have recently been developed for the marketplace, based upon nanocomposites containing silica or titania. Solution-grown nanocomposite colloid is preferred as a stating material to secure homogeneous filler dispersion. Voltron^{TM} is available for magnet wire production. Nano (silica or titania) particles are chemically linked with resin polymers such as THEIC-polyesterimide, where THEIC stands for tris(hydroxyethyl) isocyanurate [2]. PD-resistant magnet wires are also available, one of which is a nanosilica-based KMKED™ wire [3]. This wire has a two-layer structure, i.e., an inner layer coated by PD-resistant polyester-imide/silica nanocomposite enamel and then covered by a self-lubricant-type modified polyamide-imide enamel.

15.2.2 Generators

It is possible to apply nanocomposite to stator winding insulation power generators rated at 10 to 30 kV_{rms}. High-voltage insulation of this kind consists of rectangular bundled strand conductors half lapped by micaceous insulation tapes and then impregnated with resins. Mica tapes are inherently PD resistant, and therefore stator insulation is designed to accept the occurrence of PD. However, the impregnating resin, usually epoxy in modern designs, or vacuum pressure impregnated MICALASTIC™ epoxy insulation, is subject to PD, resulting in degradation and breakdown. In order to make this insulation more PD resistant, a method for reinforcing it with nanofillers is under investigation, where the idea is that the neat epoxy resin should be replaced by an epoxy nanocomposite. Figure 15.3 illustrates the possible retardation of tree propagation through the use of an impregnating resin containing nanoparticles. One trial has been made [4, 5] using surface-treated SiO_2.

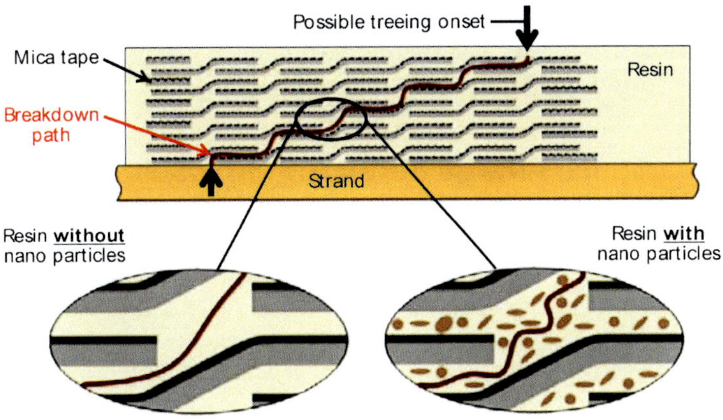

Figure 15.3 A breakdown path in resin-impregnated mica paper lapped insulation system and an enlarged path for neat resin and nanocomposite resin for comparison.

Another concern is to increase thermal conductivity. Figure 15.4 contains a schematic representation of the thermal conductivity profile across a mica-epoxy insulated stator winding in a generator slot; the thermal conductivity of the mica base insulation is around 0.3 W/m/K, which is much lower than that of surrounding copper and steel [6], and consequently limits heat dissipation through the system. Introduction of microfillers and even nanofillers can directly increase thermal conductivity to a significant degree. In addition, if thinner insulation can be used due to improved electrical performance, this will also help in thermal dissipation.

In order to make this new insulation a reality, some other requirements should be fulfilled. For instance, new stator windings insulated with nanocomposite micaceous resin should withstand repeated heat tests and bending tests.

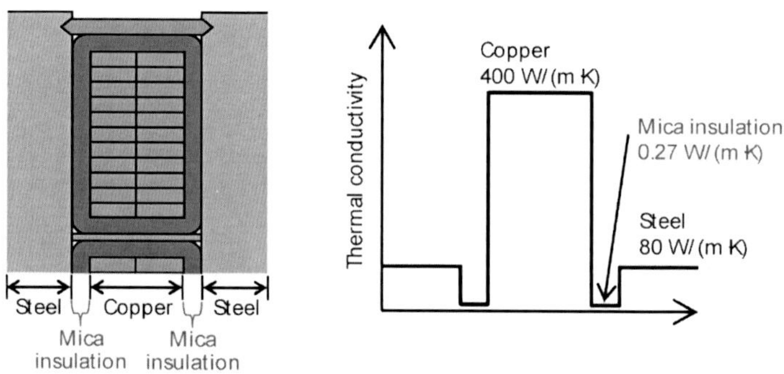

Figure 15.4 Horizontal distribution of thermal conductivity over a mica-epoxy insulated stator winding in generator slot.

15.3 Space Charge Reduction in Insulation

Space charge formation in polymers can be greatly decreased if some nanofillers are added. This is especially dramatic in polyethylene and cross-linked polyethylene (XLPE). This finding initiated the development of extra-high-voltage extruded DC cables.

500 kV and 250 kV DC XLPE submarine cables have been developed [7, 8] and the latter is now in service, as shown in Fig. 15.5. Although it is reported that the insulation used is an XLPE-based nanocomposite, no direct information is available about the type or content of nanofiller used. However, a series

of published research papers indicate that MgO nanofillers are used and that XLPE loaded with nanoparticles about 40 nm in diameter show good characteristics. This kind of cable insulation is designed to withstand the voltage polarity reversals required for line commutated converter (LCC) systems. Alternatively, voltage-source commutated systems require no voltage reversal, resulting in easier design of DC insulation. In this case, DC cable insulation is only required to withstand the high voltage for a long time.

Another cable insulation design was made public, which included a conducting inorganic filler (presumably nanocarbon) at a loading level of about 1 wt% [9]. This insulation exhibits characteristics similar to that described above, as far as space charge reduction and reduced electrical conductivity are concerned.

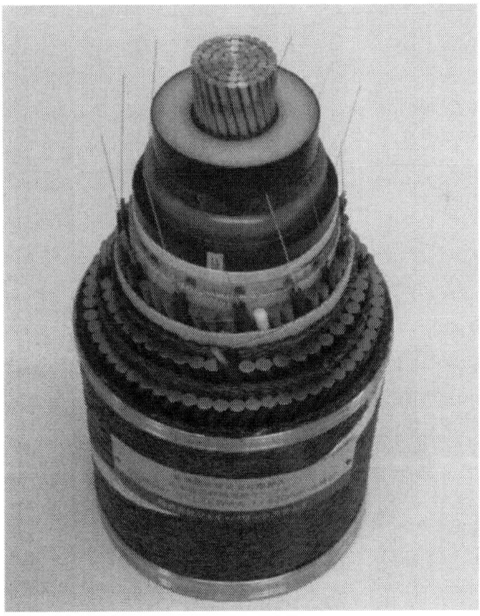

Figure 15.5 View of ±250 kV DC XLPE submarine cable in service in 2013 [8].

15.4 Increased Electric Stress Insulation

Nanocomposites can exhibit technological potential to withstand voltages, when prepared properly. What "properly" means is not, however, well understood, but it is believed to be closely related to the formation of physically and chemically sound interfaces and homogeneous filler dispersion.

15.4.1 XLPE AC Power Cables

In the case of XLPE insulation, XLPE containing vinylsilane-treated nano-silica exhibits the best ac breakdown strength performance [10] and, thereby, is intended to be used in XLPE-insulated extruded ac cables. Figure 15.6 shows possible designs for a reduced-insulation-thickness cable and an increased-ampacity cable, such that the transmittable power density is expected to increase.

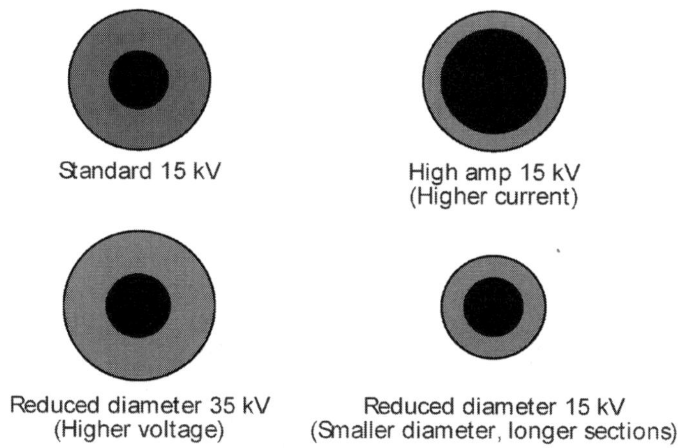

Figure 15.6 Conceptual design of ac extruded cable insulation.

15.4.2 Capacitors

The energy stored in a capacitor is proportional to the permittivity of the material used and to the square of the applied voltage. Therefore, increases in dielectric breakdown strength and/or permittivity equate to improved material performance. Polypropylene (PP) is commonly used for power capacitors and the addition of nanofillers to PP is expected to increase performance. Dielectric loss should not be increased in the course of innovation and, for power application purposes, increased dielectric breakdown strength is more important than increased permittivity. The development of such systems is under way.

PP silica nanocomposites exhibit a promising combination of dielectric properties, especially for use in ac and dc power capacitors. Indeed, it has been reported that the ac and dc breakdown strength of PP increases by 19.9% and 52.3% respectively on the addition of silica nanofillers [11, 12]. Polyaniline 1–3 wt% also increases the ac and lightning impulse breakdown strength of capacitor grade polypropylene thick films (t = 150 μm).

Figure 15.7 shows cross sections of PP capacitor structures, where the PP films are embossed for impregnation. In the case of such oil-filled types, the compatibility of the PP films with the oil should be taken into consideration. In the case of gas-filled types, PD resistance becomes an important factor in the design. It is intended to replace this PP film with a novel nanocomposite PP film.

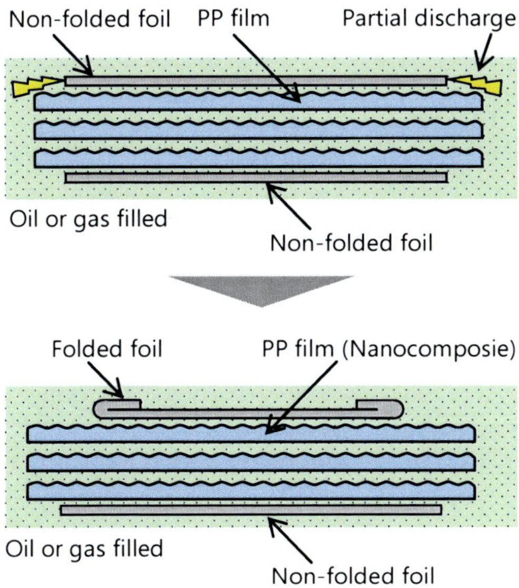

Figure 15.7 An example of PP capacitor structure—cross section.

15.5 Tracking and Erosion-Resistant Insulation

Silicone rubbers are gaining in popularity for use as polymer insulators to replace porcelain because of their performance advantages under contaminated conditions. An example of such a structure is shown in Fig. 15.8 [13]. However, the use of unfilled silicone rubber for the weathersheds will not satisfy criteria of tracking and erosion resistance and, consequently, co-loading nano- and micro silica fillers is used to improve such properties. However, this requires the effective dispersion of the nanofiller to prevent their agglomeration, which can easily take place because of their high surface energy.

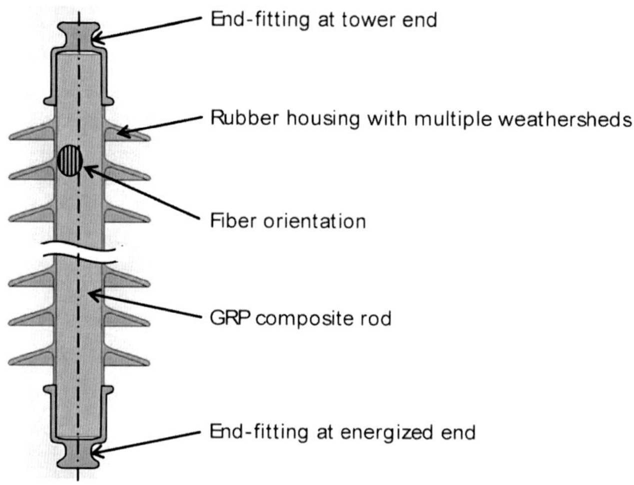

Figure 15.8 An example of polymer insulator structure—cross section [13].

One of the methods that has been suggested for better dispersion is to use surfactants such as Triton™ (X-100 ($C_{14}H_{22}O(C_2H_4O)_n$, where n = 9–10) to aid nanofiller mixing [14]. Another trial has been reported that applied an electrospinning method [15]. Although some formulations are commercially available, much work is still needed fully to characterize and to optimize their properties, to establish wide commercial application.

15.6 Nano-Micro Co-Loading for Dual Purposes

Dielectric performance can be improved by the addition of nanofillers into polymer resins of interest. However, other factors such as mechanical and thermal properties are also important in engineering design. Such dual purposes are considered to be fulfilled by co-loading of nano- and microfillers and, by way of illustration, two cases are described below.

15.6.1 Solid-State Insulated Switchgear

To avoid the use of SF_6 gas, epoxy resin solid-state insulated switchgear is attracting much attention. In order to design such new apparatus correctly, the solid epoxy resin should be characterized by a comparable thermal expansion coefficient to that of metals and, in order to meet this requirement, micro silica fillers are commonly loaded into the epoxy resins. However, microfiller loading has adverse consequences for dielectric performance, especially in connection with breakdown strength and treeing lifetime. Since nanofiller loading is one of the best ways to improve these dielectric properties, a trial development was made as shown in Fig. 15.9 [16].

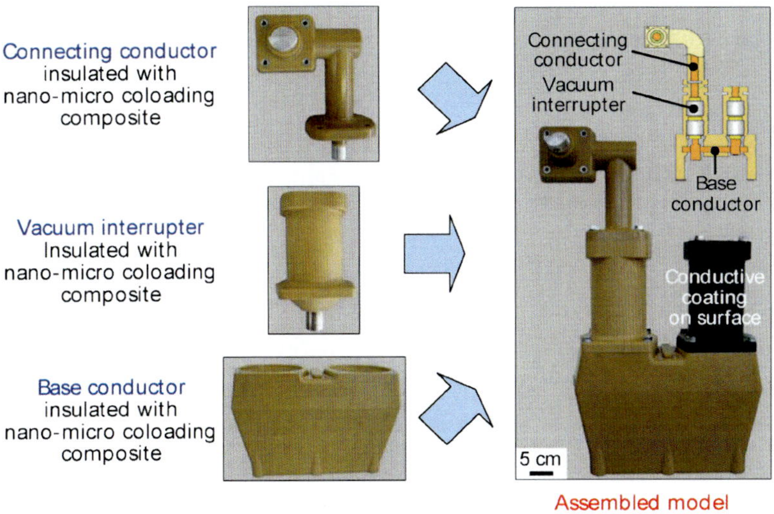

Figure 15.9 Developing component models of switchgear [16].

Figure 15.10 Developing component model of instrument transformer [17].

Nano-micro co-loading composites are candidate materials to achieve the long-life operation of instrument transformers; the instrument transformer is the one of the components that make up solid-state insulated switchgear. Since nano-micro co-loading composite would appear appropriate for use in this application, a trial development has been made, as shown in Fig. 15.10 [17].

15.6.2 Molded Transformers

Molded transformers are also a target application for nano-micro co-loading composites, since the equipment can be made more compact if the constituent insulation becomes more PD resistant and more thermally conductive. In general, polymer resins are characterized by a low thermal conductivity, which results in a significant temperature gradient from conductor side to insulation surface. This can result in incidental microcrack formation. In order to significantly increase thermal conductivity and thereby reduce the temperature difference between the inside and outside of a molded transformer, epoxy resins should be reinforced with microfillers. Then, in order to maintain sound dielectric performance, nanofillers can be used as co-fillers and a trial of this concept has been undertaken using AlN nano- and microfillers [18]. Figure 15.11 shows an example of a molded transformer by using conventional (micro-filler loading) composite [19].

Figure 15.11 Structure of a molded transformer [19].

15.7 Composite Insulation with Distributed Filler Particle size

Gas-insulated switchgear and cables involve the use of solid-state spacers, usually made of filled epoxy resins. Since the permittivity of the component material is uniform, the electric field distribution is not uniform because of the form of the structure. Since design should be made to accommodate the highest electric field, if this field can be mitigated or be made more uniform, then the apparatus can become more compact and more reliable. Mitigation of the field distributions can be achieved through the use of fillers that are distributed in size from nano- to micrometers and by controlling their local concentration to a certain extent. Such systems are termed permittivity functionally graded materials (FGM) and can be used in relevant ac applications, where fillers such as TiO_2 and SiO_2 have been considered. Although methods to achieve the required distribution remain a target of research, some trials are nevertheless shown in Fig. 15.12 [20]. For dc applications, electrical conductivity grading is needed, leading to the concept of conductivity FGM. Nanocarbon is a candidate filler in this case.

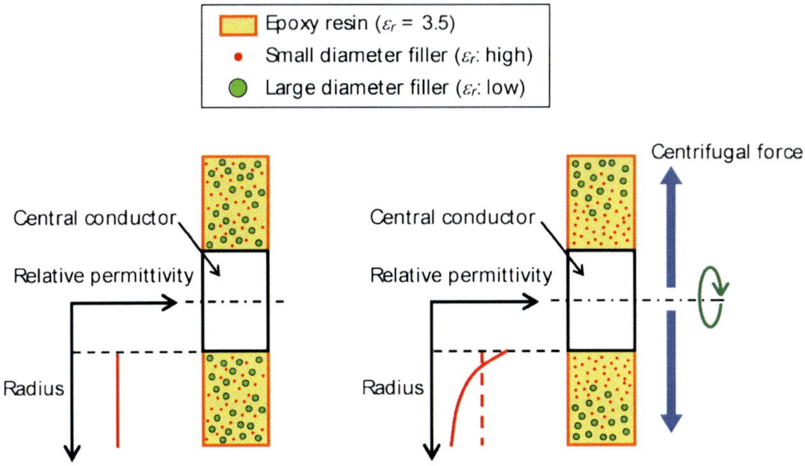

Figure 15.12 An example of a FGM formation [20].

15.8 Prospect of Power Application

The main purpose of this chapter is to illustrate the possible beneficial effects of nanocomposite dielectrics within the power sector. Plausible future applications areas are shown in Fig. 15.13 [21], while certain applications such as submarine dc cables using the XLPE nanocomposite and PD-resistant magnet wires using nanocomposite enamel resin are available for practical use.

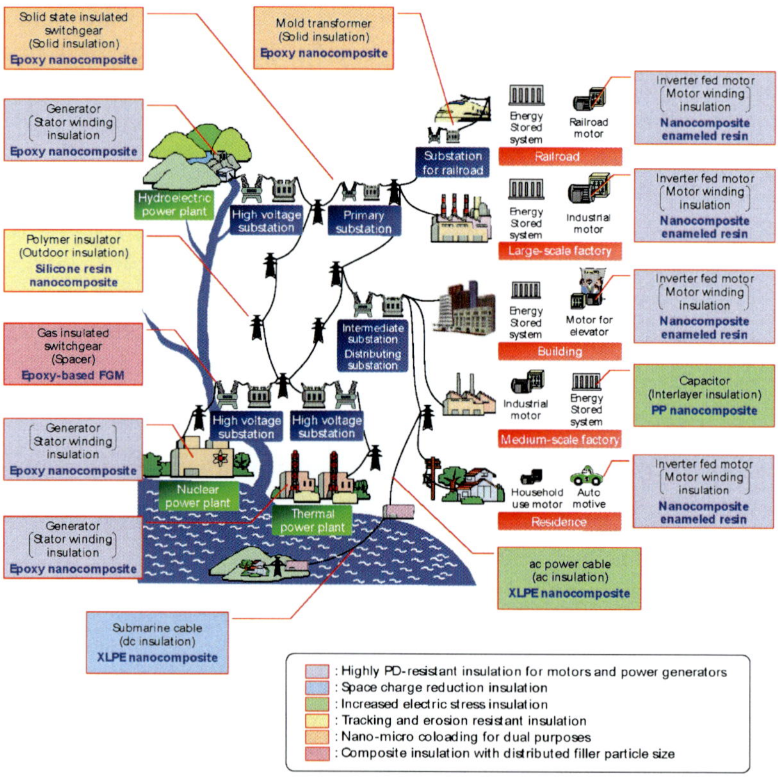

Figure 15.13 Plausible applications of nanocomposites in power sector [21].

At present, energy consumption is increasing around the world and global warming is becoming more serious every year. Energy saving and environmental friendliness are international concerns. It is often said that the twenty first century has three issues: economy, energy and environment. These issues interact and conflict with each other, as shown in Fig. 15.14. Therefore, they are termed the "Trilemma of 3E" ("Trilemma" is derived from "dilemma" which means a conflict between two issues). Balance of the 3E is very important in order to realize a sustainable society. Nanocomposites are applicable to various kinds of equipment in the power sector and can provide advantages of high efficiency, long-life operation and downsizing that will, in turn, enable energy saving and environmental friendliness. Therefore, nanocomposite dielectrics have the potential to contribute to solving a principal challenge of the current century.

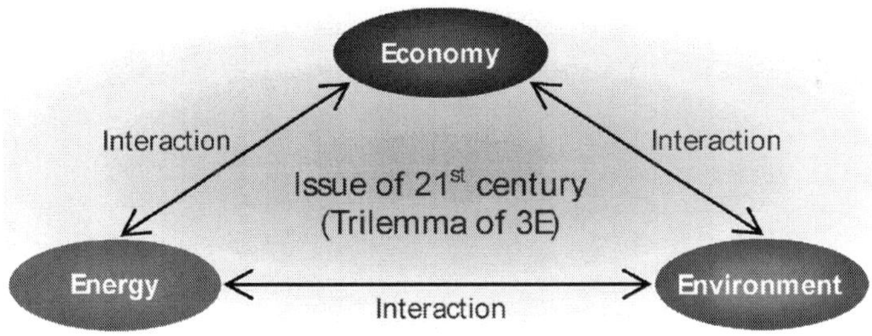

Figure 15.14 Trilemma of 3E (economy, energy, and environment).

References

1. Cherney, E. A. (2013). Nanodielectrics applications: Today and tomorrow, *IEEE Electr. Insul. Magazine*, **29**(6), pp. 59–65.
2. Böhm, F-R., Nagal, K., and Schndler, H. (2003). New generation of wire enamel for the production of magnet wires with outstanding corona resistance, *Proceedings of Electrical Insulation Conference and Electrical Manufacturing and Coil Winding Technology Conference*, pp. 109–113.
3. Kikuchi, H., and Hanawa, H. (2012). Inverter surge resistant enameled wire with nanocomposite insulating material, *IEEE Trans. Dielectr. Electr. Insul.*, **19**(1), pp. 99–106.
4. Weidner, J. R., Pohlmann, F., Gröppel, P., and Hildinger, T. (2011). Nanotechnology in high voltage insulation systems for turbine generators: First results-, *Proceedings of XVII International Symposium on High Voltage Engineering (ISH)*, no. E1-5, p. 6.
5. Gröpper, P., Hildinger, T., Pohlmann, F., and Weidner, J. R. (2012). Nanotechnology in high voltage insulation systems for large electrical machinery: First results, *CIGRE Paper*, no. A1-103, p. 15.

6. Tari, M., Yoshida, K., Sekito, S., Allison, J., Brütsch, R., and Lutz, A. (2001). A high voltage thermal conductivity for turbo generators, *Proceedings of Coil Winding, Insulation and Electrical Manufacturing Conference*, p. 5.
7. Murata, Y., and Kanaoka, Y. (2014). Development history of HVDC extruded cable with nanocomposite materials, *Proceedings of IEEE International Conference on Properties and Applications of Dielectric Materials (ICPADM)*, no. P1-29, pp. 460–463.
8. Watanabe, C., Itou, Y., Sasaki, H., Murata, Y., Suizu, M., Sakamaki, M., Watanabe, M., and Katakai, S. (2014). +/– 250 kV DC-XLPE cable for Hokkaido-Honshu HVDC link, *IEEJ Trans. Power and Energy*, **134**(1), pp. 64–75 (in Japanese).
9. Yagi, Y., Sakai, Y., Mori, H., Niinobe, H., and Tanaka, H. (2014). Development of HVDC XLPE cable and accessories, *IEEJ Trans. Power and Energy*, **134**(8), pp. 665–672 (in Japanese).
10. Roy, M., Nelson, J. K., MacCrone, R. K., Schadler, L. S., Reed, C. W., Keefe, R., and Zenger, W. (2005). Polymer nanocomposite dielectrics: The role of the interface, *IEEE Trans. Dielectr. Electr. Insul.*, **12**(4), pp. 629–643.
11. Takala, M., Ranta, H., Nevalainen, P., Pakonen, P., Pelto, J., Karttunen, M., Virtanen, S., Koivu, V., Pettersson, M., Sonerud, B., and Kannus, K. (2010). Dielectric properties and partial discharge endurance of polypropylene-silica nanocomposite, *IEEE Trans. Dielectr. Electr. Insul.*, **17**(4), pp. 1259–1267.
12. Takala, M., Sonerud, B., Ranta, H., Pelto, J., Ahonen, S., Pettersson, M., and Kannus, K. (2010). Effect of low amount of nanosilica on dielectric properties of polypropylene, *Proceedings of IEEE International Conference on Solid Dielectrics (ICSD)*, no. B2-5, pp. 187–191.
13. Kumosa, M., Kumosa, L., and Armentrout, D. (2005). Failure analyses of nonceramic insulators: Part 1: Brittle fracture characteristics, *IEEE Electr. Insul. Magazine*, **21**(3), pp. 14–27.
14. Ramirez, I., Cherney, E. A., Jayaram, S., and Gauthier, M. (2008). Nanofilled silicone dielectrics prepared with surfactant for outdoor insulation applications, *IEEE Trans. Dielectr. Electr. Insul.*, **15**(1), pp. 228–235.
15. Bian, S., Jayaram, S., and Cherney, E. A. (2013). Erosion resistance of electrospun silicone rubber nanocomposites, *IEEE Trans. Dielectr. Electr. Insul.*, **20**(1), pp. 185–193.

16. Imai, T., Yamazaki, K., Komiya, G., Murayama, K., Ozaki, T., Sawa, F., Kurosaka, K., Kitamura, H., and Shimizu, T. (2010). Component models insulated with nanocomposite material for environmentally-friendly switchgear, *Proceedings of IEEE International Symposium on Electrical Insulation (ISEI)*, no. P-52, pp. 229–232.

17. Nakamura, Y., Yamazaki, K., Imai, T., Ozaki, T., Takeuchi, M., and Maeda, T. (2014). Study on improvement of reliability of transformer using nanocomposite insulation materials, *Proceedings of IEEJ International Symposium on Electrical Insulating Materials (ISEIM)*, no. S3, p. 531.

18. Lee, T.-H., Park, J.-H., Kim, J.-U., and Cho, H.-G. (2013). Thermal properties and temperature distribution of epoxy composite with micro and nano AlN for molded transformer, *Proceedings of IEEE International Conference on Solid Dielectrics (ICSD)*, pp. 927–930.

19. Toshiba Corporation (2010). *Toshiba Review*, **65**(3), p. 80.

20. Ishiguro, J., Kurimoto, M., Kojima, H., Kato, K., Okubo, H., and Hayakawa, N. (2014). Electric field control in coaxial disk-type solid insulator by functionally graded materials (FGM), *Annual Report of IEEE Conference on Electrical Insulation and Dielectric Phenomena (CEIDP)*, no. 7–6, pp. 663–666.

21. Tanaka, T., and Imai, T. (2013). Advances in nanodielectric materials over the past 50 years. *IEEE Electr. Insul. Magazine*, **29**(1), pp. 10–23.

Index